David Colton
Heinz W. Engl
Alfred K. Louis
Joyce R. McLaughlin
William Rundell (eds.)

Surveys on Solution Methods for
Inverse Problems

Springer-Verlag Wien GmbH

Dr. David Colton

Department of Mathematical Sciences, University of Delaware,
Newark, Delaware, U.S.A.

Dr. Heinz W. Engl

Institut für Mathematik, Johannes-Kepler-Universität, Linz, Austria

Dr. Alfred K. Louis

Fachbereich Mathematik, Universität des Saarlandes, Saarbrücken,
Federal Republic of Germany

Dr. Joyce R. McLaughlin

Department of Mathematical Sciences, Rensselaer Polytechnic Institute
Troy, New York, U.S.A.

Dr. William Rundell

Department of Mathematics, Texas A & M University, College Station,
Texas, U.S.A.

© 2000 Springer-Verlag Wien
Originally published by Springer-Verlag Wien New York in 2000

Typesetting: Camera ready by authors

Graphic design: Ecke Bonk

SPIN 10732324

With 41 Figures

ISBN 978-3-211-83470-1   ISBN 978-3-7091-6296-5 (eBook)
DOI 10.1007/978-3-7091-6296-5

Contents

Contents

Introduction

D. Colton, H.W. Engl, A.K. Louis, J.R. McLaughlin and W. Rundell

It has only been since the mid-1960s that inverse problems has been identified as a proper subfield of mathematics. Prior to this conventional wisdom held it was not an area appropriate for mathematical analysis. This historical prejudice dates back to Hadamard who claimed that the only problems of physical interest were those that had a unique solution depending continuously on the given data. Such problems were *well-posed* and problems that were not well-posed were labeled *ill-posed*. In particular, ill-posed problems connected with partial differential equations of mathematical physics were considered to be of purely academic interest and not worthy of serious study. In the meantime, the success of radar and sonar during the Second World War caused scientists to ask the question if more could be determined about a scattering object than simply its location. Such problems are in the category of *inverse scattering problems* and it was slowly realised that these problems, although of obvious physical interest, were ill-posed mathematically. Similar problems began to present themselves in other areas such as geophysics, medical imaging and non-destructive testing. However, due to the lack of a mathematical theory of inverse problems together with limited computational capabilities, further progress was not possible.

Things began to change in the mid-1960s with the introduction of regularization methods for linear ill-posed problems by Tikhonov and his school in the Soviet Union. Basic to Tikhonov's theory of ill-posed problems was the careful formulation of what is meant by a "solution" to an ill-posed problem through the use of "nonstandard" information that reflects the physical situation being modeled. In particular, Morozov's method for choosing the Tikhonov regularization parameter by using an a priori knowledge of the noise level and Miller's use of an a priori bound on the norm of the solution to choose this parameter are examples of the use of such information. The subsequent extension of Tikhonov's ideas to nonlinear ill-posed problems and the rapid development of high speed computational facilities set the stage for the practical solution to inverse problems appearing in a wide variety of areas of application.

Unfortunately, during the 1970s and 1980s, developments in the mathematical theory of inverse problems and the practical problems arising in applications often appeared out of step as the techniques and problems in each area become more sophisticated. In a response to this tendency, David Colton, Heinz Engl, Alfred Louis and Bill Rundell initiated a series of conferences in the 1990s devoted to a specific application area topic (inverse problems in heat conduction, diffusion, geophysics, medical imaging and non-destructive testing) with the idea of bringing mathematicians and practitioners together to discuss problems of common interest. These conferences enjoyed the support of GAMM and SIAM and were highly successful. In 1998 it was decided to conclude this series by holding a conference on "Mathematical Methods in Inverse Problems for Partial Differential Equations", with Joyce McLaughlin as an additional member of the organising committee. The Joint Summer Research Conferences Programme Committee generously agreed to sponsor such a meeting as part of the long-standing summer research conference series held under the auspices of the AMS, IMS and SIAM.

At the conclusion of our meeting it was decided not to have a traditional conference proceedings but rather to focus on papers that we felt most typified our main theme – a new or promising method that could have impact on a wide variety of inverse problems. The organising committee asked some of the participants of the Mount Holyoke meeting to write extended surveys on the methods they presented. This book is the result of that effort, and not a proceedings volume in the usual sense.

There are three papers that deal with new issues in the regularization of nonlinear ill-posed problems. Many inverse problems can be formulated as a nonlinear operator equation $F(x) = y$ where F models the corresponding direct problem, x is the solution sought (undetermined coefficient, obstacle, boundary etc.) and y is the given data. Although there may be uniqueness for the problem, that is, there is at most one solution x for a given y, actually recovering such an x cannot be done in a stable way. Thus a simplistic approach, for example based on minimising the functional $x \to \|F(x) - y\|$, is doomed to failure. There are several ways around this dilemma and the most well known, and probably the most studied, is that due to Tikhonov: one minimises the functional $x \to \|F(x) - y\|^2 + \alpha\|x - x_0\|^2$. Here x_0 is an example of the sort of solution one might expect and the effect of the additional term is to "penalise" a large deviation away from such a solution. Another alternative is to solve the nonlinear equation by an iterative method. If one chooses certain parameters correctly, it can be possible to simply truncate the iteration procedure before noise amplification has a chance to destroy the solution. An important aspect of this idea is the need to choose a preconditioner for the problem.

In their paper, Engl and Scherzer introduce a scheme based on this last idea. They develop a convergence theory in the case where the function F is continuously differentiable, a common situation arising from problems in partial differential equations. Hanke takes a similar tactic developing a number of preconditioners that are suitable for image reconstruction problems. Not all problems are amenable to these two approaches and for others it is not an optimal approach. Tikhonov regularization works well for solving a first kind integral equation of Fredholm type, essentially converting this into a second kind equation. If the first kind integral equation is of Volterra type then the Tikhonov approach destroys this structure, and in turn loses the causal behavior of the original problem. Causality is an important feature of, for example, parabolic differential operators and giving this up is to ignore a critical aspect. In her survey, Lamm discusses discrete regularization methods specifically for first kind Volterra operators with smooth kernels.

Consider the problem of determining information about some physical object that is contained in the unit sphere from data measured on the surface of a much larger sphere. Information is transferred by the action of a partial differential operator, for example the Laplacian. This problem will be highly ill-conditioned. The effect of the object, in particular its fine structure, will be diluted in the solution as this is measured on the sphere of large radius. The further the bounding sphere is from the obstacle the more severe the ill-conditioning of the inverse problem. If somehow the data on a sphere of radius r_1 could be transferred to a sphere of radius r_2 with $r_2 < r_1$ the ill-conditioning of the problem would be reduced. Of course, by continuing the process we can extend the solution from its known values on the large sphere into the interior of the unit sphere and hence effect recovery of the object. This is the underlying principle behind layer-stripping. Although the idea is simple, making an effective inversion method out of it is not. Sylvester has developed just such an approach for the one-dimensional Helmholtz equation. It allows the stacking and splitting of layers, and when the scheme is refined, allows the use of the Born approximation for inversion of a scattering problem.

The fundamental problem of inverse obstacle scattering is to determine the shape of an object from a measurement of its far field pattern – the values of the amplitude and phase of a scattered wave. Over the years, several techniques have been developed for this problem, and a very rough classification would place these into linearised models or solving the full nonlinear problem by optimisation methods. These approaches have their advantages as well as their drawbacks. A common one is a requirement for some a-priori information such as the boundary condition at the surface of the unknown object. In all cases, traditional methods have relied on some inversion of the map that takes the boundary of the region onto the far field patterns. Recently, a new method has been developed that seeks to determine whether a given point lies inside the obstacle. Although this requires far field patterns arising from incident waves from multiple directions it has two remarkable properties. First, it requires only the inversion of linear integral equations and thus is much less computationally expensive that non-linear optimisation methods. Second, since it simply detects the boundary by checking whether a given point is inside or out, it does not require knowledge of, for example, the boundary condition. In their paper, Colton, Monk and Kirsch survey some very recent work on this subject including some of the applications of the technique to problems in medical imaging.

In 1939, Carleman developed a technique for proving uniqueness results for ill-posed Cauchy problems. The method can in fact be applied to a wide class of inverse problems in partial differential equations and the paper by Klibanov is a survey of such results that includes some future directions for growth.

A wide variety of inverse problems can be reduced, after linearisation, to the determination of a function from certain integrals over spheres. The CAT scan is perhaps the most famous of these, but the applications of such tomographic problems are far reaching. In many cases we are looking for "edges" or singularities of a quantity and must take advantage of this information. The paper by Louis and Quinto uses microlocal analysis and shows that this provides an effective technique for reconstructing object boundaries in shallow water using sonar. The paper by Köhler, Maaß and Wust takes a different approach and uses certain functionals that give much more efficient algorithms than the use of global optimisation techniques.

Perhaps the most well known inverse problem is the inverse Sturm-Liouville problem which seeks to recover a coefficient in an ordinary differential equation from knowledge of its spectrum. Historically, this was one of the first to be studied from a mathematical viewpoint and now there is a fairly complete understanding of the problem, when the coefficient is reasonably smooth, and a number of mathematically elegant ways to attack it. Notable among the unsolved problems are the ones where additional roughness is allowed for the coefficient when the mathematical model for the physical problem requires it. For many applications a more realistic mathematical model is higher dimensional. There the situation changes considerably. Eigenvalues are no longer well spaced and perturbation results, which so prominently play a part in one dimension, are much more difficult to establish. Only in the last few years has significant progress been made on these problems. In her article, McLaughlin reviews known results for two-dimensional problems. In one case data includes eigenvalues and level set information from the eigenmodes; in a second case the data is the eigenvalues and boundary measurements from the eigenfunctions; and in the third case the data is four sequences of eigenvalues from related boundary value problems. Existence, uniqueness results, error estimates and results from numerical calculations are considered; some open problems are suggested. Successful techniques that have so far yielded results are the use of perturbation theory, the use of Dirichlet to Neumann maps, and the application of the boundary control method.

One of the standard inverse problems is to recover the values of coefficients of an elliptic operator from a prescription of Cauchy data on the boundary. The "impedance tomography" problem is an example of this where one attempts to recover the value of the coefficient a in $\nabla.(a\nabla)u = 0$ within a domain Ω from a complete set of pairs $\{f_n, g_n\}$ representing the values of u and its normal derivative on the boundary of Ω. This problem is severely ill-posed and many of the standard tools, such as linearisation about a known solution, give poor results when the variation in the values of a are large. Yet, this is precisely the situation in many of the most important applications. In their paper Borcea and Papanicolaou demonstrate a new approach that works well in the high contrast case and allows effective recovery of an electrical conductivity by means of low frequency electromagnetic waves. Their technique is to use variational principles to construct discrete network approximations to the problem.

Uhlmann surveys some recent developments in the problem of determining a Riemannian metric in R^n which is Euclidean outside of a ball from certain scattering information. The idea is to use the wave front set of the scattering operator to measure the travel times of geodesics passing through the domain. Such problems are important in the recovery of anisotropic media using electromagnetic waves.

Stark points out that mathematicians, scientists, engineers and statisticians view inverse problems as having some common threads as well as some distinctly different approaches related to their own disciplines. Some common elements are that the data are finite in number and contain errors; further the data are only indirectly related to the unknown. Some mathematical tools are similar: functional analysis, convex analysis, optimization theory, nonsmooth analysis, approximation theory, harmonic analysis and measure theory. Specific to statistics, a statistician views an inverse problem as an inference or estimation problem; the data are modeled as stochastic. Standard statistical concepts, questions, and considerations such as bias, variance, mean-square error, identifiability, consistency, efficiency and various forms of optimality can be applied. Stark's article discusses inverse problems as statistical estimation and inference problems and points to the literature for a variety of techniques and results.

We thank all contributors for their efforts they put into the survey papers contained in this volume and hope that the readers will find these papers useful in their own research.

Convergence Rates Results for Iterative Methods for Solving Nonlinear Ill–Posed Problems

H. W. Engl and O. Scherzer

Industrial Mathematics Institute, Johannes Kepler Universität, Altenberger Str. 69,
A-4040 Linz, Austria
engl@indmath.uni-linz.ac.at, scherzer@indmath.uni-linz.ac.at,
WWW home page: http://www.indmath.uni-linz.ac.at

Abstract. The growth of the area of inverse problems within applied mathematics in recent years has been driven both by the needs of applications and by advances in a rigorous convergence theory of regularization methods for the solution of nonlinear ill–posed problems. There are at least two widely used approaches for solving inverse problems in a stable way: Tikhonov regularization and iterative regularization techniques. In this paper we give an overview over the latter. Moreover, we put the analysis of iterative methods for the solution of ill–posed problems into perspective with the analysis of iterative methods for the solution of well-posed problems.

1   Introduction

In the last two decades, the field of inverse problems has been one of the fastest growing areas in applied mathematics. This growth, on the one hand, has largely been driven by the needs of applications both in other sciences and in industry, and, on the other hand, by advances in a rigorous convergence theory of regularization methods for nonlinear ill–posed problems. For an account of both aspects up to 1996 see e.g. [37]; an overview over inverse problems in various applications can be gained by browsing through the proceedings of various recent conferences on inverse problems in different fields of applications, e.g., [26, 41, 42, 40, 39, 22]. A field where methods from inverse problems are of growing importance is image processing as discussed by M. Hanke in this volume, see also [8, 9].

Many inverse problems can be formulated as a *nonlinear operator equation*

$$F(x) = y \,, \tag{1}$$

with a Fréchet-differentiable operator F, where F models the corresponding *direct problem*. Usually, for an inverse problem, (1) is *ill–posed*.

There are at least two general approaches for solving (1) in a stable way. The first approach consists in approximating a solution of (1) by the minimizer of the Tikhonov functional

$$x \to \|F(x) - y\|^2 + \alpha \|x - x_0\|^2 \,. \tag{2}$$

Here α is a positive parameter, $x_0 \in X$ is an initial guess for the solution of (1). This approach was invented by Tikhonov [124, 123] originally for linear ill–posed problems, and has been applied successfully for the solution of many inverse problems. A nice account about Tikhonov regularization for linear problems is given in [48]; since the late 1980s, the theory has also been systematically developed for nonlinear problems, see e.g. [124, 123, 122, 90, 108, 83, 38, 119, 95, 80, 96, 109, 91, 116, 125, 133, 37, 120]).

As opposed to the linear case, the functional in (2) is, for nonlinear F, no longer strictly convex. This can cause severe problems due to possibly many local minima; overcoming them when minimizing (2) is not any easy task. For a class of nonlinear ill–posed problems, this problem disappears at least locally (see e.g. [17–21]), but in general, these undesirable properties of the functional (2) are a reason for looking at alternatives like iterative methods for solving (1). In such an iterative process

$$x_{n+1} := U(x_n)$$

for solving (1), the instability which stems from the ill–posedness of the problem has to be somehow controlled. While in Tikhonov regularization, this is done by appropriately choosing the regularization parameter α (see e.g. [37, 38, 116] for a detailed discussion), for iterative methods, it is decisive to stop the iteration "at the right iteration index", since after that, noise amplification will completely destroy the result. Such "stopping rules" will be a central part of discussion here.

In this paper we give an overview on iterative regularization techniques for the solution of inverse problems which can be formulated as a nonlinear operator equation (1) with a continuously differentiable operator F. We do not consider the case where the operator F is not differentiable, for this topic we refer to [23, 102, 24, 25, 60, 99, 61]. Inverse problems which have to be formulated via a nonlinear variational inequality instead of (1) will also not be addressed in this paper; for this, we refer to [78, 93] and the references therein.

2 Preconditioning as a General Concept for Constructing Iterative Regularization Algorithms

Preconditioning is a central concept for iteratively solving well–posed problems. It can also be used in the context of ill–posed problems and unifies the construction of iterative methods. Iterative preconditioning transforms the operator equation (1) into a sequence

$$B(n, x)(F(x) - y) = 0 \,, \tag{3}$$

where for $n \in \mathbb{N}_0$, $B(n,x) : Y \to X$ is a linear, bounded operator. This concept generalizes the usual preconditioning of finite-dimensional linear problems. For more background on preconditioning of linear systems we refer to [49–51]. Preconditioning of linear ill–posed problems has been considered recently in [105, 58, 59] for the linear case and in [28] for nonlinear problems.

In this paper, we consider the solution of (1) with iterative schemata of the form

$$x_{n+1} = x_n - B(n, x_n)(F(x_n) - y) . \tag{4}$$

2.1 Algorithms for the Solution of Nonlinear Operator Equations in the Framework of Preconditioning

Let F be a Fréchet-differentiable operator between two Hilbert spaces X and Y. We are concerned with solving (1); in order to develop a reasonably unified theory, we formulate iterative methods in the form (4), i.e., as fixed point iterations for the preconditioned version (3). By appropriately defining the preconditioner B, many well-known methods result in this way. In the following we indicate how it is possible to use this framework for studying convergence in a unified way. We consider three approaches:

1. Let $\{C(n,x) : n \in \mathbb{N}_0, x \in X\}$ be a family of linear bounded operator between Y and a Hilbert space Z. Then $B(n,x) := F'(x)^*C(n,x)^*C(n,x)$ are bounded operator between Y and X; here $C(n,x)^*$ is the adjoint operator to $C(n,x)$ with respect to the dual system Z and Y, i.e.,

$$\langle C(n,x)y, z \rangle_Z = \langle y, C(n,x)^*z \rangle_Y \text{ for all } z \in Z \text{, and } y \in Y \text{,}$$

and $F'(x)^*$ is the adjoint operator to $F'(x)$ with respect to the dual system X and Y, i.e,

$$\langle F'(x)v, y \rangle_Y = \langle v, F'(x)^*y \rangle_X \text{ for all } v \in X \text{, and } y \in Y .$$

$\{B(n,x) : n \in \mathbb{N}_0, x \in X\}$ constitutes a suitable class of preconditioners.
   – By letting $C(n,x) := F'(x)^{-1}$, we obtain *Newton's method*

$$x_{n+1} = x_n - F'(x_n)^{-1}(F(x_n) - y) .$$

The inverse $F'(x)^{-1}$ need not be defined everywhere; in fact, for ill–posed problems, e.g., if F is compact and $F'(x)$ is compact, $F'(x)^{-1}$ is unbounded and densely defined at best. Hence, Newton's method in its original form is inappropriate for ill–posed problems, since each iteration involves solving a linear ill–posed problem.
   – $C(n,x) = I$, i.e., $B(n,x) = F'(x)^*$ results in *Landweber's method*.
$C(n,x) = \sqrt{\alpha_n}I$, and $B(n,x) = \alpha_n F'(x)^*$, with $\alpha_n = \frac{\|s_n\|^2}{\|F'(x_n)s_n\|^2}$ and $s_n = F'(x_n)^*(F(x_n) - y)$ is a *steepest descent* method. $C(n,x) = \sqrt{\alpha_n}I$, i.e., $B(n,x) = \alpha_n F'(x)^*$, where $\alpha_n = \frac{\|F(x_n) - y\|^2}{\|s_n\|^2}$, is a *minimal error method*. See Section 4 for a further discussion of these methods.

- For $F = (F_i)_{i=0,1,\ldots,N-1} : X \to Y^N$, $y = (y_0, \ldots, y_{N-1}) \in Y^N$, and $C(i) = (0, \ldots, 0, I, 0, \ldots, 0)$, where the identity occurs at the $j = i(\text{modulo} (N))$-th vector entry, (4) becomes the *Kacmarcz method*

$$x_{n+1} = x_n - F'_j(x_n)^*(F_j(x_n) - y_j),$$
$$n = 0, 1, 2, \ldots. \text{ and } j = n(\text{modulo}(N)).$$

For well-posed problems, the Kacmarcz method has been analyzed e.g. by McCormick [86,87] and Meyn [88]. Natterer [94] has studied this method for the solution of (ill–posed) bilinear problems. A convergence analysis for nonlinear ill–posed problems will be found in [115].

2. We treat (1) in the more general form of the "output least squares minimization problem" to minimize the functional $x \to \|F(x) - y\|_Y$. If F is Fréchet-differentiable, then the minimizer $x^\dagger$ satisfies

$$F'(x^\dagger)^*(F(x^\dagger) - y) = 0. \tag{5}$$

Let for $n \in \mathbb{N}$ and $x \in X$, $A(n, x)$ be a bounded operator from X to X and define $B(n, x) = A(n, x)F'(x)^*$. If $x^\dagger$ solves (5), then

$$B(n, x^\dagger)(F(x^\dagger) - y) = 0 \text{ for all } n \in \mathbb{N}_0. \tag{6}$$

Fixed point iteration for (6) results in

$$x_{n+1} = x_n - B(n, x_n)(F(x_n) - y). \tag{7}$$

Several well known iterative methods are of this form:
- Suppose that for any $x \in X$ there exists a unique $h \in X$ satisfying

$$F'(x)^*F'(x)h = F'(x)^*(F(x) - y).$$

We use the notation $h = (F'(x)^*F'(x))^{-1} F'(x)^*(F(x) - y)$ (although this inverse need not exist everywhere), and formally define

$$B(n, x) = (F'(x)^*F'(x))^{-1} F'(x)^*.$$

We keep in mind that $A(n, x)$ has to be defined only at the single point $F'(x)^*(F(x) - y)$. This method is the *Gauß–Newton method*. For a detailed analysis of this method we refer to Deuflhard and Hohmann [30].

- If $B(n, x) = (\alpha_n I + F'(x)^*F'(x))^{-1}F'(x)^*$, then (7) is a version of the *Levenberg-Marquardt method*. For the solution of ill–posed problems, this method has been analyzed by Hanke [54].

- Let $\theta(\alpha, \lambda)$, be a real valued function defined on $]0, \infty[\times]0, \infty[$. Then

$$\theta(\alpha, F'(x)^*F'(x))$$

is well–defined via spectral theory (see e.g. [136]). If

$$\lim_{\alpha \to 0} \theta(\alpha, F'(x)^* F'(x)) = (F'(x)^* F'(x))^{-1}$$

in a sense that will be made more precise in this paper, then methods of the form

$$x_{n+1} = x_n - \theta(\alpha_n, F'(x)^* F'(x)) F'(x)^* (F(x) - y)$$

are called *inexact Gauß–Newton methods.*

3. Let $G(n, \cdot) : X \to X$ be a family of continuous (probably nonlinear) operators with $\lim_{n \to \infty} G(n, x^\dagger) \to 0$. Moreover, let $\{B(n, x) : n \in \mathbb{N}_0 \text{ and } x \in X\}$ be a family of uniformly bounded operators and let $\{F(n, \cdot) : n \in \mathbb{N}_0\}$ be a family of Fréchet-differentiable operators approximating the operator $F(x) - y$, i.e.,

$$\lim_{n \to \infty} F(n, \cdot) = F(\cdot) - y,$$

uniformly over bounded sets. Then it follows from Lebesgue's theorem on dominated convergence that

$$\lim_{n \to \infty} \left\{ B(n, x^\dagger) F(n, x^\dagger) + G(n, x^\dagger) \right\} = 0,$$

if $x^\dagger$ solves (1). With these settings, we consider the iterative method

$$x_{n+1} = x_n - (B(n, x_n) F(n, x_n) + G(n, x_n)), \tag{8}$$

For $F(n, x) = F(x) - y$ and $G(n, x) = 0$, this iteration scheme generalizes the techniques discussed above. Several well known iterative techniques can be put into this general framework:

– Let $F'(x)$ be positive definite. We set $G(n, x) = 0$, $F(n, x) := F(x) - y$, $B(n, x) := (\alpha_n I + F'(x))^{-1}$, where $\{\alpha_n\}$ is a sequence of positive numbers. Then (8) is the *Levenberg Marquardt method.*
– Let $F(n, x) := F(x) - y$, $B(n, x) := \beta_n F'(x)^*$, $G(n, x) := \alpha_n(x - x_0)$, for $\beta_n = 1$, we obtain the *modified Landweber iteration*, for $\beta_n = \frac{\|s_n\|^2}{\|F'(x_n) s_n\|^2}$ and $s_n = F'(x_n)^* (F(x_n) - y)$ the *modified steepest descent method*, and $\beta_n = \frac{\|F(x_n) - y\|^2}{\|s_n\|^2}$ in the *modified minimal error method.* The term "modified" refers to the additional terms $\alpha_n(x - x_0)$ with a positive sequence α_n, which provides additional stabilization. Note that the minimizer x_α of the Tikhonov functional (2) satisfies

$$F'(x)^* (F(x) - y) + \alpha(x - x_0) = 0.$$

Thus each iteration of the modified Landweber iteration can be considered to be descending in the direction of the gradient of the Tikhonov functional.

– Set $F(n, x) := F(x) - y$, $B(n, x) := (\alpha_n I + F'(x)^* F'(x))^{-1} F'(x)^*$, and $G(n, x) := \alpha_n(\alpha_n + F'(x)^* F'(x))^{-1}(x - x_0)$. Then (8) is called the *iteratively regularized Gauß–Newton method.*

12

- For $F : X \to X$ and $y \in X$, let $U : X \to X$ be such that $F(x) - y = U(x) - x$; then the fixed points of U satisfy (1). If we put $F(n, x) := F(x) - y$, $B(n, x) := -d_n I$ where $d_n \in]0, 1[$, and let $G(n, x) = 0$, then (8) becomes

$$x_{n+1} = x_n + d_n(F(x_n) - y) = (1 - d_n)x_n + d_n U(x_n) . \qquad (9)$$

This method is called the *segmenting Mann iteration*.
- Let P_N, P_M be the orthogonal projectors of X onto a finite dimensional subspace X_N, and from Y onto a finite dimensional subspace Y_M, respectively. Let $F(n, x) = F(P_{N(n)}(x))$ and $B(n, x) = P_{M(n)} F'(P_{N(n)}(x))^*$. Then (8) becomes the *multi-level Landweber iteration*. We note that the discretizations M and N depend on the iteration number n. The multi-level Landweber iteration has been investigated in [104, 113]. A multi-level iteratively regularized Gauß–Newton method has been studied in [28].
- If (1) is uniquely solvable, then also the artificially expanded system

$$\hat{F}(u, v) = (y, y)$$

with $\hat{F}(u, v) := (F(u), F(v))$ has a unique solution. A fixed point iteration for this system reads as

$$(u_{n+1}, v_{n+1})^t = (u_n, v_n)^t - \\ \Big(B(n, (u_n, v_n))(\hat{F}(u_n, v_n)^t - (y, y)^t) + \\ G(n, (u_n, v_n)^t)\Big) .$$

If F is a monotone operator in a partially ordered set and B is the 2×2 identity matrix operator and $G = 0$, then this scheme is the monotone iterative scheme of Vasin [133, 130]. Other variants of the monotone iteration scheme have been studied in [131]. In [134, 1] monotone iteration schemes for the solution of geophysical inverse problems were used.

By the preconditioning approach for formulating iterative methods in a unified way, we aim at a unified convergence analysis of iterative methods for the solution of nonlinear operator equations. Several methods which have not been considered in the literature so far for ill–posed problems can thus also be analyzed for nonlinear ill–posed problems – like the modified steepest descent method and the modified minimal error method. Moreover, *hybrid methods*, which combine various iterative techniques, fit into the framework of this presentation.

3 Convergence of fixed point iterations

Since all methods discussed so far are fixed point iterations in some more general sense (the operator in the iteration also depends on n), their convergence theory can be deduced from a corresponding theory of fixed point iterations, which is widely developed.

So far, there is no general convergence analysis available covering all the methods discussed in the previous section. If $B(x) := B(n, x)$ does not depend on n, then (4) can be considered as a (classical) fixed point iteration with fixed point operator $U(x) = x - B(x)(F(x) - y)$. In particular, if $F'(x)$ is self–adjoint and positive definite, then for each parameter $\lambda \in [-1, 1]$ the iterative scheme

$$x_{k+1} = x_k - F'(x_k)^\lambda (F(x_k) - y) =: U^\lambda(x_k) \tag{10}$$

can be considered as a fixed point iteration with fixed point operator U^λ. For $\lambda = 1$, (10) is Landweber's iteration, setting $\lambda = 0$ is a fixed point iteration for (1), and for $\lambda = -1$, (10) becomes Newton's method.

In the theory of nonlinear fixed point iterations, a basic assumption is that the operator $U(.)$ is of contractive type (see e.g. [14, 15, 101]). This assumption is strong enough to ensure the existence of a fixed point and to guarantee convergence of the fixed point iteration. Assuming the existence of a fixed point, convergence of a fixed point iteration follows under weaker contractivity assumptions. Under such assumptions, convergence of Newton's method was proven in [85]. This approach links fixed point theory with theoretical results on convergence of Newton's method. We discuss this point in detail in Section 5. Relations between the convergence analysis of fixed point iterations and in the convergence analysis of Landweber's method were pointed out in [111]. In [128, 129, 133, 132], fixed point iterations for the solution of ill–posed problems have been investigated. Note that for ill–posed problems, one usually has to assume the existence of a solution of (1), i.e., of a fixed point of U, anyway, which is in fact an "attainability" assumption on y.

The *segmenting Mann* iteration (9) (introduced in [100]) is another popular method for solving fixed point equations $x = U(x)$. This method is a special case of the *Mann* iteration [84] (see also [46]) which is defined as follows: let $A = (a_{n,k})$ be an infinite lower triangular matrix, and define an iteration process via

$$v_n = \sum_{k=1}^{n} a_{n,k} x_k, \quad x_{n+1} = U(v_n), \quad n = 1, 2, 3, \dots$$

Following [100] the matrix A is called *segmenting* if $a_{n+1,k} = (1 - a_{n+1,n+1})a_{n,k}$ for $k \leq n$. In this case, v_{n+1} lies in the segment joining v_n and $U(v_n)$. The Landweber iteration is a special case of the segmenting Mann iteration. To see this, set $U(x) = x - F'(x)^*(F(x) - y)$ and $v_n = x_n$. Then the segmenting Mann iteration is

$$x_{n+1} = (1 - d_n)x_n + d_n U(x_n) = x_n - d_n F'(x_n)^*(F(x_n) - y) .$$

In the case $d_n < 1$, this is the *damped Landweber iteration* and for $d_n = 1$, it is classical Landweber iteration.

Convergence result for the segmenting Mann iteration can be found in [46, 47]. For some reference on weak convergence results for the Mann iteration we refer to [98, 33, 35, 34]. In particular, (weak) convergence results for the Landweber iteration follow from general convergence results for Mann iteration.

4 Steepest Descent Algorithms for Ill–Posed Problems

In this section, we review recent convergence (rates) result for the Landweber iteration, a steepest descent method, a minimal error method, and some modifications of these methods. We recall that Landweber's iteration can be considered as a fixed point iteration with fixed point operator $U(x) = x - F'(x)^*(F(x) - y)$. This has been directly used in [111] to derive results on weak convergence of Landweber iteration. In many inverse and ill–posed problems the operator F has some structure, which is lost by passing to U. Thus, a theory based on assumptions involving the operator F rather than U is desirable. A convergence analysis of the Landweber iteration based on assumptions on the operator F has been developed in [57]. There, the essential ingredient is that F satisfies

$$\|F(x) - F(\tilde{x}) - F'(x)(x - \tilde{x})\| \le \eta \|F(x) - F(\tilde{x})\| , \qquad \eta < 1/2 ,$$
$$x, \tilde{x} \in \overline{B_\rho(x_0)} \subset \mathcal{D}(F) , \tag{11}$$

in a ball $\overline{B_\rho(x_0)}$ of radius ρ around the starting point x_0 of the iteration process containing the desired solution $x^\dagger$. (11) together with

$$\|F'(x)\| \le \psi < 1, \text{ for } x \in \overline{B_\rho(x_0)} \subset \mathcal{D}(F) \tag{12}$$

is strong enough to ensure convergence of the Landweber iteration to a solution $x^\dagger$ of (1) (cf. [57], see also [10, 37]).

A fundamental problem in solving ill–posed problems is the treatment of data errors, since it is nearly synonymous with ill–posedness that such data errors can have severe effects (see e.g. [37]). Thus, the result of Landweber iteration

$$x_{k+1}^\delta = x_k^\delta - F'(x_k^\delta)^*(F(x_k^\delta) - y^\delta) \tag{13}$$

with $\|y^\delta - y\| \le \delta$ has in general nothing to do with the corresponding result for data y even if δ is arbitrarily small; if y^δ is not in the range of F, (13) will not converge at all. However, a stable approximation of $x^\dagger$ can still be obtained by stopping the iteration "at the right iteration index", where the stopping index N_0 either depends on the noise level δ alone ($N_0 = N_0(\delta)$, "a priori stopping rule") or the stopping index k_* also depends on the data and hence on the results of the iteration itself ($k_* = k_*(\delta, y^\delta)$, "a posteriori stopping rule"). A popular stopping rule of the latter kind is the discrepancy principle, where the iteration is stopped after k_* steps with k_* determined by

$$\|y^\delta - F(x_{k_*}^\delta)\| \le \tau\delta < \|y^\delta - F(x_k^\delta)\| , \qquad 0 \le k < k_* , \tag{14}$$

where τ is a positive number depending on η from (11) satisfying

$$\tau > 2\frac{1+\eta}{1-2\eta} > 2 . \tag{15}$$

The implementation of (14) requires that an estimates of the data error is available, i.e., that the measured data $y^\delta \in Y$ satisfies

$$\|y^\delta - y\| \le \delta . \tag{16}$$

Note that a quite general result due to Bakushinskii essentially says that there can be no stable and convergent method for solving an ill–posed problem that does not use the estimate of the data error (cf. [37, Theorem 3.3] for the linear case).

In [57] it was proven that

$$x^\delta_{k_*} \to x^\dagger , \qquad \delta \to 0 ,$$

which shows that the Landweber iteration is a *regularization method*, i.e., convergent for exact data and stable with respect to data errors (for a precise definition see [37]). We note that Morozov's discrepancy principle [89], i.e., (14) with $\tau > 1$, has been applied by Vainikko [127] for the regularization of linear ill-posed problems with the Landweber iteration.

For ill–posed problems, convergence of iterative methods may be arbitrarily slow (cf. [118]). It is therefore of utmost importance to give conditions under which one can derive convergence rates. In contrast to Tikhonov regularization (see e.g. [38, 95, 37]) the *source condition*

$$x^\dagger - x_0 = \left(F'(x^\dagger)^* F'(x^\dagger)\right)^\nu f , \qquad \nu > 0 , \quad f \in \mathcal{N}(F'(x^\dagger))^\perp \backslash \{0\} \subseteq X , \tag{17}$$

(with $\|f\|$ sufficiently small) is not enough to obtain convergence rates for the Landweber iteration. See [37] for a discussion of such source conditions in the linear and nonlinear case. Some further properties of F are required:

$$F'(x) = R_x F'(x^\dagger) , \qquad x \in \overline{B_\rho(x_0)} , \tag{18}$$

where $\{R_x : x \in B_\rho(x_0)\}$ is a family of bounded linear operators $R_x : Y \to Y$ with

$$\|R_x - I\| \le C\|x - x^\dagger\| , \qquad x \in \overline{B_\rho(x_0)} , \tag{19}$$

and C is a positive constant. Note that in the linear case $R_x \equiv I$; therefore, (18) may be interpreted as a further restriction of the "nonlinearity" of F (in addition to (11)). In particular, (18) implies that

$$\mathcal{N}(F'(x^\dagger)) \subset \mathcal{N}(F'(x)) , \qquad x \in \overline{B_\rho(x_0)} .$$

It is not difficult to see that (18), (19) imply (11) with $\tilde{x} = x^\dagger$ for ρ sufficiently small (cf. [57]), thus giving rise to local convergence results. We now summarize results on convergence rates for Landweber iteration based on such considerations:

Theorem 1. *[57] Assume that problem (1) has a solution $x^\dagger$ in $\mathcal{B}_{\rho/2}$, that y^δ satisfies (16), and that F fulfills (11), (12), (18) and (19). If $x^\dagger - x_0$ satisfies (17) with some $0 < \nu \le 1/2$ and $\|f\|$ sufficiently small, then*

$$\|x^\dagger - x_{k_*}^\delta\| \le c_2 \|f\|^{1/(2\nu+1)} \delta^{2\nu/(2\nu+1)}, \tag{20}$$

with some constant $c_2 > 0$, depending on ν only. Here, x_k^δ is defined by (13), and the stopping index k_* by (14).

In [28], stability and convergence (rates) of the Landweber iteration have been derived under the assumption that the operator F satisfies a *Newton–Mysovskii condition*

$$\|(F'(x) - F'(x^\dagger))F'(x^\dagger)^\#\| \le C_{NM}\|x - x^\dagger\|, \qquad x \in D(F), \tag{21}$$

where $F'(x)^\#$ a *left inverse* of $F'(x)$; recall that an operator B is called a left inverse to $F'(x^\dagger)$ if

$$BF'(x^\dagger) = I.$$

For a survey on left, outer, and inner inverses we refer to Nashed [92]. As we will point out in Sections 5 and 6, Newton-Mysovskii conditions play an important role in the analysis of Newton-type methods. For more background on Newton-Mysovskii conditions we refer to [30, 31, 77].

Source conditions of the type (17) are too strong if the operator $F'(x)$ is strongly smoothing, e.g., if its range consists of analytic functions. Then, (17) is an a priori smoothness condition on $x^\dagger - x_0$ which is very strong. For such *severely ill–posed* problems, Hölder convergence rates in the noise level like in (20) cannot be expected under reasonable assumptions. The next best thing to look for are convergence rates which are logarithmic in δ. We now present such a result for Landweber's iteration based on Newton-Mysovskii conditions; logarithmic rates can be obtained under much weaker a–priori smoothness conditions than (17) (cf. (22) below).

Theorem 2. *[28] Let $p \ge 1$. Assume that problem (1) has a solution $x^\dagger$ in $\mathcal{B}_{\frac{\rho}{2}}(x_0)$, that y^δ satisfies (16), and that F fulfills (12), (21), with $\|F'(x^\dagger)^* F'(x^\dagger)\| < 1$. Moreover, let τ be chosen appropriately. If*

$$x^\dagger - x_0 = g_p(F'(x^\dagger)^* F'(x^\dagger))f, \tag{22}$$

with

$$g_p(\lambda) := \begin{cases} \left(\ln \frac{\exp(1)}{\lambda}\right)^{-p} & \text{for } 0 < \lambda \le 1 \\ 0 & \text{else.} \end{cases} \tag{23}$$

and $\|f\|$ is sufficiently small, then there exists a constant c_1, depending on p and $\|f\|$ only, such that

$$\|x^\dagger - x_{k_*}^\delta\| = c_1(-\ln\delta)^{-p}.$$

Condition (22) with (23) has been used first in [68] (for an iteratively regularized Gauß-Newton method), and interpretations for (22), (23) and (17) in *inverse scattering problems* have been given in [68–70]. For inverse scattering, a severely ill-posed problem, condition (17) is much too restrictive.

At the end of this section we summarize the convergence (rates) results for the Landweber iteration:

Table 1. Overview on Convergence Rates Results for the Landweber Iteration

Assumptions on F	Smoothness assumptions	Convergence Rates	Reference
(11)	–	strong convergence, stability	[57]
(18), (19)	$\nu > 0$ in (17)	Hölder rates	Theorem 1
(21)	$\nu > 0$ in (17)	Hölder rates	[28]
(21)	(22), (23)	Logarithmic rates	Theorem 2, [28]

A different approach to prove convergence rates results for regularization methods, in particular Tikhonov regularization, Landweber iteration, and the iteratively regularized Gauß- Newton method was developed in [67, 110]. The basic idea of this approach relies on the fact that under similar assumptions on the operator F as above, if (17) or (22) are satisfied, the regularized solutions again satisfy conditions similar to (17), (22), respectively, where of course now x_0 has to be replaced by the regularized solution. Then, the fact that for all elements satisfying source conditions *a posteriori estimates* hold is used to prove convergence–rates results. A posteriori estimates (which are also of interest by themselves) allow to estimate the norm of the difference to $x^\dagger$ as a function of the norm of the residual. Exemplarily we outline the idea of a posteriori error estimates for Tikhonov regularization in the case that (17) with $\eta = 1/2$ holds. In [110] convergence rates results for Tikhonov regularization for (17) with $0 < \eta < 1/2$ and under logarithmic source conditions (23) have been proven using a posteriori error estimates. For iterative regularization techniques convergence rates results using a posteriori estimates have been proven in [67].

For Tikhonov regularization and $\eta = 1/2$ it can be shown that

$$x_\alpha - x^\dagger = (F'(x^\dagger)^* F'(x^*))^\nu f_\alpha + g_\alpha \,,$$

where $\|g_\alpha\|$ is relatively small compared with $\|(F'(x^\dagger)^* F'(x^*))^\nu f_\alpha\|$. Moreover, on the set

$$B := \{x : x - x^\dagger = (F'(x^\dagger)^* F'(x^*))^\nu f + g, \text{ with } \|f\| \text{ sufficiently small}$$
$$\text{and } \|g\| \text{ relatively small compared with } (F'(x^\dagger)^* F'(x^*))^\nu f\} \,,$$

a posteriori estimates of the form

$$\|x - x^\dagger\|^2 \le C \|F(x) - y\|$$

hold. The combination of these two results gives again a classical convergence rates result for Tikhonov regularization.

The condition (11) has been verified for some parameter estimation problems in linear ordinary differential equations (cf. [57, 10]). For an inverse problem in a hyperbolic partial differential equation, (11) has been verified in [72]. The Landweber iteration has been successfully implemented for the solution of *inverse potential problems* [64], for the *recovery of the support of a source term* in an elliptic differential equation [66] (a similar topic is treated also in [63]), and for the determination of discontinuity sets of a source term and a conductivity parameter [65, 66]. For *inverse scattering problems*, Landweber iteration has been studied in [56, 62]. In [16], Landweber iteration has been used for the numerical solution of an inverse problem in polymer processing. Tautenhahn [121] considered asymptotic regularization of nonlinear ill–posed problems by a continuous version of the Landweber iteration

$$\frac{\partial x}{\partial t}(t) = F'(x(t))^*(y - F(x(t))), \text{ for } t \in [0, T], \text{ and } x(0) = x_0 .$$

For linear problems, this method is called *asymptotic regularization* (see [37, Example 4.7]). The role of a stopping rule is played by the "time" t_* up to which the initial value problem is integrated. A forward Euler discretization gives back a damped Landweber iteration.

We now turn to the modified Landweber iteration

$$x_{k+1}^\delta = x_k^\delta - \left(F'(x_k^\delta)^*(F(x_k^\delta) - y^\delta) + \alpha_k(x_k^\delta - x_0)\right);$$

it has been shown in [114] that for appropriately chosen $\{\alpha_k\} \to 0$, it is convergent and stable (in an analogous sense as the Landweber iteration) if it is stopped according the a generalized discrepancy principle (14), (15), or if the iteration is terminated a-priori when during the iteration for the first time

$$\frac{\delta}{\alpha_k} > C . \tag{24}$$

The latter termination index will be denoted by N_0. For convergence rates results, this method does not require assumptions on the nonlinearity of the operator F like (18), (19): let the operator F be Fréchet-differentiable with Lipschitz-continuous derivative in a sufficiently large neighborhood of the solution. If for fixed $0 < \psi < 1$ and $l_0 \in \mathbb{N}$ sufficiently large, $\{\alpha_k\}$ is chosen as

$$\alpha_k = (k + l_0)^{-\psi}, \quad k \in \mathbb{N}_0,$$

if (17) holds with $\nu = 1/2$, and if the modified Landweber iteration is terminated with the a–priori stopping criterion (24), then, for $\delta > 0$,

$$\|x_{N_0+1}^\delta - x^\dagger\| = O(\sqrt{\delta}) .$$

If it is terminated by the a–posteriori stopping principle (14), with τ appropriately chosen, then [1]

$$\|x_{k_*}^\delta - x^\dagger\| = O(\sqrt{\alpha_{k_*}}) \,. \tag{25}$$

For linear problems, the *steepest descent method* and the *minimal error method* are well-known known to be regularization methods, cf. [13, 75, 76, 79, 82, 83]. An analysis of the steepest descent iteration and the minimal error method for the solution of nonlinear (ill–posed) operator equations has been developed in [112, 97] (cf. 1 in Section 2.1). If (11) holds, then both methods are convergent and stable if the iteration is terminated by a discrepancy principle (cf. [112]). A convergence rates result for the steepest descent method and the minimal error method has been proven in [97] using Gilyazov's ideas [43] for proving convergence rates results of α-processes for linear ill–posed problems.

5 Links between Gradient Methods and Newton's Method

For simplicity of exposition, in this section, we assume that F is an operator on a Hilbert space X over the reals and that for each $x \in X$, $F'(x)$ is self–adjoint and positive definite. A unified convergence theory for the class of iterative schemata (10) requires generalized Newton–Mysovskii conditions:

Definition 1. *F satisfies a Newton-Mysovskii condition with index μ and exponent ϕ if for any $w, x, y \in X$ and $t \in [0, 1]$*

$$\|F'(w)^\mu(F'(x + t(y - x)) - F'(x))z\| \le \eta t \|F'(w)^{\mu+1} z\|^\phi \,. \tag{26}$$

For each $\lambda \in \mathbb{R}$, (10) characterizes an iterative method; e.g. $\lambda = -1$ is Newton's method, $\lambda = 1$ is Landweber's method, and $\lambda = 0$ is a fixed point iteration. We show below that (26) with $\mu = (\lambda - 1)/2$ is a natural condition for convergence of the iterative method with parameter λ. I.e., we have a *scale of methods* (with scale parameter λ) and a *scale of convergence conditions* (with scale parameter μ). We note that for $\mu = -1$, $w = y$, $z = y - x$, and $\phi = 1$, (26) has been introduced in [29] to prove convergence of Newton's method. If (26) holds with $\phi = 1$, $\mu = 0$, $w = x$, $z = y - x$, and η sufficiently small, then (11) holds. To see this we note that from (26) it follows that

$$\|F(y) - F(x) - F'(x)(y - x)\| = \left\| \int_0^1 (F'(x + t(y - x)) - F'(x))\, dt\, (y - x) \right\|$$
$$\le \int_0^1 \|(F'(x + t(y - x)) - F'(x))(y - x)\|\, dt$$
$$\le \tfrac{\eta}{2}\|F'(x)(y - x)\| \,,$$

which shows that

$$\|F(y) - F(x) - F'(x)(y - x)\| \le \frac{\frac{\eta}{2}}{1 - \frac{\eta}{2}} \|F(y) - F(x)\| \,,$$

---

[1] Prof. Jin Qi-Nian, University of Nanjing, China, pointed out to the second author, that there is a mistake in [114], where it was wrongly stated that $\|x_{k_*(\delta)}^\delta - x^\dagger\| = O(\sqrt{\delta})$ instead of (25).

and thus (11) holds as long as η is sufficiently small. This shows that the Newton–Mysovskii condition with $\mu = 0$ provides a natural convergence condition for the Landweber iteration. If $\mu = -1/2$, $\phi = 1$, and $z = y - x$, then (26) guarantees that the operator F is monotone. To see this we note that

$$
\begin{aligned}
0 &\le \int_0^1 \|F'(x + t(y - x))^{1/2}(y - x)\|^2 \, dt \\
&= \int_0^1 \langle y - x, F'(x + t(y - x))(y - x) \rangle \, dt \\
&= \langle y - x, F(y) - F(x) \rangle .
\end{aligned}
$$

This shows that the Newton-Mysovskii condition with $\mu = -1/2$ provides monotonicity of F, which is a typically condition in the theory of fixed point iterations. In our scale of iterative methods (10) at $\lambda = 0$ we have a fixed point iteration.

We note that Newton-Mysovskii conditions of the form (26) have been used in the literature in a different context to derive convergence (rates) results for Newton's method in various norms:

1. Hohmann [71] showed that the Newton iterates are quadratically convergent in the image space, i.e.,

$$
\|F(x_{k+1}) - F(x^\dagger)\| \le \frac{w}{2}\|F(x_k) - F(x^\dagger)\|^2 ,
$$

if F satisfies the *affine contravariant condition*

$$
\|(F'(x + th) - F'(x))h\| \le tw\|F'(x)h\|^2 .
$$

2. Deuflhard and Weiser [32] proved convergence of Newton's method in the *energy space*, i.e.,

$$
\|F'(x_{k+1})^{1/2}(x_{k+1} - x_k)\| \le \frac{w}{2}\|F'(x_k)^{1/2}(x_k - x_{k-1})\|^2 ,
$$

if the *affine conjugate Newton–Mysovskii condition*

$$
\|F'(z)^{1/2}(F'(y) - F'(x))(y - x)\| \le \frac{w}{2}\|F'(x_k)^{1/2}(y - x)\|^2
$$

holds.

To explain the formal coherence with the above considerations, we highlight Hohmann's result in the context of this paper. The results of Deuflhard and Weiser can be interpreted analogously.

Suppose for the sake of simplicity that $F'(x^\dagger)$ is invertible; this assumption is not essential, actually we only require that $v^\dagger := F'(x^\dagger)x^\dagger$, and $v_k := F'(x^\dagger)x_k$ satisfy

$$
x_k = F'(x^\dagger)^{-1}v_k , \text{ and } x^\dagger = F'(x^\dagger)^{-1}v^\dagger ,
$$

and that the Newton iteration is well-defined.

Under these assumptions we can rewrite the Newton iteration as

$$v_{k+1} = v_k - F'(x^\dagger)F'(x_k)^{-1}\left(F(x_k) - y\right) . \qquad (27)$$

Noting that $x_k = F'(x^\dagger)^{-1}v_k$ this shows that this iteration is a fixed point iteration with fixed point operator

$$U(v) := v - F'(x^\dagger)F'(F'(x^\dagger)^{-1}v)^{-1}\left(F(F'(x^\dagger)^{-1}v) - y\right) .$$

Now, we show that the affine contravariant condition implies that the fixed point iteration is quadratically convergent. Let us assume for the sake of simplicity that

$$\|(F'(x_k) - F'(x^\dagger))F'(x_k)^{-1}\| \leq C\|x^\dagger - x_k\| ; \qquad (28)$$

Under our general assumption that $F'(x^\dagger)$ is invertible and that F is uniformly Lipschitz continuous in a neighborhood of $x^\dagger$ this condition is always locally satisfied.

From the definition of Newton's method and the affine contravariant condition it immediatelly follows that

$$\|F'(x_k)(x_{k+1} - x^\dagger)\| \leq \frac{w}{2}\|F'(x_k)(x_k - x^\dagger)\|^2 .$$

Then, from (28) it follows that there exists a generic constant C such that

$$\|v_{k+1} - v^\dagger\| = \|F'(x^\dagger)(x_{k+1} - x^\dagger)\| \leq C\frac{w}{2}\|F'(x^\dagger)(x_k - x^\dagger)\|^2 = C\frac{w}{2}\|v_k - v^\dagger\|^2$$

for x_k in a sufficiently small neighborhood of $x^\dagger$. This shows that the fixed point iteration is quadratically convergent. Moreover, from the affine contravariant condition it follows that there exist positive constants a, b such that in a sufficiently small neighborhood of $x^\dagger$

$$a\|F'(x^\dagger)(x - x^\dagger)\| \leq \|F(x) - F(x^\dagger)\| \leq b\|F'(x^\dagger)(x - x^\dagger)\| . \qquad (29)$$

Condition (29) actually implies that the ill–posedness of the nonlinear ill–posed problem is as severe as the ill–posedness of the linearized problem.

Consequently it follows that

$$\|F(x_{k+1}) - F(x^\dagger)\| \leq C\|F(x_k) - F(x^\dagger)\|^2 .$$

This is Hohmann's result.

We note that we studied quadratic convergence of the Newton iteration in the image space. The imposed conditions are such that the fixed point iteration (27) is convergent. If (29) holds, then instead of the affine contravariant condition we could have used any condition which guarantees quadratic convergence of the fixed point iteration $U(v)$ to derive analogous results as Hohmann. Again this shows the strong link between convergence results in fixed point theory and in the convergence analysis of Newton's method.

Convergence of Newton's iteration in the image space is equivalent to considering convergence (rates) of Newton iteration in the original space with a semi-norm induced by the operator $F'(x^\dagger)$, i.e., in the semi-norm $\|.\|_{F'(x^\dagger)} = \|F'(x^\dagger).\|$.

The equivalence relation (29) shows that the affine contravariant condition guarantees that the operator is not "too" nonlinear.

6 Inexact Newton Methods for the Solution of Ill–Posed Problems

Newton's method is based on the Taylor expansion of the operator F. Assuming that $x^\dagger$ is the solution of the nonlinear problem (1), and x_n is an approximation of $x^\dagger$, then

$$F(x^\dagger) - F(x_n) = F'(x_n)(x^\dagger - x_n) + R(x^\dagger; x_n), \tag{30}$$

where $R(x^\dagger; x_n)$ is the Taylor remainder term. From (15) it follows that

$$F'(x_n)(x^\dagger - x_n) = \left(y^\delta - F(x_n)\right) + \left(y - y^\delta - R(x^\dagger; x_n)\right) =: y_n. \tag{31}$$

The "ideal" update $h = x^\dagger - x_n$ would allow to calculate the solution of (1) in one step. However, y_n consist of the term $y - y^\delta - R(x^\dagger; x_n)$ which is not computable. This motivates to neglect all terms on the right hand side of (31) which are not computable. Therefore we get

$$F'(x_n)h = y^\delta - F(x_n). \tag{32}$$

The error of the right hand sides between (31) and (32) can be estimated as

$$\|(y^\delta - F(x_n)) - y_n\| \le \delta + \|R(x_n, x^\dagger)\|.$$

In general, solving (32) is still ill–posed. This is e.g. the case in the practical important situation of ill-posed problems when the family $\{F'(v) : v \in D(F)\}$ is *collectively compact* (see [2]). There is a well-developed theory on how to regularize *linear* ill-posed problems with inexact data, cf., e.g., [48, 83, 37]. In the following we summarize several regularization techniques for the solution of (32), thus arriving at iterative methods for nonlinear problems where the iterates can be computed in a stable way.

Linear filtering techniques: Bakushinskii [3, 4], Kaltenbacher [74], Rieder [107, 106], propose to use parametric approximations of the equation (32):

$$(\theta(F'(x_n)^* F'(x_n), \alpha_n))^{-1} F'(x_n)^* h = y^\delta - F(x_n), \tag{33}$$

when $\theta(\lambda, \alpha)$ is a functional defined on the product space of the spectral values of $F'(x_n)^* F'(x_n)$ and the set of positive numbers; for all λ, $\lim_{\alpha \to 0} \theta(\lambda, \alpha) = \frac{1}{\lambda}$

has to hold to guarantee that $\theta(F'(x_n)^* F'(x_n), \alpha_n) F'(x_n)^*$ indeed approximates $F'(x_n)^{-1}$. This approach results in the method

$$x_{n+1} = x_n - \theta(F'(x_n)^* F'(x_n), \alpha_n) \left(F(x_n) - y^\delta \right) .$$

For the further presentation of this paper it is convenient to derive an equivalent formulation to (32): for $\zeta \in D(F)$ and the "ideal update" $h = x^\dagger - x_n$ we obtain

$$F'(x_n)(x^\dagger - \zeta) = - \left(F(x_n) - y^\delta - F'(x_n)(x_n - \zeta) \right) . \tag{34}$$

Using the above discussed parametric approximation for (34) results in the method

$$x_{n+1} = \zeta - \theta(F'(x_n)^* F'(x_n), \alpha_n) \left(F(x_n) - y^\delta - F'(x_n)(x_n - \zeta) \right) . \tag{35}$$

Methods of this type have been investigated by Bakushinskii [3], Bakushinskii and Goncharskii [5, 6], Blaschke-Kaltenbacher [73, 74], see also [11].

In the following we summarize methods which fit into this general framework:

Levenberg and Marquardt: Here $\theta(\lambda, \alpha) = \frac{1}{\alpha + \lambda}$ in (33). This corresponds to the method

$$x_{k+1} = x_k - \left(\alpha_k I + F'(x_k)^* F'(x_k) \right)^{-1} F'(x_k)^* \left(F(x_k) - y \right) .$$

We note that x_{k+1} has the variational characterization

$$\|F(x_k) - y + F'(x_k)(x - x_k)\|^2 + \alpha_k \|x - x_k\|^2 . \tag{36}$$

Essentially, under the assumption that

$$\|F(x) - F(z) - F'(z)(x - z)\| \le C\|x - z\|\|F(x) - F(z)\| ,$$

Hanke [54] was able to prove convergence and stability of the Levenberg–Marquardt scheme if the parameters α_k are such that

$$\|y - F(x_k) - F'(x_k)(x_{k+1} - x_k)\| \le \rho \|F(x_k) - y\| . \tag{37}$$

The combination of an inexact Newton's method with (37) is sometimes referred to *residual inexact Newton method* in the literature (see e.g.,[27, 7, 71]).

Iteratively Regularized Gauß–Newton method: This method is based on the Levenberg–Marquardt algorithm, but augmented by a term $-(\alpha_k I + F'(x_k)^* F'(x_k))^{-1} \alpha_k (x_k - \zeta)$ for additional stabilization (see below):

$$x_{k+1} = x_k - \left(\alpha_k I + F'(x_k)^* F'(x_k) \right)^{-1}$$
$$\left[F'(x_k)^* \left(F(x_k) - y \right) + \alpha_k \left(x_k - \zeta \right) \right] . \tag{38}$$

The iteratively regularized Gauß-Newton method fits into the general framework of this paper by using $\theta(\lambda, \alpha) = \frac{1}{\alpha + \lambda}$ in (35). Usually, ζ is taken as x_0, but this is not necessary. Leaving ζ arbitrary gives additional freedom. This is especially useful in a multi-level versions of (38) (cf. [28]). The iterates of the iteratively regularized Gauß–Newton technique x_{k+1} have the variational characterization

$$\|F(x_k) - y + F'(x_k)(x - x_k)\|^2 + \alpha_k \|x - \zeta\|^2 .$$

A comparison with (36) shows that as long as α_k is not too small, the last term has a tendency to keep the iterates close to the point ζ which explains why we spoke of "additional stabilization" above. In the following, we summarize some convergence (rates) results for the iteratively regularized Gauß-Newton method. In order to become a regularization method, the iteratively regularized Gauß-Newton technique must be either terminated with the *a posteriori* stopping criterion (14) (with an appropriately chosen parameter τ) or using the following a priori stopping criterion (with ν as in the source condition (17)):
if $0 < \nu \leq 1$: $N_0 = N_0(\delta)$ is chosen such that

$$\eta \alpha_{N_0}^{\nu + 1/2} < \delta \leq \eta \alpha_n^{\nu + 1/2} \quad \text{for } 0 \leq n < N_0 , \tag{39}$$

if $\nu = 0$: $N_0 = N_0(\delta)$ is chosen such that

$$N_0 \to \infty \text{ and } \eta \geq \frac{\delta}{\sqrt{\alpha_{N_0}}} \to 0 \text{ as } \delta \to 0 . \tag{40}$$

The convergence results for the iteratively regularized Gauß–Newton technique presented below are (if not stated otherwise) based on the assumption that the operator F satisfies the following condition in a sufficiently large neighborhood of the starting point x_0 containing the solution $x^\dagger$:

$$F'(\bar{x}) = R(\bar{x}, x)F'(x) + Q(\bar{x}, x)$$
$$\|I - R(\bar{x}, x)\| \leq C_R$$
$$\|Q(\bar{x}, x)\| \leq C_Q \|F'(x^\dagger)(\bar{x} - x)\|^\beta \|\bar{x} - x\|^\gamma \tag{41}$$
$$\text{with } \beta \left(\tfrac{1}{2} + \nu \right) + \gamma \nu \geq \tfrac{1}{2} .$$

Note that for $\nu \geq 1/2$ (41) is satisfied if F is Fréchet-differentiable with Lipschitz-continuous derivative, since in this case we can take $\beta = 0, \gamma = 2$.

The following result is taken from [12]; see there for the precise assumptions in the conditions where we use "sufficiently small" and "appropriately" here for simplicity.

Theorem 3. *Let*

$$\alpha_n > 0, \quad 1 \leq \frac{\alpha_n}{\alpha_{n+1}} \leq r, \quad \lim_{n \to \infty} \alpha_n = 0 . \tag{42}$$

Let x_n be as in (38), and let x_n^δ be the iterates when in (38) y is replaced by y^δ. Let the condition (17) be satisfied and let $x^\dagger \in B_\rho(x_0)$. Let $N_0 = N_0(\delta)$ be chosen according to (39) if $0 < \nu \le 1$ and (40) if $\nu = 0$. Moreover, we assume that $\|f\|$ in (17), η in (39), (40), and α_0 are sufficiently small.

Let either one of the following two assumptions hold:
1. $0 \le \nu < \frac{1}{2}$ and (41) hold with C_R and C_Q sufficiently small.
2. $\frac{1}{2} \le \nu \le 1$, and let F be appropriately scaled.
Then

$$\|x_{N_0}^\delta - x^\dagger\| = \begin{cases} o(1) & , \text{if } \nu = 0 , \\ O(\delta^{\frac{2\nu}{2\nu+1}}) & , \text{if } 0 < \nu \le 1 , \end{cases}$$

where N_0 is the stopping index determined by (39), (40), respectively. For the noise free case ($\delta = 0$, $\eta = 0$) we obtain that for all $n \in \mathbb{N}$

$$\|x_n - x^\dagger\| = \begin{cases} o(\alpha_n^\nu) & , \text{if } 0 \le \nu < 1 , \\ O(\alpha_n) & , \text{if } \nu = 1 , \end{cases}$$

and that

$$\|F(x_n) - y\| = \begin{cases} o(\alpha_n^{\nu+\frac{1}{2}}) & , \text{if } 0 \le \nu < \frac{1}{2} , \\ O(\alpha_n) & , \text{if } \frac{1}{2} \le \nu \le 1 . \end{cases}$$

If the iteration is stopped according to a generalized discrepancy (14) with $\tau > 1$ sufficiently large instead of the a priori stopping criterion (39) and (40), then it was proven in [12] under essentially the same assumptions as in Theorem 3 that

$$\|x_{k_*}^\delta - x^\dagger\| = \begin{cases} o(\delta^{\frac{2\nu}{2\nu+1}}) & , \text{ if } 0 < \nu \le 1/2 \\ O(\sqrt{\delta}) & , \text{ if } \nu \ge 1/2 . \end{cases}$$

For $\nu = 1$ the above result was proven by Bakushinskii [3]. Alternative stopping criteria for the iteratively regularized Gauß–Newton method have been investigated in [11]. For other work related to the iteratively regularized Gauß-Newton technique we refer to [4, 6, 37, 103]. In [68], a convergence rates result was proven which is applicable to severely ill-posed problems:

Theorem 4. *[68] Let x_k be as in (38), and let x_k^δ be the iterates when in (38) y is replaced by y^δ. Let F satisfy (41) with $\beta = 1$, $\gamma = 0$, where the constants C_R is sufficiently small, and let $\|F'(x^\dagger)^* F'(x^\dagger)\| < 1$. Let (22), (23) be satisfied, with g_p and f as in Theorem 2, where $\|f\|$ is sufficiently small. Moreover, let $\{\alpha_k\}$ satisfy (42) and let τ as in (14) be sufficiently large, and let ρ be sufficiently small. Let k_* be determined by the discrepancy principle (14), then*

$$\|x_{k_*}^\delta - x^\dagger\| = O\left((-\ln \delta)^{-p}\right) .$$

In the case of exact data we have

$$\|x_k - x^\dagger\| = O\left(\left(-\ln\frac{\alpha_k}{\exp(1)\alpha_0}\right)^{-p}\right),$$

with α_0 sufficiently large.

In the case of exact data a similar convergence rates result for the iteratively regularized Gauß-Newton method was proved in [28] for the case that the operator F satisfies a Newton-Mysovskii condition (24).

In the results summarized above, the parameters $\{\alpha_k\}$ are chosen to satisfy (42) and are therefore chosen problem-independent. In [117], a strategy for choosing these parameters which uses information on the residual error was investigated.

In the following we give an overview on convergence (rates) results of the iteratively regularized Gauß-Newton method.

Table 2. Overview on convergence (rates) results for the iteratively regularized Gauß-Newton method

Assumptions on F	Smoothness assumptions	Convergence rates	Reference
(41), (42)	$0 \le \nu \le 1/2$ in (17)	$O\left(\delta^{\frac{2\nu}{2\nu+1}}\right)$ A posteriori stop	Theorem 3
		A priori stop	
Lipschitz-continuous	$1/2 \le \nu$ in (17)	$O(\sqrt{\delta})$ A posteriori stop	
derivative, (42)	$1/2 \le \nu \le 1$ in (17)	$O\left(\delta^{\frac{2\nu}{2\nu+1}}\right)$ A priori stop	Theorem 3
(41) with $\gamma = 0$, (42)	(22), (23)	$O\left((\ln(-\delta))^{-p}\right)$ A posteriori stop	Theorem 4

<u>The truncated Newton CG-method:</u> Hanke [55] analyzed another variant of an inexact Newton method for the solution of nonlinear ill–posed problems, where in each Newton step an approximate solution of the linearized problem is computed with a truncated conjugate gradient method. The conjugate gradient method is terminated when the residual has been reduced significantly. Hanke proved that this method is stable and convergent provided that the conjugate gradient iteration is appropriately terminated. For some literature on the conjugate gradient method for the solution of linear problems we refer to [45] and the references therein. For the solution of linear ill–posed problems with the conjugate gradient technique we refer to [53, 43, 44].

<u>Quasi-Newton methods:</u> Kaltenbacher [74] studied a Broyden's method for the solution of nonlinear ill–posed problems and proved convergence of this method under Newton-Mysovskii conditions (21).

Second degree method: Hettlich and Rundell [66] have studied an inexact Newton method which utilizes an approximation of the second derivative. They suggest to use the following two-step algorithm:

$$(\alpha_2 I + S(x_n)^* S(x_n))\,(x_{n+1} - x_n) = F'(x_n)^*(y^\delta - F(x_n))$$

with

$$S(x_n)h = F'(x_n)h + \frac{1}{2}F''(x_n)(\tilde{h}, h)$$

and

$$\tilde{h} = (F'(x_n)^* F'(x_n) + \alpha_1 I)^{-1} F'(x_n)^*(y^\delta - F(x_n))\,.$$

Hettlich and Rundell showed that this method is stable and convergent if the operator F satisfies (38) and the parameters α_1 and α_2 are chosen appropriately. In comparison with the iteratively regularized Gauß-Newton method, the constants α_1 and α_2 are not changed during the the iteration. In the notation of Section 2 we can consider $(\alpha_2 I + S(x_n)^* S(x_n))^{-1} F'(x_n)^*$ as a preconditioner $B(n, x_n)$. Assuming the convergence of x_n to $x^\dagger$ we realize that the fixed point iteration for the preconditioned equation

$$0 = (\alpha_2 I + S(x^\dagger)^* S(x^\dagger))^{-1} F'(x^\dagger)^* (F(x^\dagger) - y)$$

is the method proposed by Hettlich and Rundell, i.e., this method is also contained in the general setup of this paper. It might be worthwhile to analyze monotonicity assumptions of the fixed point operator defined by this equation, and deduce convergence results from general fixed point theory.

We have summarized convergence rates results for iterative methods for the solution of nonlinear ill–posed problem and we have shown that general Newton-Mysovskii conditions of the form (26) are the basic ingredients to prove convergence rates results of iterative regularization schemes. Iterative methods, like Landweber's method, fixed point iteration, and Newton's method can be considered part of a scale of iterative methods, which defines via (26) a convergence condition that can be used for a convergence analysis. This fundamental link relates the convergence analysis of classical proofs of the Newton process with the convergence analysis of Landweber iteration, and of fixed point iterations.

Acknowledgement

This work is supported by the Austrian Fonds zur Förderung der wissenschaftlichen Forschung, grants F–1308 (H.W.E) and F–1310 (O.S).

References

1. A.L. Ageev, T.V. Bolotova, and V.V. Vasın. Solution to the inverse gravity problem for two interfaces in a medium. *Izvestiya, Physics of the Solid Earth.* 34:225 227, 1998.

2. P. Anselone. *Collectively Compact Operator Approximation Theory.* Prentice-Hall, Englewood Cliffs, New Jersey, 1971.

3. A.B. Bakushinskii. The problem of the convergence of the iteratively regularized Gauß–Newton method. *Comput. Maths. Math. Phys.*, 32:1353–1359, 1992.

4. A.B. Bakushinskii. Universal linear approximations of solutions to nonlinear operator equations and their application. *J. Inv. Ill-Posed Problems*, 5:501–521, 1997.

5. A.B. Bakushinskii and A.V. Goncharskii. *Iterative Methods for the Solution of Incorrect Problems.* Nauka, Moscow, 1989. in Russian.

6. A.B. Bakushinskii and A.V. Goncharskii. *Ill–Posed Problems: Theory and Applications.* Kluwer Academic Publishers, Dordrecht, Boston, London, 1994.

7. R.E. Bank and D.J. Rose. Analysis of a multilevel iterative method for nonlinear finite element equations. *Math. Comput.*, 39:453–465, 1982.

8. M. Bertero and P. Boccacci. *Introduction to Inverse Problems in Imaging.* IOP Publishing, London, 1998.

9. M. Bertero, T.A. Poggio, and V. Torre. Ill-posed problems in early vision. *Proc. IEEE*, 76:869–889, 1988.

10. A. Binder, M. Hanke, and O. Scherzer. On the Landweber iteration for nonlinear ill-posed problems. *J. Inverse Ill-Posed Probl.*, 4:381–389, 1996.

11. B. Blaschke. *Some Newton Type Methods for the Regularization of Nonlinear Ill-Posed Problems.* Universitätsverlag Rudolf Trauner, Linz, Austria, 1996. PhD–Thesis, Schriften der Johannes–Kepler–Universität.

12. B. Blaschke, A. Neubauer, and O. Scherzer. On convergence rates for the iteratively regularized Gauss-Newton method. *IMA J. Numer. Anal.*, 17:421–436, 1997.

13. H. Brakhage. On ill–posed problems and the method of conjugate gradients. In *[36]*, pages 177–185, 1987.

14. F.E. Browder, editor. *Nonlinear Functional Analysis*, volume 18. Amer. Math. Soc., Providence, 1970. Proc. Symposia in Pure Math.

15. F.E. Browder and W.V. Petryshyn. Construction of fixed points of nonlinear mappings in Hilbert space. *J. Math. Anal. Appl.*, 20:197–228, 1967.

16. M. Burger, V. Capasso, and H.W. Engl. Inverse problems related to crystallization of polymers. *Inverse Probl.*, 15:155–173, 1999.

17. G. Chavent. New size × curvature conditions for strict quasiconvexity of sets. *SIAM J. Control and Optimiz.*, 29:1348–1372, 1991.

18. G. Chavent. Quasi-convex sets and size × curvature condition, applications to nonlinear inversion. *Appl. Math. Optimiz.*, 24:129–169, 1991.

19. G. Chavent and K. Kunisch. A geometric theory for l^2-stabilization of the inverse problem in a one-dimensional elliptic equation from an h^1-observation. *Appl. Math. Optimiz.*, 27:231–260, 1993.

20. G. Chavent and K. Kunisch. Regularization in state space. *J. Numer. Anal.*, 27:535–564, 1993.

21. G. Chavent and K. Kunisch. State space regularization: geometric theory. *Appl. Math. Optimiz.*, 37:243–267, 1998.

22. G. Chavent and P. C. Sabatier, editors. *Inverse Problems of Wave Propagation and Diffraction.* Springer, Berlin, 1997.

23. X. Chen, M.Z. Nashed, and L. Qi. Convergence of Newton's method for singular smooth and nonsmooth equations using adaptive outer inverses. *SIAM J. Optim.*, 7:445–462, 1997.

24. X. Chen and L. Qi. A parameterized Newton method and a quasi-Newton method for nonsmooth equations. *Comput. Optim. Appl.*, 3:157–179, 1994.

25. X. Chen and T. Yamamoto. On the convergence of some quasi-Newton methods for nonlinear equations with nondifferentiable operators. *Computing*, 49:87–94, 1992.

26. D. Colton, R. Ewing, and W. Rundell, editors. *Inverse Problems in Partial Differential Equations*. SIAM, Philadelphia, 1990.

27. R.S. Dembo, S.C. Eisenstat, and T. Steihaug. Inexact Newton methods. *SIAM J. Numer. Anal.*, 19:400–408, 1982.

28. P. Deuflhard, H.W. Engl, and O. Scherzer. A convergence analysis of iterative methods for the solution of nonlinear ill–posed problems under affinely invariant conditions. *Inverse Probl.*, 14:1081–1106, 1998.

29. P. Deuflhard and G. Heindl. Affine invariant convergence theorems for Newton's method and extensions to related methods. *SIAM J. Numer. Anal.*, 16:1–10, 1979.

30. P. Deuflhard and A. Hohmann. *Numerical Analysis. A First Course in Scientific Computation*. De Gruyter, Berlin, 1991. Transl. from the German by F.A. Potra and F. Schulz.

31. P. Deuflhard and A. Hohmann. *Numerische Mathematik I. Eine algorithmisch orientierte Einführung (Numerical Mathematics I. An algorithmically oriented Introduction)*. De Gruyter, Berlin, 1993. 2., überarb. Aufl. (German).

32. P. Deuflhard and M. Weiser. Local inexact Newton multilevel FEM for nonlinear elliptic problems. In M-O. Bristeau, G. Etgen, W. Fitzgibbon, J.-L. Lions, J. Periaux, and M. Wheeler, editors, *Computational Science for the 21st Century, Tours, France*, pages 129–138. Wiley-Interscience-Europe, 1997.

33. W.G. Dotson jr. On the Mann iterative process. *Trans. Amer. Math. Soc.*, 149:65–73, 1970.

34. H.W. Engl. Weak convergence of asymptotically regular sequences for nonexpansive mappings and connectors with certain Chebyshef-centers. *Nonlinear Analysis, Theory & Applications*, 1:495–501, 1977.

35. H.W. Engl. Weak convergence of Mann iteration for nonexpansive mappings without convexity assumptions. *Bollettino U.M.I*, 14:471–475, 1977.

36. H.W. Engl and C.W. Groetsch, editors. *Inverse and Ill-Posed Problems*. Academic Press, Boston, 1987.

37. H.W. Engl, M. Hanke, and A. Neubauer. *Regularization of Inverse Problems*. Kluwer Academic Publishers, Dordrecht, 1996.

38. H.W. Engl, K. Kunisch, and A. Neubauer. Convergence rates for Tikhonov regularization of nonlinear ill–posed problems. *Inverse Probl.*, 5:523–540, 1989.

39. H.W. Engl, A.K. Louis, and W. Rundell, editors. *Inverse Problems in Geophysical Applications*. SIAM, Philadelphia, 1996.

40. H.W. Engl, A.K. Louis, and W. Rundell, editors. *Inverse Problems in Medical Imaging and Nondestructive Testing*. Springer, Wien, New York, 1996.

41. H.W. Engl and J. McLaughlin, editors. *Inverse Problems and Optimal Design in Industry*. B.G. Teubner, Stuttgart, 1994.

42. H.W. Engl and W. Rundell, editors. *Inverse Problems in Diffusion Processes*. SIAM, Philadelphia, 1995.

43. S.F. Gilyazov. Iterative solution methods for inconsistent linear equations with non self–adjoint operators. *Moscov University Computational Mathematics and Cybernetics*, pages 8–13, 1977.

44. S.F. Gilyazov. Regularizing algorithms based on the conjugate-gradient method. *U.S.S.R. Comput. Math. Math. Phys.*, 26:8–13, 1986.

45. G.H. Golub and D.P. O'Leary. Some history of the conjugate gradient method and Lanczos algorithms:1948–1976. *SIAM Review*, 31:50–102, 1989.

46. C.W. Groetsch. A note on segmenting Mann iterates. *J. Math. Anal. Appl.*, 40:369–372, 1972.

47. C.W. Groetsch. A nonstationary iterative process for nonexpansive mappings. *Proc. Am. Math. Soc.*, 43:155–158, 1974.

48. C.W. Groetsch. *The Theory of Tikhonov Regularization for Fredholm Equations of the First Kind.* Pitman, Boston, 1984.

49. W. Hackbusch. *Multi-Grid Methods and Applications.* Springer-Verlag, Berlin, Heidelberg, New York, 1985.

50. W. Hackbusch. *Iterative Solution of Large Sparse Systems of Equations.* Springer-Verlag, New York, 1994. Applied Mathematical Sciences 95.

51. W. Hackbusch. *Integral Equations. Theory and Numerical Treatment.* Birkhäuser, Basel, 1995.

52. G. Hämmerlein and K.-H. Hoffmann, editors. *Constructive Methods for the Practical Treatment of Integral Equations.* Birkhäuser Verlag, Basel, Boston, Stuttgart, 1985. Proceedings of the Conference at the Mathematisches Forschungsinstitut Oberwolfach, June 24-30, 1984. International Series of Numerical Mathematics, Vol. 73.

53. M. Hanke. *Conjugate Gradient Type Methods for Ill-Posed Problems.* Longman Scientific & Technical, Harlow, 1995. Pitman Research Notes in Mathematics Series.

54. M. Hanke. A regularizing Levenberg-Marquardt scheme, with applications to inverse groundwater filtration problems. *Inverse Probl.*, 13:79–95, 1997.

55. M. Hanke. Regularizing properties of a truncated Newton-CG algorithm for nonlinear inverse problems. *Numer. Funct. Anal. Optimiz.*, 18:971–993, 1997.

56. M. Hanke, F. Hettlich, and O. Scherzer. The Landweber iteration for an inverse scattering problem. In *[135]*, pages 909–915, 1995.

57. M. Hanke, A. Neubauer, and O. Scherzer. A convergence analysis of Landweber iteration for nonlinear ill-posed problems. *Numer. Math.*, 72:21–37, 1995.

58. M. Hanke and C. Vogel. Two-level preconditioners for regularized inverse problems I: theory. *Numer. Math.*, 1999. to appear.

59. M. Hanke and C. Vogel. Two-level preconditioners for regularized inverse problems II: implementation and numerical results. submitted.

60. W.M. Häussler. A Kantorovich-type convergence analysis for the Gauss-Newton-method. *Numer. Math.*, 48:119–125, 1986.

61. M. Heinkenschloss, C.T. Kelley, and H.T. Tran. Fast algorithms for nonsmooth compact fixed-point problems. *SIAM J. Numer. Anal.*, 29:1769–1792, 1992.

62. F. Hettlich. An iterative method for the inverse scattering problem from sound-hard obstacles. *Z. Angew. Math. Mech.*, 76:165–168, 1996.

63. F. Hettlich, J. Morgan, and O. Scherzer. On the estimation of interfaces from boundary measurements. In *[40]*, pages 163 – 178, 1996.

64. F. Hettlich and W. Rundell. Iterative methods for the reconstruction of an inverse potential problem. *Inverse Probl.*, 12:251–266, 1996.

65. F. Hettlich and W. Rundell. Recovery of the support of a source term in an elliptic differential equation. *Inverse Probl.*, 13:959–976, 1997.

66. F. Hettlich and W. Rundell. The determination of a discontinuity in a conductivity from a single boundary measurement. *Inverse Probl.*, 14:67–82, 1998.

67. B. Hofmann and O. Scherzer. Local ill-posedness and source conditions of operator equations in Hilbert spaces. *Inverse Probl.*, 14:1189–1206, 1998.

68. T. Hohage. Logarithmic convergence rates of the iteratively regularized Gauß-Newton method for an inverse potential and an inverse scattering problem. *Inverse Probl.*, 13:1279–1300, 1997.

69. T. Hohage. Convergence rates of a regularized Newton method in sound-hard inverse scattering. *SIAM Numer. Anal.*, 36:125–142, 1999.

70. T. Hohage and C. Schormann. A Newton-type method for a transmission problem in inverse scattering. *Inverse Probl.*, 14:1207–1228, 1998.

71. A. Hohmann. *Inexact Gauss Newton methods for parameter dependent nonlinear problems*. Shaker, Aachen, 1994. Berichte aus der Mathematik.

72. S. Kabanikhin, R. Kowar, and O. Scherzer. On the Landweber iteration for the solution of a parameter identification problem in a hyperbolic partial differential equation of second order. *J. Inv. Ill-Posed Problems*, 6:403–430, 1998.

73. B. Kaltenbacher. Some Newton-type methods for the regularization of nonlinear ill-posed problems. *Inverse Probl.*, 13:729 – 753, 1997.

74. B. Kaltenbacher. A posteriori parameter choice strategies for some Newton type methods for the regularization of nonlinear ill–posed problems. *Numer. Math.*, 79:501 – 528, 1998.

75. W.J. Kammerer and M.Z. Nashed. Steepest descent for singular linear operators with nonclosed range. *Applicable Analysis*, 1:143–159, 1971.

76. W.J. Kammerer and M.Z. Nashed. On the convergence of the conjugate gradient method for singular linear operator equations. *SIAM J. Numer. Anal.*, 9:165–181, 1972.

77. L.W. Kantorowitsch and G.P. Akilow. *Funktionalanalysis in normierten Räumen*. Akademie Verlag, Berlin, 1964.

78. A. Kaplan and R. Tichatschke. *Stable Methods for Ill–Posed Problems*. Akademie Verlag, Berlin, 1994.

79. J.T. King. A minimal error conjugate gradient method for ill–posed problems. *Journal of Optimization Theory and Applications*, 60:297–304, 1989.

80. K. Kunisch and G. Geymayer. Convergence rates for regularized nonlinear illposed problem. In *[81]*, pages 81–92, 1991.

81. A. Kurzhanski and I. Lasiecka, editors. *Modelling and Inverse Problems of Control for Distributed Parameter Systems*. Springer, Berlin, 1991. Lecture Notes in Control and Information Sciences, Vol. 154.

82. A.K. Louis. Convergence of the conjugate gradient method for compact operators. In H.W. Engl and C.W. Groetsch, editors, *[36]*, pages 177–185, 1987.

83. A.K. Louis. *Inverse und Schlecht Gestellte Probleme*. Teubner, Stuttgart, 1989.

84. W.R. Mann. Mean value methods in iteration. *Proc. Am. Math. Soc.*, 4:506–510, 1953.

85. St. Măruster. Quasi-nonexpansity and two classical methods for solving nonlinear equations. *Proc. Amer. Math. Soc.*, 62:119–123, 1977.

86. S.F. McCormick. An iterative procedure for the solution of constrained nonlinear equations with applications to optimization problems. *Numer. Math.*, 23:371–385, 1975.

87. S.F. McCormick. The methods of Kacmarcz and row orthogonalization for solving linear equations and least squares problems in Hilbert space. *Indiana University Mathematics Journal*, 26:1137–1150, 1977.

88. K.H. Meyn. Solution of underdetermined nonlinear equations by stationary iteration methods. *Numer. Math.*, 42:161–172, 1983.

89. V.A. Morozov. On the solution of functional equations by the method of regularization. *Soviet Math. Dokl.*, 7:414–417, 1966.

90. V.A. Morozov. *Methods for Solving Incorrectly Posed Problems*. Springer Verlag, New York, Berlin, Heidelberg, 1984.

91. V.A. Morozov. *Regularization Methods for Ill-Posed Problems*. CRC Press, Boca Raton, 1993.

92. M. Z. Nashed, editor. *Generalized Inverses and Applications*. Academic Press, New York, 1976.

93. M.Z. Nashed and O. Scherzer. Least squares and bounded variation regularization with nondifferentiable functional. *Num. Funct. Anal. and Optimiz.*, 19:873–901, 1998.

94. F. Natterer. Numerical solution of bilinear inverse problems. *Universtät Münster, Germany*, 1998. preprint.

95. A. Neubauer. Tikhonov regularization for non–linear ill–posed problems: optimal convergence rates and finite–dimensional approximation. *Inverse Probl.*, 5:541–557, 1989.

96. A. Neubauer and O. Scherzer. Finite–dimensional approximation of Tikhonov regularized solutions of non–linear ill–posed problems. *Numer. Funct. Anal. and Optimiz.*, 11:85–99, 1990.

97. A. Neubauer and O. Scherzer. A convergence rate result for a steepest descent method and a minimal error method for the solution of nonlinear ill–posed problems. *Z. Anal. Anwend.*, 14:369–377, 1995.

98. Z. Opial. Weak convergence of the sequence of successive approximations of nonexpansive mappings. *Bull. Amer. Math. Soc.*, 67:591–597, 1967.

99. J.M. Ortega and W.C. Rheinboldt. *Iterative Solution of Nonlinear Equations in Several Variables*. Academic Press, New York, 1970.

100. C. Outlaw and C.W. Groetsch. Averaging iteration in a Banach space. *Bull. Am. Math. Soc.*, 75:430–432, 1969.

101. W.V. Petryshyn and T.E. Williamson. Strong and weak convergence of the sequence of successive approximations for quasi–nonexpansive mappings. *J. Math. Anal. Appl.*, 43:459–497, 1973.

102. L. Qi and X. Chen. A preconditioning proximal Newton method for nondifferentiable convex optimization. *Math. Program.*, 76B:411–429, 1997.

103. J. Qi-Nian. On the iteratively regularized Gauss-Newton method for solving nonlinear ill–posed problems. 1998. Preprint, University of Nanjing, China.

104. R. Ramlau. A modified Landweber-method for inverse problems. *Num. Funct. Anal. Opt.*, 20:79–98, 1999.

105. A. Rieder. A wavelet multilevel method for ill-posed problems stabilized by Tikhonov regularization. *Numer. Math.*, 75:501–522, 1997.

106. A. Rieder. On convergence rates of inexact Newton regularizations. *Preprint, Universität Saarbrücken*, 1998.

107. A. Rieder. On the regularization of nonlinear ill–posed problems via inexact Newton iterations. *Inverse Probl.*, pages 309–327, 1999.

108. P.C. Sabatier (ed.). *Some Topics in Inverse Problems*. World Scientific, Singapore, 1988.

109. P.C. Sabatier (ed.). *Inverse Methods in Action*. Spinger, Berlin, Heidelberg, New York, 1990.

110. O. Scherzer. A posteriori error estimates for nonlinear ill–posed problems. submitted.

111. O. Scherzer. Convergence criteria of iterative methods based on Landweber iteration for solving nonlinear problems. *J. Math. Anal. Appl.*, 194:911–933, 1995.

112. O. Scherzer. A convergence analysis of a method of steepest descent and a two-step algorithm for nonlinear ill-posed problems. *Numer. Funct. Anal. Optimization*, 17:197–214, 1996.

113. O. Scherzer. An iterative multi level algorithm for solving nonlinear ill–posed problems. *Numer. Math.*, 80:579–600, 1998.

114. O. Scherzer. A modified Landweber iteration for solving parameter estimation problems. *Appl. Math. Optimiz.*, 38:45–68, 1998.

115. O. Scherzer. An note on Kacmarz's method for the solution of ill posed problems. *in preparation*, 1999.

116. O. Scherzer, H.W. Engl, and K. Kunisch. Optimal a-posteriori parameter choice for Tikhonov regularization for solving nonlinear ill-posed problems. *SIAM J. Numer. Anal.*, 30:1796–1838, 1993.

117. O. Scherzer and M. Gullikson. An adaptive strategy for updating the damping parameters in an iteratively regularized Gauss-Newton method. *JOTA*, 100, 1999. to appear.

118. E. Schock. Approximate solution of ill-posed equations: Arbitrarily slow convergence vs. superconvergence. In *[52]*, pages 234–243, 1985.

119. T.I. Seidman and C.R. Vogel. Well posedness and convergence of some regularisation methods for non–linear ill posed problems. *Inverse Probl.*, 5:227–238, 1989.

120. V.P. Tanana. *Methods for Solution of Nonlinear Operator Equations.* VSP, Utrecht, 1997.

121. U. Tautenhahn. On the asymptotical regularization of nonlinear ill-posed problems. *Inverse Probl.*, 10:1405–1418, 1994.

122. A. N. Tikhonov and V. Y. Arsenin. *Solutions of Ill-Posed Problems.* John Wiley & Sons, Washington, D.C., 1977. Translation editor: Fritz John.

123. A.N. Tikhonov. Regularization of incorrectly posed problems. *Soviet Math. Dokl.*, 4:1624–1627, 1963.

124. A.N. Tikhonov. Solution of incorrectly formulated problems and the regularization methods. *Soviet Math. Dokl.*, 4:1035–1038, 1963.

125. A.N. Tikhonov, A. Goncharsky, V. Stepanov, and A. Yagola. *Numerical Methods for the Solution of Ill-Posed Problems.* Kluwer, Dordrecht, 1995.

126. A.N. Tikhonov, A.S. Leonov, A.I. Prilepko, I.A. Vasin, V.A. Vatutin, and A.G. Yagola, editors. *Ill-Posed Problems in Natural Sciences.* VSP, Utrecht, 1992.

127. G.M. Vainikko. Error estimates of the successive approximation method for ill-posed problems. *Automat. Remote Control*, 40:356–363, 1980.

128. V.V. Vasin. Iterative methods for the approximate solution of ill-posed problems with a priori informations and their applications. In *[36]*, pages 211–229, 1987.

129. V.V. Vasin. Ill–posed problems and iterative approximation of fixed points of pseudo–contractive mappings. In *[126]*, pages 214–223, 1992.

130. V.V. Vasin. Monotone iterative processes for nonlinear operator equations and their applications to volterra equations. *J. Inverse Ill-Posed Probl.*, 4:331–340, 1996.

131. V.V. Vasin. Monotonic iterative processes for operator equations in semiordered spaces. *Dokl. Math. 54*, 53:487–489, 1996. Translation from Dokl. Akad. Nauk, Ross. Akad. Nauk 349, No.1, 7-9.

132. V.V. Vasin. On the convergence of gradient-type methods for nonlinear equations. *Doklady Mathematics*, 57:173–175, 1998. Translated from Doklady Akademii Nauk Vol. 359 (1998), pp. 7 - 9.

133. V.V. Vasin and A.L. Ageev. *Ill–Posed Problems with A-Priori Information.* VSP, Utrecht, 1995.

134. V.V. Vasin, I.L. Prutkin, and L.Yu Timerkhanova. Retrieval of a three-dimensional relief of geological boundary from gravity data. *Izvestiya, Physics of the Solid Earth*, 32:901–905, 1996.

34

135 K.-W. Wang, B. Yang, J.Q. Sun, K. Seto, K. Nonami, H.-S. Tzou, S.S. Rao, G.R. Tomlinson, B. Yang, H.T. Banks, G.M.L. Gladwell, M. Link, G. Lallement, T.E. Alberts, C.-A. Tan, and Y.Y. Hung, editors. *Proceedings of the 1995 Design Engineering Technical Conferences*. The American Society of Mechanical Engineers, New York, 1995. Vibration Control, Analysis, and Identification, Vol. 3, Part C.

136. J. Weidmann. *Lineare Operatoren in Hilberträumen*. Teubner, Stuttgart, 1976.

Iterative Regularization Techniques in Image Reconstruction

Martin Hanke

Fachbereich Mathematik, Johannes-Gutenberg-Universität Mainz, D–55099 Mainz, Germany.

Abstract. In this survey we review recent developments concerning the efficient iterative regularization of image reconstruction problems in atmospheric imaging. We present a number of preconditioners for the minimization of the corresponding Tikhonov functional, and discuss the alternative of terminating the iteration early, rather than adding a stabilizing term in the Tikhonov functional. The methods are examplified for a (synthetic) model problem.

1 Introduction

Atmospheric turbulences are the reason for severe problems in ground based astronomical imaging. On the passage through the atmosphere, light waves are scattered because of temperature fluctuations both in space and time, which lead to strong aberrations of astronomical images taken by a telescope on the surface of the Earth.

In principle, if a sophisticated model of the scattering process is available, the true image can be reconstructed from the photo by solving the associated inverse problem. Such models, however, are very difficult to derive, because atmospheric turbulences are hard to predict and can currently only be accessed via stochastical processes.

In this survey we shall focus on the inverse problem using a very simple model according to which the observed image, y, is a linear convolution of the true image (a nonnegative function x of two variables) with a certain convolution kernel k,

$$y(\xi,\eta) = \int k(\xi-\xi',\eta-\eta')x(\xi',\eta')\,d(\xi',\eta')\,. \qquad (1)$$

The function k is known as *point spread function*: it is nonnegative and its $\mathcal{L}^1$-norm equals one; this refers to conservation of energy in the imaging process. The model (1) is quite appropriate for a long-time exposure of incoherent light waves. It is based on the assumption that the way a point source in the sky is mapped onto its image point and the neighbouring points on the photo is space-invariant. More sophisticated models also take space dependency into account. Long-time exposures usually lead to rather wide-spread point spread functions and thus to a significant loss of high-frequent and small detail information. As a consequence, imaging models for coherent light waves are currently under development in order to deal with short-time exposures, cf., Roggeman and Welsh [21]. We shall not consider these models in the present survey.

Part of the modeling process is the selection of a realistic point spread function k to be used in (1). Based on stochastic reasoning, simple Gaussian kernels were a common choice for k in the early days of ground-based astronomical imaging, cf., e.g., Lagendijk and Biemond [17]. More recently, a method known as guide star imaging has become popular: This refers to a photo of a bright light source, which can be a known star or a so-called artificial beacon, i.e., the backscatter from a laser beam. According to the space-invariance of the imaging process a guide star image is essentially the convolution of k with a delta distribution, and therefore provides an approximation of the values of k. Such a (simulated) guide star image is shown in Fig. 1.

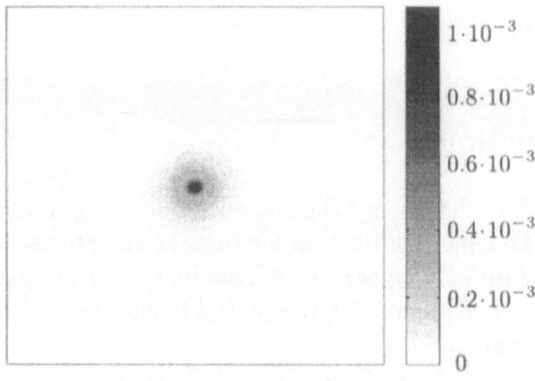

Fig. 1. Point spread function

Another option, which is currently under investigation, consists in reconstructing the point spread function *and* the original image simultaneously. This problem is known as *blind deconvolution* and is formally strongly underdetermined; to improve the setting a series of images can be taken within a short time interval, or additional physics and known a priori constraints can be incorporated to make the problem better determined.

2   Tikhonov Regularization

In the hardware (gray-scale) photos are encoded as two-dimensional arrays of pixel values, i.e., integers between 0 and 255 describing the darkness of the corresponding pixel. For numerical computations the integer assumption is usually dropped, and pixel values are allowed to take any real value, preferrably nonnegative reals. We denote by x and y the vector of all pixel values (in a row-wise ordering) corresponding to the functions x and y of the continuous model (1); given $N \times N$ pixels for each image, the vectors $\boldsymbol{x}$ and $\boldsymbol{y}$ have dimension N^2.

Using the midpoint rule for the discretization of the convolution (1) we then end up with a finite dimensional linear system of equations,

$$T\boldsymbol{x} = \boldsymbol{y}. \tag{2}$$

The matrix T is an $N \times N$ block matrix, with each block being itself an $N \times N$ matrix corresponding to one pair of pixel rows of the two images encoded in $\boldsymbol{x}$ and $\boldsymbol{y}$, respectively. A careful inspection of the quadrature process reveals that the matrix T has additional structure in that, first of all, each of its N^2 blocks is a Toeplitz matrix, i.e., its entries do not change along each individual diagonal, and second, the blocks on each block-diagonal of the entire matrix are all the same. We therefore call T a block Toeplitz matrix with Toeplitz blocks (BTTB). We mention that for current images N ranges from 256, say, up to 1024 and more; already for $N = 256$ this yields a dimension of 65536×65536 for matrix T. This is also the size of our numerical test problem which is used as example throughout this survey: This is a test problem from the Phillips Laboratory at Kirtland US Air Force Base, New Mexico (see [21]). The corresponding point spread function k is the one from Fig. 1; the test image and its blurred photo are plotted in Fig. 2.

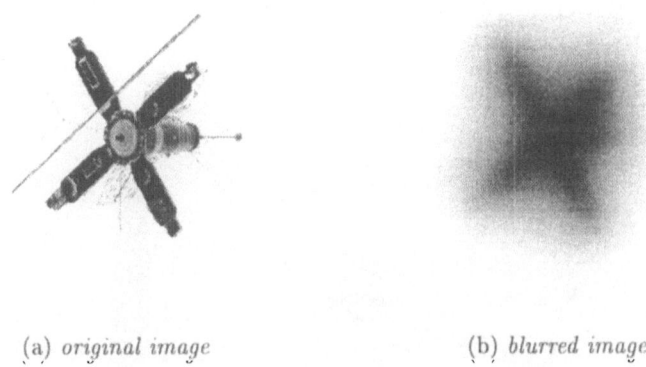

(a) *original image*   (b) *blurred image*

Fig. 2. Model problem.

In many cases of interest, in particular for the aforementioned model problem the matrix T is symmetric or close to a symmetric matrix. A symmetric matrix would correspond to a convolution kernel which is symmetric with respect to the origin, i.e.,

$$k(s,t) = k(-s,-t),$$

an assumption which is likely to hold (at least approximately) if the light waves propagate in a normal direction to the surface of the Earth. Whenever appropri-

38

ate we will limitate our discussion to this symmetric case for simplicity, although
similar considerations apply to the general case.

A symmetric BTTB matrix is completely determined by its first column (just
as a self adjoint convolution operator (1) is determined by the values of the
kernel function $k(\xi, \eta)$ for all nonnegative arguments ξ and η). We denote the
entries of T by $t_{\nu,\mu}$ where the first index counts the block and the second one the
index relative to the first entry of this block; it is convenient to start counting
by zero so that $0 \leq \nu, \mu < N$. In this case it follows for the midpoint quadrature
rule that the $(\nu N + \mu, \nu' N + \mu')$ index of the symmetric BTTB matrix T is given
by

$$t_{|\nu-\nu'|,|\mu-\mu'|} = \Delta\, k\big((\nu - \nu')\Delta, (\mu - \mu')\Delta\big)\,,$$

where Δ is the mesh width.

To analyze the spectrum of T the functions

$$f_N(\phi,\theta) = \sum_{1-N}^{N-1} t_{\nu,\mu}e^{i(\nu\phi+\mu\theta)}\,, \qquad -\pi \leq \phi,\theta \leq \pi\,, \tag{3}$$

play a prominent role. In general, f_N is a smooth real-valued function which is
essentially zero except for a neighborhood of the origin. Moreover, the distribution of the eigenvalues of T is related to the distribution of the values of f_N,
which implies that the spectrum of T usually clusters at the origin. The function
f_N corresponding to the point spread function of Fig. 1 is shown as a logarithmic
gray scale image in Fig. 3; it is obvious that f_N is essentially zero for all angles
ϕ and θ with $|\phi|, |\theta| > \pi/3$.

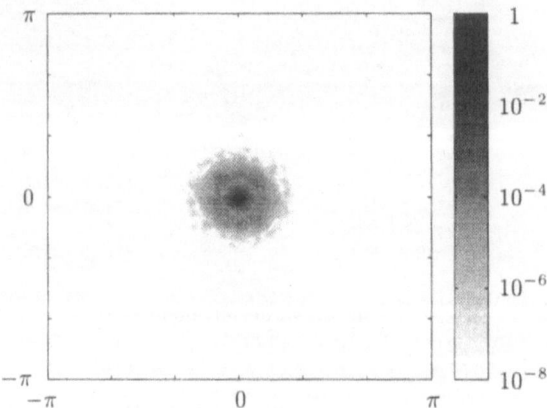

Fig. 3. A logarithmic plot of f_N.

As we have seen, the eigenvalues of T cluster at the origin so that T has a really large condition number in general. As a matter of fact, the solution x of the

linear system (2) is very sensitive to measurement errors in the right-hand side y resulting from the imaging process. To overcome this ill-conditioning regularization techniques have to be employed, among which Tikhonov regularization has outstanding popularity, cf. Groetsch [9]. In Tikhonov regularization, the goal is to minimize $\|y - Tx\|_2$ subject to a constraint on the size or the smoothness of x, i.e., a bound for the norm $\|Lx\|_2$ for some given matrix L. This leads to the minimization problem

$$\|y - Tx\|_2^2 + \alpha \|Lx\|_2^2 \longrightarrow \min . \qquad (4)$$

Here α is some positive parameter, the *regularization parameter*, and the matrix L is often chosen to be the identity matrix, either for simplicity, or for the lack of more sophisticated alternatives. In image restoration, penalty terms $\|Lx\|_2^2$ approximating a total variation functional of x have also received increasing interest recently, cf., e.g., Vogel and Oman [24]. The regularization parameter α can be viewed as an a posteriori tuning parameter: theoretically, decreasing α should give higher resolution, but in practice rather leads to increasingly strong artefacts because of the influence of high-frequent noise in the data; with increasing α such artefacts can be reduced but then the details of the reconstructions get smeared. The optimal balance between the two extremes is a very delicate issue, and the costs for tuning the parameter are so high (usually, the code is restarted for every new value α) that in practice α is chosen a priori, on the basis of preliminary experiments and experience.

Since the above Tikhonov functional is quadratic, the minimization process is equivalent to solving a linear system, namely

$$(T^*T + \alpha L^*L)x = T^*y . \qquad (5)$$

Because of the tremendous size of matrix T this system cannot be solved by direct methods. In the engineering literature it is therefore commonly recommended to replace the doubly Toeplitz matrix T by a doubly circulant one, i.e., a block circulant matrix S with circulant blocks (BCCB), which coincides with T in all central block diagonals and the central diagonals of all blocks. In the important case $L = I$ the resulting linear system

$$(S^*S + \alpha I)\tilde{x} = S^*y \qquad (6)$$

has again a BCCB coefficient matrix and can explicitly be solved with only two 2D-FFTs. Moreover, if the original image x is only nonzero in its inner quarter and the point spread function is sufficiently narrow, then the reconstructions x and $\tilde{x}$ coincide. This assumption is essentially satisfied for the satellite image in Fig. 2 (a) because there is a zero boundary layer of roughly 64 pixels width around the satellite. As a consequence, the reconstruction $\tilde{x}$ obtained from (6) is pretty good for this particular model problem. We refer to Figs. 8 (b) and (c) for the two reconstructions obtained from (5) and (6)*.

---

* The reconstruction for (5) corresponds to the case $L = I$. The optimal regularization parameters have been determined to be $\alpha = 2.3 \cdot 10^{-4}$ and $\alpha = 2.2 \cdot 10^{-4}$ for problems (5) and (6), respectively.

The approximation $T \approx S$ can also be interpreted in terms of the function f_N of (3). In fact, the eigenvalues of a BCCB matrix are given by a 2D-FFT of its first column, and the eigenvectors are the two-dimensional Fourier vectors. By construction of S the eigenvalues $\lambda_{\nu,\mu}$ of S corresponding to the trigonometric monomials

$$p_{\nu,\mu}(\phi,\theta) = e^{i(\nu\phi+\mu\theta)}, \qquad -N/2 < \nu, \mu \leq N/2,$$

are the values of $f_{N/2}$ at equidistant mesh points,

$$\lambda_{\nu,\mu} = f_{N/2}(2\nu\pi/N, 2\mu\pi/N), \qquad -N/2 < \nu, \mu \leq N/2.$$

In particular, high frequent monomials correspond to eigenvalues close to $f_{N/2}(\pm\pi, \pm\pi)$ and are therefore close to zero, cf. Fig. 3.

3   Iterative Minimization of the Tikhonov Functional

3.1   The Conjugate Gradient Iteration

Aside of using the BCCB approximation S instead of T, one can use iterative methods rather than direct methods to minimize (4), most notably by applying the well-known *conjugate gradient method* (CG) to (5). The CG-method is certainly one of the most efficient methods to solve large-scale linear systems of equations with a positive definite coefficient matrix (cf.,e.g., Golub and Van Loan [7]). What is important to mention is that although BTTB matrices cannot be inverted by use of FFTs, a matrix-vector multiplication with T can be implemented using FFTs by imbedding T into a BCCB matrix of four times the size of T. This well-known fact leads to an operation count of roughly 16 2D-FFTs (of length $N \times N$) per iteration of the conjugate gradient method.

It follows that the CG iteration for solving (5) will be a competitive algorithm provided that the number of required iterations is small. Unfortunately, the rate of convergence of the CG iteration depends significantly on the condition number of the linear system (5), which is still pretty large despite the regularizing term: To illustrate this fact we fix without loss of generality $\|T\|_2 = 1$ and $\|x\|_2 = 1$, in which case it is also reasonable to expect $\|y\|_2 \approx 1$; given a small norm $\delta \ll 1$ for the data error in y we quote from [5] that a reasonable regularization parameter α will typically be of the order of δ^2 but not less – although it might be as large as $O(\delta^{2/3})$ in very special circumstances. From this we conclude the bound

$$\mathrm{cond}(T^*T + \alpha I) \leq \frac{1+\alpha}{\alpha} = O(\delta^{-2}),$$

which turns out pretty sharp in practice. Since the spectrum of T^*T is fairly densely distributed in the interval $[0, 1]$ there are no or little additional spectral properties that the CG-method can use with advantage. As a consequence the CG iteration will only be moderately fast in general, requiring about a hundred iterations or so to converge**. This is illustrated in Fig. 4 which shows the iteration history of CG applied to (5) with $L = I$ and with the optimal regularization parameter $\alpha = 2.3 \cdot 10^{-4}$: the left-hand plot (a) shows the norm of the relative residual of (5) versus the iteration count; the solid line in the right-hand side plot (b) corresponds to the relative errors of the reconstructions as compared to the true image; the dashed and dash-dotted lines in plot (b) will be referred to in Sect. 4.

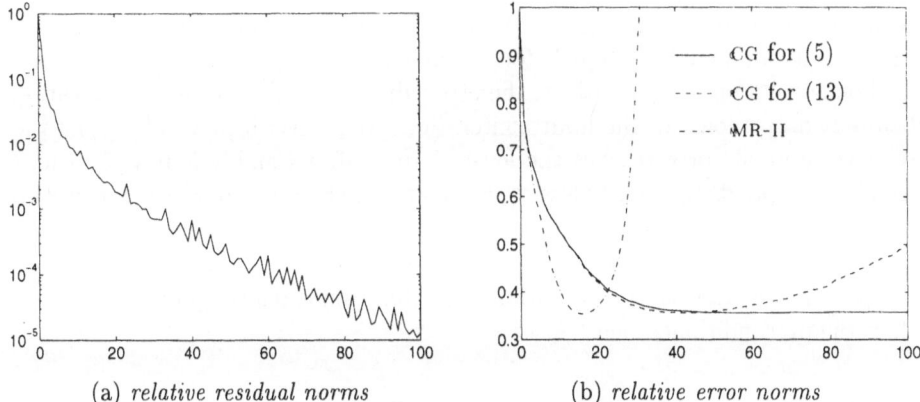

(a) *relative residual norms*   (b) *relative error norms*

Fig. 4. Convergence history: CG applied to problem (5) (solod lines).

3.2   Circulant Preconditioners

On the other hand there are good news in that several preconditioners are meanwhile available for our problem. In fact, whereas the solution of the BCCB system (6) may often not be an acceptable reconstruction in itself, the matrix S provides a useful approximation of T which can be used for preconditioning the system (5). It has been shown by R. Chan and others (see [2] for a survey of these results) that a variety of BCCB matrices C as well as some other structured matrices approximate T in that

$$C - T = R + E,\qquad (7)$$

where R is a matrix of small rank and E is a matrix of small norm (depending on the rank of R). From this follows, cf. Chan, Nagy, and Plemmons [3], that the preconditioned matrix

$$(C^*C + \alpha I)^{-1}(T^*T + \alpha I)\qquad (8)$$

** It must be mentioned, though, that high accuracy of the solution of (5) is typically not an important requirement because the optimal regularization parameter α is not exactly known anyway.

contains relatively few eigenvalues which do not cluster around $\lambda = 1$. This is a situation which is ideally suited for the conjugate gradient method, and which often reduces the number of CG iterations by one order of magnitude.

For our test problem we have used as C in (8) the BCCB matrix which is closest to T in the Frobenius norm (called Level-2 preconditioner in [3]), i.e.,

$$\|C - T\|_F \longrightarrow \min_{\mathrm{BCCB}\ C};$$

the results (see the dashed line in Fig. 5) are somewhat better than using $C = S$ in (8). We observe, cf. Fig. 5 (b), that the preconditioning deteriorates the iteration in the first few steps before (a much more rapid) convergence to the Tikhonov solution sets in; in fact, the error almost doubles in the first iteration before it drops down to the limit at iteration twenty. With preconditioning the relative residual norm reaches the level of 10^{-5} after roughly fourty iterations which corresponds to slightly less than about half the iterations as compared to the unpreconditioned scheme. It should be mentioned that the implementation of this preconditioner requires only two 2D-FFTs, i.e., less than 15% extrawork per iteration, so that the preconditioned CG iteration is really significantly faster than the preconditioned one.

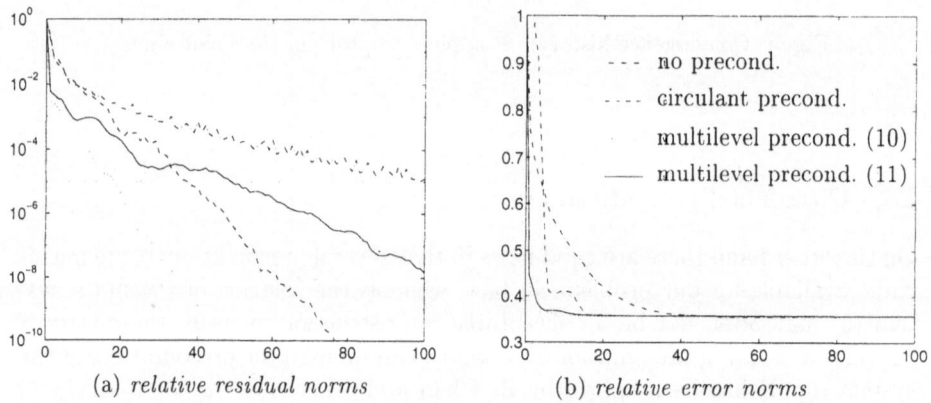

(a) *relative residual norms*   (b) *relative error norms*

Fig. 5. Convergence history of CG for (5) with various preconditioners.

3.3   Multilevel Preconditioners

Another preconditioner has recently been suggested in [23]. It is based on certain multigrid ideas first proposed by Hackbusch [10] for general integral equations of the second kind. For this preconditioner we introduce a second (coarse) grid of 16 × 16 pixels with each pixel representing itself 16 × 16 pixels from the original fine grid; we denote by h and $H = 16h$ the mesh-widths of the fine and coarse grid, respectively. We require the usual intergrid transfer operations, i.e., orthogonal restriction (I_h^H) and associated prolongation operators $(I_H^h = I_h^{H*})$ such that $I_H^h I_h^H$ is the orthogonal projector onto the coarse grid functions. The corresponding projector onto the orthogonal complement is denoted by $Q = I - I_H^h I_h^H$. We use the coarse grid piecewise constant basis functions and the Haar tensor-wavelet basis for the orthogonal complement on the fine grid to represent $A = T^*T + \alpha I$ as a two by two block matrix

$$A \sim \begin{bmatrix} A_{11} & A_{12} \\ A_{21} & A_{22} \end{bmatrix},$$

where, for example, $A_{11} = I_h^H A I_H^h$ is a representation for the restriction A_H of A to the coarse grid (more details are provided in [23]).

After approximating $A_{22} \approx \alpha I$ the preconditioner is defined as the symmetric Gauß-Seidel preconditioner with respect to this block representation/approximation of A. With $M = -\alpha^{-1/2} A_{11}^{-1} A_{12}$ the preconditioned matrix WA has the block representation

$$WA \sim \begin{bmatrix} A_{11}^{-1/2} & M \\ 0 & \alpha^{-1/2} I \end{bmatrix} \begin{bmatrix} A_{11}^{-1/2} & 0 \\ M^* & \alpha^{-1/2} I \end{bmatrix} \begin{bmatrix} A_{11} & A_{12} \\ A_{21} & A_{22} \end{bmatrix}.$$

After some further manipulations we obtain a representation W of the preconditioner in the original fine grid basis, namely

$$W = A_H^\dagger + \frac{1}{\alpha} (A_H^\dagger T^*T - I) Q (T^*T A_H^\dagger - I).$$

Here, $A_H^\dagger$ is defined to be the generalized inverse of the coarse grid blurring matrix, i.e.,

$$A_H^\dagger = I_H^h A_{11}^{-1} I_h^H = I_H^h (I_h^H (T^*T + \alpha I) I_H^h)^{-1} I_h^H. \tag{9}$$

Even for the 16×16 dimensional coarse grid the (exact) construction of $A_{11} \in \mathbb{R}^{N \times N}$ is prohibitive because this would require the amount of work of approximately $N = 256$ iterations of the unpreconditioned CG iteration. Instead, Riley and Vogel [20] approximate A_{11} by a BTTB matrix B_{11} of size 256×256 (the blocks have size 16×16) with a kernel function obtained from a restriction of the original kernel k to the coarse grid. The Cholesky factor of B_{11} can be computed with approximately $2/3\,N^3$ operations prior to the iteration so that matrix vector multiplications with

$$B_H^\dagger = I_H^h B_{11}^{-1} I_h^H$$

using this Cholesky factorization are negligible extrawork. We denote the corresponding preconditioner by

$$W_1 = B_H^\dagger + \frac{1}{\alpha} (B_H^\dagger T^*T - I) Q (T^*T B_H^\dagger - I). \tag{10}$$

As shown in [20] the preconditioner W_1 outperforms the BCCB preconditioners for image restoration problems in terms of iteration counts. However, the preconditioner W_1 as it stands requires two further multiplications with T^*T in each iteration, so that the total amount of work per iteration is *tripled* as compared to the unpreconditioned or the circulant preconditioned iteration. We therefore proceed with a second level of approximation in that we replace T in (10) once again by its optimal BCCB approximation C; by this we finally end up with the preconditioner

$$W_2 = B_H^\dagger + \frac{1}{\alpha}(B_H^\dagger C^* C - I) Q (C^* C B_H^\dagger - I).$$ (11)

The implementation of W_2 takes four 2D-FFTs and four triangular solves with the Cholesky factors of B_{11}. It is therefore only marginally more expensive than the implementation of a circulant preconditioner.

The solid and dotted lines in Fig. 5 illustrate the performance of the two multilevel preconditioners W_1 and W_2 for our model problem. In agreement with [20] we observe that W_1 is most effective in speeding up the asymptotic rate of convergence; on the other hand, the relatively high costs per iteration almost compensate for this advantage. The approximation W_2 of (11) is inferior as far as the asymptotic rate of convergence is concerned, but the initial error history is almost the same with W_1 and W_2 as preconditioner. Because of its low cost, W_2 therefore seems to be the most efficient preconditioner for this particular problem.

In summary, there are quite a few alternative preconditioners for the Tikhonov regularized problem (5). With any of these it should be possible to minimize the Tikhonov functional in a fairly small number of iterations, i.e., in only $O(N^2 \log N)$ operations. But still, even under the assumption that it is possible to solve the regularized problem in only $O(N^2 \log N)$ operations, the aforementioned open problem of choosing an appropriate regularization parameter α remains a crucial issue to deal with. Some attempts to solve this problem can be found in the literature (cf., e.g., [1, 6, 8, 15, 18]) but more research is still necessary.

4 Regularization by Iteration

As an alternative we advocate the possibility of *regularizing by iteration*. Historically, this technique originated with the Landweber iteration, which can be viewed as a variant of the steepest-descent method applied to the least-squares functional $\|y - Tx\|_2^2$. An intuitive understanding of its performance can be summarized as follows: As long as the current approximation x_k is not too close to the true solution, the residual $y - Tx_k$ of the linear system is quite large and the data error in the right-hand side is negligible as compared to the size of the residual. It follows that the negative gradient of the least squares functional, i.e., $T^*(y - Tx_k)$, essentially points to the right descent direction. As the residual shrinks, the data error component becomes increasingly important before it eventually dominates the objective function. This is the turning point at which the iterates start to lose orientation and eventually diverge, a phenomenon entitled *semiconvergence* in the literature.

It should be clear from this introduction that the regularization parameter (the tuning parameter) of the Landweber iteration is the iteration index, which should be chosen in such a way that the the residual and the errors in the data are essentially of the same size: a very nice proof of this conclusion was given by Defrise and de Mol [4]. The choice of an optimal regularization parameter therefore amounts to choosing an appropriate stopping rule, and the stopping rule that we have outlined above is the so-called *discrepancy principle*: Find the smallest nonnegative integer k for which

$$\|y - Tx_k\|_2 \approx \delta. \tag{12}$$

Instead of the Landweber iteration one can again apply the CG-method to the normal equation system

$$T^*Tx = T^*y. \tag{13}$$

When T is symmetric (not necessarily positive definite) then it is also possible to apply a variant of the CG method to the minimization of $\|y - Tx\|_2$ without forming the normal equation system. This latter scheme, called MR-II in [11], has been applied successfully in [12] to our atmospheric imaging model problem after a suitable approximation of the point spread function in Fig. 1 by a symmetric one.

Although much more difficult to establish rigorously, the results for the Landweber iteration extend to these much faster iteration schemes, cf. [5, 11]: It turns out that one can "naively" apply the CG or MR-II iteration, i.e., without any prior regularization as in (5), provided that the iteration is ultimately stopped as soon as the residual first satisfies the discrepancy principle (12).

Interestingly, for the CG method it is *not* the condition number (which is huge) but only the noise level that determines the number of iterations to satisfy this stopping criterion. In fact, as shown by Plato [19] the total number k of CG iterations to meet (12) is always bounded by

$$k = O(\delta^{-1}), \tag{14}$$

and this bound is the sharper the more dense the spectrum of T^*T is. For symmetric and positive definite matrices T the MR-II iteration requires at most $O(\delta^{-1/2})$ iterations.

The iteration histories for (plain) CG and MR-II when applied to our model problem are included in Fig. 4 (b): the dashed line refers to CG, the dash-dotted line shows the performance of MR-II. Note that one iteration of MR-II is only half as expensive as one CG iteration because MR-II does not refer to the normal equation system. From this follows that the sixteen MR-II iterations to find the optimal reconstruction of x require less than a sixth of the total work for the fifty or so iterations of the conjugate gradient method applied to (5) to obtain the same accuracy. Furthermore, the MR-II iteration requires no a posteriori selection of a suitable regularization parameter α. As can be seen from Figs. 8 (b) and (d) the reconstructions with and without prior regularization are very similar.

As compared to (14), there are significantly smaller bounds for the number of iterations to achieve (12) when the eigenvalues of T^*T decay rapidly to zero. One of these improved bounds is stated in [5, Theorem 7.15] and concerns the case when the spectrum of the normal equation operator clusters around $\lambda = 1$ *and* $\lambda = 0$, except for only a few eigenvalues outside these two clusters. Then, as shown in [5], it is this number of outlying eigenvalues which essentially determines the number of CG iterations. This result makes use of two major assumptions:

- The first assumption relates the size of the eigenvalue cluster at the origin to the noise level in the data. The less noise there is, the smaller this cluster should be. In fact, during the early stage of the CG iteration this eigenvalue cluster is pretty much ignored – which is good because the corresponding eigenvector components of the right-hand side can be considered hidden by noise. We therefore call this span of eigenvectors the *noise subspace*. If, for some reason, this eigenvalue cluster is larger, then the CG iteration needs to pick up information from the corresponding eigenvectors. This slows down the convergence in a second stage of the iteration (we call this the *transient phase*) before the iteration eventually diverges.
- The other assumption is more significant: it requires that the true image is essentially a linear combination of the eigenvectors which do *not* belong to the noise subspace, i.e., the eigenvectors corresponding to the eigenvalues near $\lambda = 1$ and the eigenvalues which do not belong to either cluster. We call this the *signal subspace*.

It is instructive to interpret these remarks in the context of the approximating BCCB system

$$Cx = y \,.$$

To this end, we plot in Fig. 6 (left) the eigenvalues of C (smooth line) and the absolute values of the corresponding Fourier coefficients of y versus the eigenvalue count. The right-hand side plot is a zoom onto the interesting part above the noise level. Under the assumption that the noise level in the Fourier coefficients is about the same for all eigenvalues, this plot leads to the dashed line as a first guess on how to separate signal and noise subspace (more sophisticated algorithms for a separation of the two subspaces are given in [12]). Since, as we have seen in the previous section, the very small eigenvalues of C correspond to high-frequent basis images it follows that this distinction between signal and noise agrees well with the intuition that low and high frequencies are associated with signal and noise, respectively.

5 Preconditioning

We stress that the introduction of signal and noise subspace (and their separation) is merely a means for understanding the different stages of the CG iteration. There is no need to explicitly determine appropriate subspaces in order to run the MR-II iteration. This gets different, though, if one is interested in speeding up the iteration by means of preconditioning.

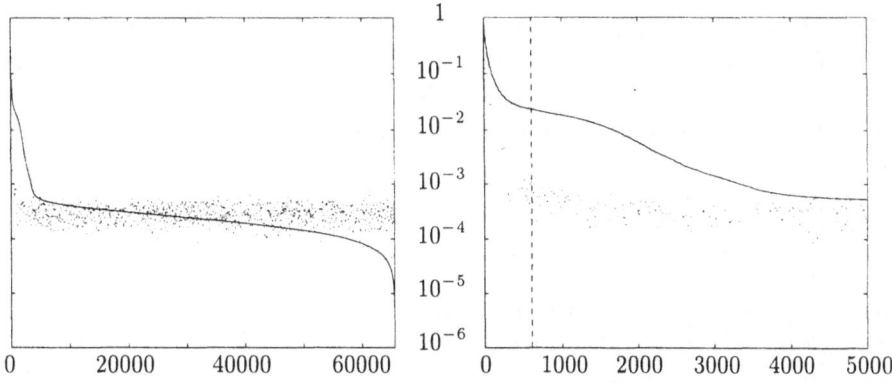

Fig. 6. Eigenvalues of C and Fourier coefficients of y.

To this end we turn back to the aforementioned theorem ([5, Theorem 7.15]) on the number of iterations of the conjugate gradient iteration. The two assumptions of this theorem are satisfied, for example, for the Gerchberg-Papoulis algorithm from signal processing, cf. [13]. This is a method to extend a given band-limited signal from its values in the time interval $[-T, T]$ to all times $t \in \mathbb{R}$. As was shown by Landau, Pollak, and Slepian in a sequel of papers in the early sixties (cf. [22] for a survey), the operator T of the underlying linear equation $Tx = y$ has a spectrum which consists of an eigenvalue cluster at $\lambda = 1$ corresponding to low-frequent eigenfunctions (the signal subspace), and another cluster at $\lambda = 0$ with high-frequent eigenfunctions (the noise subspace). In image deblurring problems, on the other hand, such a clustering will hardly occur; in Fig. 6, for instance, one can see that the signal subspace corresponds to some hundred eigenvalues which are fairly well distributed in an interval $[0.02, 1]$.

A preconditioning of the original problem may change the situation, though. Limiting our discussion once again to symmetric problems and the MR-II iteration only, we refer by preconditioning to a transformation of the linear system $Tx = y$ into

$$M^{1/2} T M^{1/2} z = M^{1/2} y, \qquad x = M^{1/2} z, \tag{15}$$

for some symmetric and positive definite matrix M. While in well-posed problems, the preconditioner is chosen so as to reduce the condition number of (15), the aim in ill-posed problems – according to the above discussion – is to cluster the eigenvalues corresponding to the signal subspace.

In other words, we are searching a matrix M such that $M^{1/2} T M^{1/2}$ has an eigenvalue cluster at $\lambda = 1$ (and possibly another cluster at $\lambda = -1$ if T is indefinite) in the signal subspace, and an eigenvalue cluster at the orgin in the noise subspace. Because of this we restrict our attention to matrices M with

$$M \approx |T|^{-1} \text{ on the signal subspace}, \text{ and}$$
$$M \approx I \qquad \text{on the noise subspace}.$$

Here, $|T| = (T^*T)^{1/2}$ is the modulus of T. Since the Level-2 BCCB approximation C of T is a useful approximation of T throughout, cf. (7), and since the eigenstructure of C is easily available and allows a plausible separation into signal and noise subspaces of 2D Fourier vectors, cf. Fig. 6, the choice

$$M = |C|^{-1} \text{ on the signal subspace, and}$$
$$M = I \qquad \text{on the noise subspace} \tag{16}$$

becomes a natural candidate for M. More precisely, if P_S and P_N are the orthogonal projectors onto the 2D-Fourier vectors corresponding to signal and noise subspaces, respectively, then we take

$$M = P_S|C|^{-1}P_S + P_N.$$

It was shown in [12] that the approximation property (7) implies that

$$M^{1/2}TM^{1/2} - P_{S+} + P_{S-} = E + F + R, \tag{17}$$

where $P_{S\pm}$ are the orthoprojectors onto signal vectors corresponding to positive and negative eigenvalues of C; furthermore, E and F are matrices of small norm, and R is a matrix of small rank. (17) implies that the eigenvalues of $M^{1/2}TM^{1/2}$ cluster around $\lambda = \pm 1$ and at the origin, except for a few outliers.

In this approach the distinction between signal and noise subspace can be based on the magnitude of the eigenvalues $\lambda_{\nu,\mu}$ of C: the signal subspace consists of all eigenvectors of C corresponding to eigenvalues $|\lambda_{\nu,\mu}| > \tau$ (low frequencies, cf. Fig. 3); the noise subspace is the complementary space.

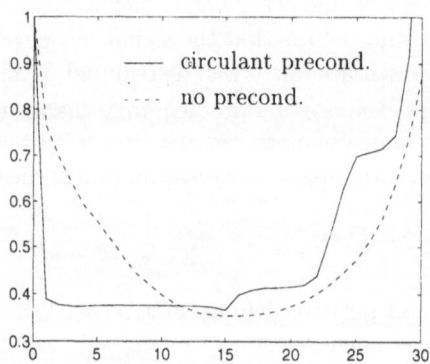

Fig. 7. Convergence history of MR-II with/without preconditioner.

For the performance of this preconditioner (with $\tau = 0.0137$) we refer to Fig. 7. Note that the error of the preconditioned MR-II iteration drops rapidly (in three iterations) down to a level which is only marginally above the best possible error attained in iteration fifteen. Still, this reconstruction (shown in Fig. 8 (e)) is slightly worse than what is obtained in sixteen iterations without preconditioner. The reason is that it is not possible to *completely* separate noise and signal in the preconditioning process: Some information in the right-hand side must have been destroyed by the preconditioner.

Better results might be achieved with a different choice of the tolerance parameter τ that is used to separate noise and signal subspace. In this example τ has been selected with the L-curve criterion as described in detail in [12]. This can be considered as a black box criterion for constructing the preconditioner and, as every black box criterion, might not be the most sophisticated.

The above choice for M is not the only possible, although there are certainly not many strategies for preconditioning T such that all aforementioned requirements on M are satisfied. For example, in [12] the preconditioner M was chosen as a modification of the BCCB extension S of T and not as a modification of the optimal BCCB approximation C. Meanwhile, other useful choices for M have been suggested, cf. [13, 16].

6 Summary

We have outlined a number of iteration methods to regularize 2D convolution problems. Our presentation is focusing on a particular model problem arising in astronomical imaging. Those images are affected by the scattering of light due to atmospheric turbulences on their passage to the surface of the Earth.

The methods that we have presented differ in the way regularization is incorporated into the scheme. One option is to first regularize (e.g., by using Tikhonov regularization) before starting the conjugate gradient method to compute the regularized approximation; another option consists in using CG for the 'plain' original problem, with early termination of the iteration to incorporate a different type of regularization. Whereas the former approach requires an a posteriori selection of the regularization parameter (and thus, in principle, multiple restarts), the latter approach appeals because the optimal iterate can be selected interactively, e.g., by monitoring the iterates on the screen.

Either iterative scheme can be accelerated by appropriate preconditioning, but this is more subtle in the non-regularized case. A couple of preconditioners and their performance on a model problem have been investigated above. For a comparison of the various algorithms we refer to Fig. 8 for the corresponding reconstructions of the satellite image. In these images negative pixel values have been set to zero, and the gray levels in all images are the same as in the original image (top left); error numbers refer to the Euclidean norm, which — as is well known — may not to be the optimal measure for a comparison of images.

We remark that for the regularized problem (5) the preconditioner (if any) does not affect the final reconstruction (Fig. 8 b) but only the costs for its computation: reconstruction (c) is the corresponding solution of problem (6) after replacing the BTTB matrix T by its BCCB approximation S. For the iterative schemes based on the plain, unregularized equation (2) the situation is somewhat different in that the quality of the reconstruction *does* depend on the preconditioner. This can be seen by comparing the reconstructions (e) and (d) ob-

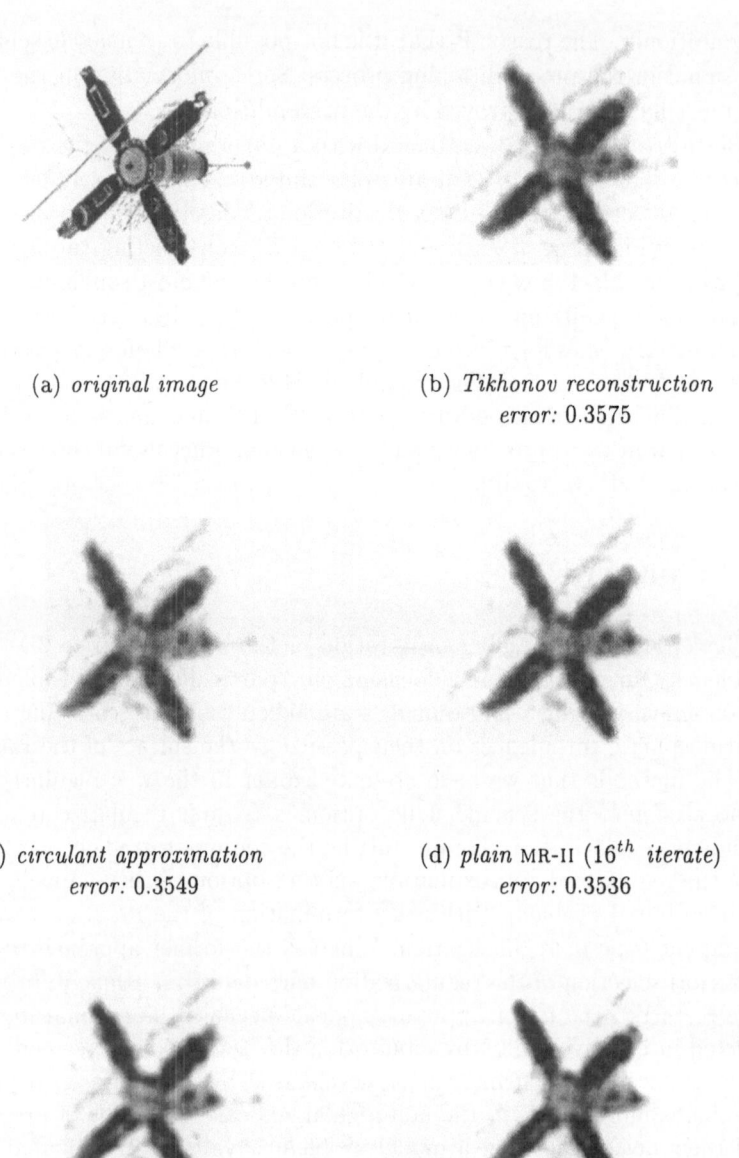

(a) *original image*

(b) *Tikhonov reconstruction*
error: 0.3575

(c) *circulant approximation*
error: 0.3549

(d) *plain* MR-II (16th *iterate*)
error: 0.3536

(e) *prec.* MR-II (3rd *iterate*)
error: 0.3736

(f) *nonneg. Tikhonov reconstruction*
error: 0.3340

Fig. 8. Various reconstructions.

tained with MR-II, with and without preconditioner: The preconditioner reduces the computational costs, but the reconstruction is slightly inferior. It should be emphasized that those two reconstructions use a modified, symmetrized point spread function because the original one from Fig. 1 is not symmetric. Nonetheless, the reconstruction obtained after 44 CG iterations using the normal equation system (13) without preconditioner and with the original point spread function (compare the dashed line in Fig. 4 (b)) is not much different from the image shown in Fig. 8 (d).

It should be noted that the reconstructions in (b) and (d), and to some minor extent also reconstruction (e) reveal distinct artefacts near the edges of the image. Similar artefacts are missing for the circulant reconstruction (c) because the true image is zero near its boundary; if this were not the case the circulant reconstruction would probably suffer much stronger under artificial boundary artefacts.

We finally mention that in astronomical imaging it is fairly common to incorporate known a priori constraints on the reconstruction, e.g., nonnegativity. In the reconstructions (b) through (e) the nonnegativity constraint only enters via the plot routine where all negative entries are set to zero. A more sophisticated use of this constraint is very difficult without losing the advantages of the CG iteration.

Still, to illustrate its (potential) advantage we also computed the solution of the nonnegatively constrained Tikhonov minimization problem (4), i.e.,

$$\text{minimize} \quad \|y - Tx\|_2^2 + \alpha \, \|x\|_2^2 \qquad \text{subject to } x \geq 0 \, .$$

The corresponding image is shown in Fig. 8 (f); it differs distinctively from the other reconstructions. While the other reconstructions suffer from speckles in those areas where they should be identically zero, the constrained reconstruction (f) does not exhibit such artefacts. This may also be the reason why the Euclidean error of the constrained reconstruction is somewhat smaller than the errors of the other reconstructions although it is not necessarily better from a visual point of view: for example, the details of the antenna of the satellite are more smeared than in the unconstrained reconstructions.

References

1. A. BJÖRCK, *A bidiagonalization algorithm for solving large and sparse ill-posed systems of linear equations*, BIT, 28 (1988), pp. 659-670.
2. R. CHAN AND M. K. NG, *Conjugate gradient methods for Toeplitz systems*, SIAM Rev., 38 (1996), pp. 427-482.
3. R. H. Chan, J. G. Nagy and R. J. Plemmons, *FFT-based preconditioners for Toeplitz-block least squares problems*, SIAM J. Numer. Anal., 30 (1993), pp. 1740–1768.
4. M. DEFRISE AND C. DE MOL, *A note on stopping rules for iterative regularization methods and filtered SVD*, in: P. C. Sabatier, ed., Inverse Problems: An Interdisciplinary Study, Academic Press, London, Orlando, San Diego, New York, 1987, 261–268.

5. H. W. ENGL, M. HANKE, AND A. NEUBAUER, *Regularization of Inverse Problems*, Kluwer, Dordrecht, 1996.

6. A. FROMMER AND P. MAASS, *Fast CG-based methods for Tikhonov regularization*, SIAM J. Sci. Comput. (1999), to appear.

7. G. H. GOLUB AND C. F. VAN LOAN, *Matrix Computations*, 3rd. ed., John Hopkins Univ. Press, Baltimore, 1996.

8. G. H. GOLUB AND U. VON MATT, *Generalized Cross-Validation for large-scale problems*, J. Comput. Graph. Statist., 6 (1997), pp. 1-34.

9. C. W. GROETSCH, *The Theory of Tikhonov Regularization for Fredholm Equations of the First Kind*, Pitman, Boston, 1984.

10. W. HACKBUSCH, *Integral Equations. Theory and Numerical Treatment*, Birkhäuser Verlag, Basel, 1995.

11. M. HANKE, *Conjugate Gradient Type Methods for Ill-Posed Problems*, Longman Scientific & Technical, Harlow, Essex, 1995.

12. M. HANKE AND J. G. NAGY, *Restoration of atmospherically blurred images by symmetric indefinite conjugate gradient techniques*, Inverse Problems, 12 (1996), pp. 157-173.

13. ——, *Inverse Toeplitz preconditioners for ill-posed problems*, Linear Algebra Appl., 284 (1998), pp. 137-156.

14. M. HANKE, J. G. NAGY, AND R. J. PLEMMONS, *Preconditioned iterative regularization*, in *Numerical Linear Algebra*, L. Reichel, A. Ruttan, and R. S. Varga, eds., de Gruyter, Berlin, 1993, pp. 141-163.

15. L. KAUFMAN AND A. NEUMAIER, *Regularization of ill-posed problems by envelope guided conjugate gradients*, J. Comput. Graph. Statist., 6 (1997), pp. 451-463.

16. M. E. KILMER, *Cauchy-like preconditioners for 2-dimensional ill-posed problems*, manuscript, 1997.

17. R. L. LAGENDIJK AND J. BIEMOND, *Iterative Identification and Restoration of Images*, Kluwer, Boston, 1991.

18. D. P. O'LEARY AND J. A. SIMMONS, *A bidiagonalization-regularization procedure for large scale discretizations of ill-posed problems*, SIAM J. Sci. Statist. Comput., 2 (1981), pp. 474-489.

19. R. PLATO, *Die Effizienz des Diskrepanzverfahrens für Verfahren vom Typ der konjugierten Gradienten*, in: E. Schock, ed., Beiträge zur Angewandten Analysis und Informatik, Shaker Verlag, Aachen, 1994, pp. 288-297. In German.

20. K. RILEY AND C. R. VOGEL, *Preconditioners for linear systems arising in image reconstruction*, in: F. T. Luk, ed., Advanced Signal Processing Algorithms, Architectures, and Implementations VIII, Proceedings of SPIE Vol. 3461 (1998), pp. 372-380.

21. M. C. ROGGEMANN AND B. WELSH, *Imaging Through Turbulence*, CRC Press, Boca Raton, Florida, 1996.

22. D. SLEPIAN, *Some comments on Fourier analysis, uncertainty and modelling*, SIAM Rev., 25 (1985), pp. 379-393.

23. C. R. VOGEL AND M. HANKE, *Two-level preconditioners for regularized inverse problems II: Implementation and numerical results*, manuscript, 1998.

24. C. R. VOGEL AND M. E. OMAN, *Fast, robust total variation-based reconstruction of noisy, blurred images*, IEEE Trans. Image Proc., 7 (1998), pp. 813-824.

A Survey of Regularization Methods for First-Kind Volterra Equations

Patricia K. Lamm

Mathematics Dept., Michigan State University, E. Lansing, MI 48824-1027 USA
lamm@math.msu.edu
WWW home page: http://www.mth.msu.edu/~lamm

Abstract. We survey continuous and discrete regularization methods for first-kind Volterra problems with continuous kernels. Classical regularization methods tend to destroy the non-anticipatory (or causal) nature of the original Volterra problem because such methods typically rely on computation of the Volterra adjoint operator, an anticipatory operator. In this survey we pay special attention to particular regularization methods, both classical and nontraditional, which tend to retain the Volterra structure of the original problem. Our attention will primarily be focused on linear problems, although extensions of methods to nonlinear and integro-operator Volterra equations are mentioned when known.

1 Introduction

Consider the Volterra equation of first kind

$$\int_0^t k(t,s)u(s)\,ds = f(t), \quad t \in [0,T], \tag{1}$$

where the kernel k is a continuous function on $[0,T] \times [0,T]$. If k is a convolution kernel, then $k(t,s) = \kappa(t-s)$ for some continuous function κ on $[0,T]$.

We will assume throughout that the data f is such that there exists a unique solution $\bar{u} \in U \equiv L_2(0,T)$ of equation (1), and, in particular, we require that $f(0) = 0$. For simplicity we will consider all quantities to be real-valued. Additional regularity on f, k, and $\bar{u}$ may be required in the results which follow.

It will often be useful to write equation (1) in the form

$$Au = f, \tag{2}$$

where the operator A is defined for $u \in U$ by

$$Au(t) := \int_0^t k(t,s)u(s)\,ds, \quad \text{a.e. } t \in [0,T]. \tag{3}$$

Then $A \in \mathcal{L}(U)$, the space of continuous linear operators from U to U.

If the kernel k is non-degenerate, then the range $\mathcal{R}(A)$ of A is non-closed in U. This means that equation (1) is *ill-posed*, which has serious implications in the usual case where we only have available an approximation $f^\delta \in U$ of f; here f^δ satisfies $\|f - f^\delta\| \leq \delta$ for some $\delta > 0$, where throughout $\|\cdot\|$ will denote the usual norm on $U = L_2(0,T)$. The ill-posedness means that the solution u^δ of $Au = f^\delta$ (when such a solution exists) may be arbitrarily far from the solution $\bar{u}$ of the unperturbed problem (1). Therefore, some kind of regularization procedure will be needed to solve the problem in the case of perturbed data f^δ.

1.1   Degree of Ill-Posedness of First-Kind Volterra Problems

For sufficiently smooth f and k, we may differentiate equation (1) with respect to t to obtain

$$k(t,t)u(t) + \int_0^t \frac{\partial}{\partial t}k(t,s)u(s)\,ds = f'(t), \quad t \in [0,T]. \tag{4}$$

If $k(t,t) \neq 0$, for $t \in [0,T]$, division of equation (4) by $k(t,t)$ yields a standard Volterra equation of the second kind which is known to be a *well-posed* problem. In particular, the solution of equation (4) depends continuously on the "data" in (4), which (for the new equation) is the function f'.

If $k(t,t) = 0$, $t \in [0,T]$, then we may repeat the process by differentiating the equation once again, this time resulting in a second-kind Volterra equation (with "data" f'') provided $\partial k(t,t)/\partial t \neq 0$ for $t \in [0,T]$. We generalize this idea with the following where, without loss of generality, we will normalize k.

Definition 1. *We will say that the Volterra operator A is ν-smoothing for integer $\nu \geq 1$ if the kernel k is such that $\left(\partial^\ell k/\partial t^\ell\right)(t,t) = 0$, for $0 \leq t \leq T$ and $\ell = 0,\ldots,\nu - 2$, and such that $\left(\partial^{\nu-1}k/\partial t^{\nu-1}\right)(t,t) = 1$, for $0 \leq t \leq T$, with $\partial^\nu k/\partial t^\nu$ continuous on $[0,T] \times [0,T]$.*

We will say that the Volterra problem $Au = f$ is a ν-smoothing problem if the operator A is a ν-smoothing operator and $f \in C^\nu[0,T]$.

Thus, if (1) is a ν-smoothing problem, we may differentiate equation (1) ν times to obtain

$$u(t) + \int_0^t \frac{\partial^\nu}{\partial t^\nu}k(t,s)u(s)\,ds = f^{(\nu)}(t), \quad t \in [0,T], \tag{5}$$

an equivalent problem to equation (1) in this case (provided $f(0) = f'(0) = \cdots = f^{(\nu-1)}(0) = 0$).

Definition 2. *We will say that the Volterra operator A is* infinitely-smoothing *if $\left(\partial^\ell k/\partial t^\ell\right)(t,t) = 0$, for $0 \leq t \leq T$ and all $\ell = 0,1,2,\ldots$.*

We will say that the Volterra equation $Au = f$ is an infinitely-smoothing *problem if the operator A is a infinitely-smoothing operator.*

Of course, not all equations of the form (1) fall into precisely one of the above classes of problems. Nevertheless, these definitions serve to characterize a wide class of Volterra problems and begin a discussion on the "degree" of ill-posedness of particular first-kind problems.

Among the class of problems defined above, a *one-smoothing problem* is the least ill-posed since it requires only one differentiation of the data (which, in general, is an ill-posed operation itself since perturbations f^δ of f need not be differentiable). In fact, the canonical one-smoothing convolution problem is the "differentiation problem" $u = f'$, or the problem of finding u solving the equation

$$\int_0^t u(s)\, ds = f(t), \quad t \in [0, T]. \tag{6}$$

for sufficiently smooth f. The jth singular value σ_j of the operator A in this case satisfies $\sigma_j = \mathcal{O}(1/j)$ as $j \to \infty$.

Similarly, the canonical ν-smoothing convolution problem is associated with the convolution kernel $\kappa(t) = t^{\nu-1}/(\nu - 1)!$, for which the associated Volterra operator has singular values $\sigma_j = \mathcal{O}(1/j^\nu)$ as $j \to \infty$. Thus, the degree of ill-posedness of ν-smoothing problems increases with increasing ν. See [4] for general information about degree of ill-posedness. See also [5,6] for a discussion of singular values for ν-smoothing Volterra operators and the way in which these singular values are representative (asymptotically) of a class of more general ν-smoothing problems of convolution type.

A classic example of an infinitely-smoothing problem is the inverse heat conduction problem (IHCP), or sideways heat equation, one formulation of which is as follows. Given an insulated semi-infinite bar on the non-negative x-axis, an unknown heat source $u = u(t)$ is applied to the end of the bar at $x = 0$. The goal is to determine the value of u, given measurements $f = f(t)$, $0 \le t \le T$, of the temperature at position $x = 1$ of the bar. The unknown heat source u is the solution of the first-kind equation (1), where the kernel k is of convolution type, $k(t, s) = \kappa(t - s)$, with

$$\kappa(t) = \frac{1}{2\sqrt{\pi}\, t^{3/2}} e^{-1/4t}, \quad t \in [0, T].$$

Because we can rarely obtain f exactly, the solution of the IHCP with perturbed data f^δ is a severely ill-posed problem.

Remark 1. A number of the regularization methods we describe below will be suitable for application to the IHCP. However, the specialized nature of the problem also suggests other particularly effective methods, some taking advantage of its equivalent formulation as an partial differential equation problem with overspecified data. As our focus here will be on methods applicable to more general Volterra equations of the form (1), a discussion of methods specific to the IHCP is beyond the scope of this paper. Instead, the reader is encouraged to consult surveys by Beck, Blackwell, and St. Clair, Jr. [16], Eldén [17] (as well as other papers in the collection [19]), Murio, [81], Alifanov [15], Hào and

Reinhardt [22], Kurpisz and Nowak [25], and the many references found therein. More recent contributions (not likely to be found in the above) include, for example, wavelet-based methods and analyses [27, 31, 32], updates on mollification methods [28–30, 35, 36], as well as the recent papers [18, 20, 21, 23, 24, 26, 33, 34].

In this paper we survey continuous and discrete regularization methods for first-kind Volterra problems of the form of (1). In particular, we pay specific attention to regularization methods which tend to retain the Volterra structure of the original problem. We have tried to pick methods which are especially representative of this idea and which also have a substantial theoretical basis. Certainly there will be errors in omission, particularly so since there is a large body of untranslated Russian literature on this subject.

In addition, although we will primarily be concerned with the linear problem, some of what will be said below also applies to nonlinear equations of the form

$$\int_0^t k(t, s, u(s)) \, ds = f(t), \quad t \in [0, T], \tag{7}$$

and to Volterra operator equations of the form (1) where in this case $u(t) \in H$, for some Hilbert space H, and $k(t, s)$ is a two-parameter family of continuous operators from H to H (see, for example, the discussion of Volterra integro-operator equations in Section 6.8 of [9]).

Finally, a discussion of first-kind Volterra problems with weakly singular kernels is beyond the scope of this paper. This is an important class of problems which includes the Abel integral equation as a special case (and for which the Abel integral operator might be considered "half-smoothing" in the terminology of this section). Although a number of the methods we discuss are applicable to the Abel equation, we will not in general attempt a comprehensive study of regularization methods for this problem.

2   Classical Regularization and Volterra Problems

The classical theory of regularization is well-developed for linear ill-posed problems. For example, given the equation $Au = f$ on a Hilbert space U, with $A \in \mathcal{L}(U)$, and given a perturbation f^δ of f, the method of *Tikhonov regularization* determines u_α^δ solving

$$\min_{u \in U} \|Au - f^\delta\|^2 + \alpha \|u\|^2,$$

or, equivalently, as the solution of the normal equations on U,

$$(A^*A + \alpha I)u = f^\delta. \tag{8}$$

where $A^* \in \mathcal{L}(U)$ is the (Hilbert) adjoint operator associated with A. Standard Tikhonov regularization theory (which is applicable to first-kind Volterra problems) gives well-known conditions on the selection of $\alpha = \alpha(\delta)$ so that $u_{\alpha(\delta)}^\delta \to \bar{u}$

in U as $\delta \to 0$. *A posteriori* discrepancy principles facilitate the selection of α for a particular perturbation f^δ and given value of δ. For more information on these topics, see, for example, [4, 8]. We note that the first application of Tikhonov regularization specifically to general first-kind Volterra problems was evidently due to Schmaedeke in 1968 [14].

2.1 Regularization Methods of "Volterra Type"

An observation relevant to the objectives of this paper is the following. The original Volterra operator A in (3) is *non-anticipatory*, or *causal*; as a consequence, if we wish to determine the values of the solution $\bar{u}$ of the original Volterra problem (1) on the interval $[0, \tau]$ (for any $\tau > 0$), we need only make use of the values of data f on the same interval. In contrast, the adjoint operator A^* of the Volterra operator is given by

$$A^* u(t) = \int_t^T k(s, t) u(s)\, ds, \quad \text{a.e. } t \in [0, T], \tag{9}$$

which is an *anticipatory* operator. So if, for any $\tau > 0$, we wish to determine the values of the solution u_α^δ of the regularized equation (8) on the interval $[0, \tau]$, we will need to make use of both past and future values of the data f^δ, i.e., we require knowledge of f^δ on all of $[0, T]$.

This point is especially clear when the two equations (1) and (8) are discretized. Typical numerical realizations of the Volterra operator A lead to a lower-triangular (or nearly lower-triangular) matrix $\mathbf{A}^N$, so that the solution of a discretization of (1) may be handled by efficient, sequential methods (often in near real time). In contrast, $(\mathbf{A}^N)^{\mathsf{T}} \mathbf{A}^N$ is typically a full matrix and, in general, more poorly conditioned than the original matrix $\mathbf{A}^N$.

Like Tikhonov regularization, numerous classical regularization methods are based on the computation of $g_\alpha(A^* A)$, where g_α is appropriately defined [4]. As a consequence, such methods do not generally retain the Volterra structure of the original problem. In this paper we will focus on regularization methods which specifically preserve the non-anticipatory nature of the original problem, referring to such approaches as *regularization methods of "Volterra type"*. Where possible, we'll indicate convergence results and state whether rates of convergence are known to be order-optimal (see, for example, [4]).

There is, unfortunately, a price associated with limiting our discussion to regularization methods which avoid use of the adjoint operator A^*. While methods based on the operator $g_\alpha(A^* A)$ are generally associated with well-developed convergence theories (because such theories can make use of the spectral properties of the operator $A^* A$), the same is not true in general for methods based on the Volterra operator A. Quite often this means that theoretical results for regularization methods of "Volterra type" are limited to one-smoothing problems, or to ν-smoothing problems for ν small; other methods may require the assumption that A have special properties (such as the assumption that A is accretive). These are often only theoretical limitations, however, and do not in general mean that the method may not be applicable to a larger class of Volterra problems.

3 Continuous Regularization Methods of "Volterra Type"

3.1 A Singular Perturbation Approach

The early theoretical development of a singular perturbation approach for regularizing first-kind Volterra problems is generally attributed to Sergeev [53] and Denisov [41] in the early 1970's, following the ideas of Lavrent'ev [51]. For this reason, the method is often referred to as *Lavrent'ev's classical method*, or the *small parameter method*. The ideas have seen numerous extensions to vector, nonlinear, integro-operator, and other types of Volterra equations, with contributions made by Asanov, Imanaliev, Imomnazarov, and Magnickiĭ, [39, 44–47, 52], to mention just a few of the names important in the historical development of this approach. More recent applications of this method may be found in [40, 42, 43, 48, 74].

As motivation for the method, we recall from Section 2 that most reasonable finite-dimensional approximations of equation (1) lead to a linear system governed by a lower-triangular matrix $\mathbf{A}^N$. Generally, such a matrix has very small entries along the diagonal (due to the ill-posedness of the original problem), and thus a natural way to stabilize such a system would be to augment the values on the diagonal. In the infinite-dimensional setting, the analog of this process is to add a term of the form $\alpha u(t)$ to the values of $Au(t)$, for $\alpha > 0$ small. Thus, when noisy data f^δ is used in place of f, the approach is to consider a perturbed version of equation (2), namely,

$$(\alpha I + A)u = f^\delta, \tag{10}$$

where I is the identity operator on U. That is, we consider

$$\alpha u(t) + \int_0^t k(t, s)u(s)\, ds = f^\delta(t), \quad t \in [0, T], \tag{11}$$

which is a (well-posed) second-kind Volterra equation and, as such, has a unique solution u_α^δ depending continuously on data.

The regularized convergence theory for this problem is well-understood in the case of one-smoothing kernels. For example, the following result is obtained after making a slight variation in the theoretical arguments found in Section 3.4 of [1].

Theorem 1. [1] **(Lavrent'ev's classical method)** *Let $\bar{u} \in C[0, T]$ be the solution of the original problem (1) associated with data f, and assume that (1) is a one-smoothing problem. Let u_α^δ be the solution of (11) associated with data f^δ, where $|f(t) - f^\delta(t)| < \delta$, $t \in [0, T]$.*

Then if $\alpha = \alpha(\delta)$ is selected satisfying $\delta/\alpha(\delta) \to 0$ as $\delta \to 0$, it follows that

$$u^\delta_{\alpha(\delta)} \to \bar{u} \quad \text{as} \quad \delta \to 0,$$

uniformly on $[0, T]$ provided $\bar{u}(0) = 0$; the convergence is uniform on $[a, T]$, for any $a > 0$, in the case of $\bar{u}(0) \neq 0$.

The lack of uniform convergence near $t = 0$ in the case of $\bar{u}(0) \neq 0$ is unfortunate, but not surprising if one observes that equation (11) gives that $u^\delta_\alpha(0) = f^\delta(0)/\alpha = (f^\delta(0) - f(0))/\alpha$, so that the selection of $\alpha = \alpha(\delta)$ as prescribed in Theorem 1 guarantees $u^\delta_\alpha(0) \to 0 \neq \bar{u}(0)$ as $\delta \to 0$. Thus, in the case of $\bar{u}(0) \neq 0$, there is a boundary layer near $t = 0$ where the solution u^δ_α of (11) must exhibit rapid change for α small [37]. This means that if (11) is to form the basis of a regularization method for the stable solution of (1), one must employ numerical methods for *singularly perturbed* Volterra equations, a class of "stiff" Volterra second-kind equations that has received little attention by the numerical analysis community to date [49]. (Asymptotic analysis of such singularly perturbed equations may be found, for example, in [37, 38, 49], and the papers cited therein.)

To correct for the difficulty of the rapidly-varying nature of the solution u^δ_α of (11), several authors have suggested a modification of the regularization equation as follows (e.g., see [44, 52], and the references therein):

$$\alpha[u(t) - \bar{u}(0)] + \int_0^t k(t, s)u(s)\, ds = f^\delta(t), \quad t \in [0, T]. \tag{12}$$

Although this formulation avoids the singularly perturbed nature of (11) when $\bar{u}(0) \neq 0$, the main drawback is that one must know the value of the true solution $\bar{u}$ at $t = 0$. We note that for one-smoothing problems, equation (4) gives that $\bar{u}(0) = f'(0)$, so we must either have knowledge of the exact value of f' at 0, or else we must perform a differentiation of the perturbed data f^δ (a process itself requiring regularization). In a related paper by Sergeev [53], similar ideas were put forward for general ν-smoothing kernels, but, using the approach taken there, one must have knowledge of $\bar{u}(0)$ as well as higher order derivatives of $\bar{u}(t)$ at $t = 0$ in order to avoid facing the boundary layer phenomenon [50].

We note that although Theorem 1 tells how to select $\alpha = \alpha(\delta)$ asymptotically as $\delta \to 0$, it does not provide a principle for selecting α when we are only given one value of $\delta > 0$ and a particular perturbation f^δ of f. In fact, discrepancy principles do exist for this method in the case of a particular class of Volterra problems, although they are not the classical (Morozov) discrepancy principles. We will postpone a discussion of these modified principles until Section 3.3 where Lavrent'ev's m-times iterated method (a generalization of Lavrent'ev's classical method) is considered; see, in particular, Remark 2 of that section.

3.2 "Local Regularization" Methods

Local regularization methods for Volterra problems share common features with the singular perturbation approach described above in that a second-kind equation similar to (11) is constructed; here, however, equation (11) takes the special form

$$\alpha(t;r)u(t) + \int_0^t \tilde{k}(t,s;r)u(s)\,ds = \tilde{f}^\delta(t;r), \quad t \in [0,T], \tag{13}$$

where $\alpha(\cdot;r)$ is now a prescribed function involving a (new) regularization parameter $r > 0$, and $\tilde{k}(\cdot;r)$, $\tilde{f}(\cdot;r)$ are given r-dependent approximations of k and f^δ (all of which will be defined shortly).

To motivate equation (13), we let $r > 0$ be a small fixed constant and assume that equation (1) holds on an extended domain $[0, T + r]$ (which can always be accomplished by simply decreasing the size of T). Then $\bar{u}$ satisfies

$$\int_0^{t+\rho} k(t+\rho,s)u(s)\,ds = f(t+\rho), \quad t \in [0,T], \ \rho \in [0,r],$$

or, splitting the integral at t and making a change of integration variable,

$$\int_0^t k(t+\rho,s)u(s)\,ds + \int_0^\rho k(t+\rho,t+s)u(t+s)\,ds \tag{14}$$
$$= f(t+\rho), \quad t \in [0,T], \ \rho \in [0,r].$$

For each $t \in [0,T]$, the ρ variable serves to advance the equation slightly into the future. One way to consolidate this future information is to integrate both sides of the equation with respect to $\rho \in [0,r]$, i.e.,

$$\int_0^t \int_0^r k(t+\rho,s)\,d\rho\,u(s)\,ds + \int_0^r \int_0^\rho k(t+\rho,t+s)u(t+s)\,ds\,d\rho \tag{15}$$
$$= \int_0^r f(t+\rho)\,d\rho, \quad t \in [0,T],$$

where we have made a change of order of integration in the first integral above.

We note that the true solution $\bar{u}$ satisfies equation (15); when f is replaced by f^δ, a regularized form of this equation is needed and we obtain this new equation by replacing $u(t+s)$ by $u(t)$ in the second integral term. That is, for fixed t, it is as if u is (temporarily) assumed to be constant on the small *local* interval $[t, t+r]$; the length r of this local interval becomes the regularization parameter. The "local regularization equation" which results is given by (13), where in that equation we make the definitions, for $0 \le s \le t \le T$,

$$\alpha(t;r) := \int_0^r \int_0^\rho k(t+\rho,t+s)\,ds\,d\rho, \tag{16}$$

$$\tilde{k}(t,s;r) := \int_0^r k(t+\rho,s)\,d\rho, \quad \tilde{f}^\delta(t;r) := \int_0^r f^\delta(t+\rho)\,d\rho. \tag{17}$$

It is clear that this method is similar to the singular perturbation approach discussed earlier because the coefficient $\alpha(t;r)$ of $u(t)$ in (13) can be made small by decreasing r. However, in contrast to the singular perturbation approach, there is never a boundary layer phenomenon for the local regularization equation. That is, one does not require $\bar{u}(0) = 0$ (or knowledge of $\bar{u}(0)$, if nonzero) in order to obtain uniform convergence on $[0, T]$ of the solution $u^\delta(\cdot; r)$ of (13) to $\bar{u}$ as $r, \delta \to 0$ in a coordinated way. Thus, one may apply standard numerical schemes to the solution of the local regularization equation (13) without having to worry about the difficulty of resolving rapidly varying solutions near $t = 0$ (such as was necessary for the singular perturbation approach in the absence of information about $\bar{u}(0)$).

A regularization theory for this method was developed in [56, 57] (for convolution kernels) and [61] (for nonconvolution kernels), and is summarized in the next theorem. An open problem concerns the development of a discrepancy principle for the selection of $r = r(\delta, f^\delta)$ in the case of fixed f^δ, δ.

Theorem 2. [61] **(Local regularization method)** *Let $\bar{r} > 0$ be small and assume that $\bar{u} \in C^1[0, T + \bar{r}]$ satisfies (1) on $[0, T + \bar{r}]$ where it is assumed that (1) is a one-smoothing problem. For $0 < r \leq \bar{r}$, let $u^\delta(\cdot; r)$ denote the solution of (13) associated with f^δ, where $|f(t) - f^\delta(t)| \leq \delta$ for $t \in [0, T + \bar{r}]$.*

Then if $r = r(\delta)$ is selected satisfying $\delta/r(\delta) \to 0$ as $\delta \to 0$, we have

$$u^\delta(\cdot; r(\delta)) \to \bar{u} \text{ as } \delta \to 0$$

uniformly on $[0, T]$, with optimal convergence rate $\delta^{1/2}$ for $r(\delta) = C\delta^{1/2}$, $C > 0$.

The local regularization theory described above may be generalized so that integration with respect to ρ in (16)–(17) is replaced by integration with respect to a suitable Borel measure η_r on $[0, r]$ [56, 57, 61]. In particular, a choice of a discrete η_r (such as $\int_0^r g(\rho) \, d\eta_r(\rho) = \sum_{i=1}^K s_i g(\tau_i)$, for $s_i > 0$, $i = 1, \ldots, K$, and $0 \leq \tau_1 < \tau_2 < \ldots < \tau_K = r$) is particularly useful in numerical implementations of this method. This idea will be revisited in Section 4.3.

We note that, although the above theorem is stated for one-smoothing problems, conditions guaranteeing convergence for general ν-smoothing problems may be found in [57] for convolution kernels. Verification of these conditions is not easy, but convergence has been demonstrated in the case of $1 \leq \nu \leq 4$ for classes of Borel measures η_r of practical use [57]. The theory for the local regularization method described above does not yet extend to infinitely-smoothing problems, but, as will be discussed in Section 4.3, a particular discretized version of (13) (with a fixed discrete measure η_r) has been used for over 30 years in practical solution of the infinitely-smoothing IHCP. Indeed, it was this numerical method due to Beck [16] for the IHCP that motivated the development of the local regularization methods described in this section and in Section 4.3. (As a final comment regarding local regularization and infinitely smoothing problems, we note that it was shown in [58] that a particular variation of the method described above can be shown to converge for infinitely-smoothing problems, however this alternate approach can no longer be considered a regularization method of "Volterra type".)

Theorem 2 was extended in [61] to include the possibility of a variable regularization parameter ($r = r(t)$), a generalization which allows for *variable* local regularization of equation (1) (i.e., more regularization in some regions of $[0,T]$ and less in others). Variable regularization has been shown to be helpful in numerical examples where the true solution $\bar{u}$ is not smooth.

A disadvantage of the local regularization method described above is the need for data slightly beyond the initial interval $[0,T]$, or, equivalently, one must be satisfied with approximating $\bar{u}$ on a slightly smaller interval $[0, T - \bar{r}]$. However, this is not a serious shortcoming as it is well-known that one cannot expect to adequately determine solutions of Volterra problems near the end of the interval $[0,T]$ when using data on the same interval.

3.3   Lavrent'ev's m-times iterated method.

For fixed integer $m \geq 1$ and given regularization parameter $\alpha > 0$, Lavrent'ev's m-times iterated method determines $u_\alpha^\delta \in U$ via

$$(\alpha I + A)u_n = \alpha u_{n-1} + f^\delta, \quad n = 1, 2, \ldots m, \tag{18}$$

$$u_\alpha^\delta := u_m, \tag{19}$$

starting from $u_0 = 0$. For $m = 1$, it is clear that the method (18)–(19) reduces to Lavrent'ev's classical method of Section 3.1; for $m > 1$, corrections are applied to further stabilize the problem. Because typical discretizations of the operator $(\alpha I + A)$ lead to lower triangular matrix representations, the calculation of u_α^δ in (18)–(19) is easily accomplished via backward substitution in (18); as before, the addition of the term αu_n (for $\alpha > 0$ small) serves to stabilize the numerical process. We note that because $m \geq 1$ is fixed, the approach is considered a parametric regularization method and not an iterative one.

In contrast to Lavrent'ev's classical method for first-kind Volterra equations, where the theoretical development is available for one-smoothing problems (and for general ν-smoothing problems using the approach of Sergeev [53]), the theoretical analysis of Lavrent'ev's m-times iterated method for Volterra problems appears to be limited at present to *accretive* operators, i.e., to those operators $A \in \mathcal{L}(U)$ (for a general Hilbert space U, here over a complex or real scalar field) satisfying

$$\Re\langle Au, u \rangle \geq 0, \quad u \in U,$$

where $\langle \cdot, \cdot \rangle$ denotes the usual inner product on U, and $\Re z$ denotes the real part of $z \in \mathbb{C}$.

Accretive Volterra operators include those with convolution kernels $k(t,s) = \kappa(t - s)$ that are completely monotone, i.e.,

$$\Re L[\kappa](z) \geq 0, \quad \Re z > 0,$$

where $L[\kappa]$ denotes the Laplace transform of κ. (For more details about completely monotone kernels, see, for example, Gripenberg, Londen, and Staffans [7], or Nohel and Shea [12].) Among this class of convolution kernels are κ which are positive, decreasing, and convex [12]. Another example of an accretive Volterra operator is the generalized Abel integral operator (where U must be interpreted to denote $L_2(0, T)$ with an associated weighted inner product). Further, the Volterra operator associated with the IHCP is evidently not accretive in $U = L_2(0, T)$, without the use of a weighted inner product [67].

We describe here some of the basic theoretical results for Lavrent'ev's m-times iterated method, assuming throughout the remainder of this section that A is an accretive Volterra operator. In [63], Plato showed that the method is convergent in the case of exact data f in (18)–(19), i.e.,

$$u_\alpha \to \bar{u} \text{ as } \alpha \to 0.$$

In the case of perturbed data f^δ, the definition of a classical discrepancy principle provides a way of selecting $\alpha = \alpha(\delta)$ so that regularized convergence is obtained as $\delta \to 0$. The following theorem involves the use of a classical discrepancy principle, but is valid only for the case of $m \geq 2$.

Theorem 3. [64, 65] **(Lavrent'ev's m-times iterated method, I)**
Let $A \in \mathcal{L}(\mathcal{U})$ be an accretive Volterra operator and let $\bar{u} \in \mathcal{R}(A)$ solve the original equation $Au = f$. Let u_α^δ denote the solution of the Lavrent'ev's m-times iterated method (18)–(19) for $m \geq 2$, with f^δ in place of f and $\|f - f^\delta\| < \delta$ for some $\delta > 0$.
Then if $\alpha = \alpha(\delta) > 0$ is selected satisfying the discrepancy principle,

$$b_0 \delta \leq \|Au_\alpha^\delta - f^\delta\| \leq b_1 \delta \qquad (20)$$

(for fixed constants b_0, b_1, with $b_1 \geq b_0 > 1$, $b_1 \delta < \|f^\delta\|$), it follows that $\delta/\alpha(\delta) \to 0$, and

$$\|u_{\alpha(\delta)}^\delta - \bar{u}\| \to 0,$$

as $\delta \to 0$. Moreover, under additional smoothness assumptions on $\bar{u}$ (and with some restrictions), order-optimal convergence rates are obtained.

Remark 2. Plato showed that the (classical) discrepancy method given by (20) in Theorem 3 is *divergent* for the case of $m = 1$ (i.e., for Lavrent'ev's classical method considered Section 3.1) when applied to accretive Volterra operators with nondegenerate kernels (cf. Prop. 3.2.4 of [63]). Although this result is restricted to accretive operators, it matches a similar finding for symmetric, positive definite operators in $\mathcal{L}(U)$ and thus does not give much hope that a discrepancy principle of this type will be useful for general ν-smoothing Volterra problems.

Fortunately, in the accretive operator case there are modified discrepancy principles that are convergent (and lead to "pseudo-optimal" choices of $\alpha = \alpha(\delta)$, in the terminology of [69]) for $m \geq 1$. Thus, provided that A is accretive, these modified discrepancy principles apply to Lavrent'ev's classical method described in Section 3.1. One such modified discrepancy principle is given below.

Theorem 4. [64, 69] **(Lavrent'ev's m-times iterated method, II)**
Let $A \in \mathcal{L}(\mathcal{U})$ be an accretive Volterra operator and let $\bar{u} \in \overline{\mathcal{R}(A)}$ solve the original equation $Au = f$. Let u_α^δ denote the solution of the Lavrent'ev's m-times iterated method (18)–(19) for any fixed $m \geq 1$, with f^δ in place of f, $\|f - f^\delta\| < \delta$.

Then if $\alpha = \alpha(\delta) > 0$ is selected satisfying the discrepancy principle,

$$b_0\delta \leq \|\alpha(A + \alpha I)^{-1}(Au_\alpha^\delta - f^\delta)\| \leq b_1\delta \qquad (21)$$

(for fixed constants b_0, b_1, with $b_1\delta < \|f^\delta\|$, $b_1 \geq b_0 > \tau_0$, for suitable $\tau_0 > 0$), it follows that $\delta/\alpha(\delta) \to 0$, and

$$\|u_{\alpha(\delta)}^\delta - \bar{u}\| \to 0,$$

as $\delta \to 0$, with order-optimal convergence rates under additional smoothness assumptions on $\bar{u}$.

Plato and Hämerik considered a second convergent parameter choice in [69] for the m-times iterated method in the case of $m \geq 1$. Both modified discrepancy principles extend (to accretive Volterra operators) analogous results for symmetric operators on a Hilbert space [13] and similar results for normal equations [3]. It was also shown in [69] that an alternate parameter selection method based on an extrapolation strategy is convergent for $m = 1$ and $m = 2$; the question is unresolved for $m > 2$.

3.4   Iterative Methods of "Volterra Type"

Classical iterations such as Landweber, conjugate gradient, ν-methods, and iterated Tikhonov regularization, generally rely on the computation of the anticipatory operator A^* and thus will not be considered here. For a discussion of classical iterative regularization, see, for example, [4].

A few simple iterations do, however, fit our notion of a regularization method of "Volterra type".

Richardson Iteration. Starting from an initial guess of u_0, the Richardson iteration is a simple explicit method which defines a sequence $\{u_n^\delta\}$ of functions satisfying

$$u_n = u_{n-1} - \beta_n(Au_{n-1} - f^\delta), \quad n = 1, 2 \ldots, \qquad (22)$$

starting from a suitable initial guess $u_0 \in U$.

In the case of noise-free data ($f^\delta = f$, $u_n^\delta = u_n$), Vasin obtained the following result concerning convergence of the iteration under quite general conditions on the original problem (1) (in particular, equation (1) need not be ν-smoothing), with results applicable to certain nonlinear Volterra problems [70, 71].

Theorem 5. [70, 71] **(Richardson iteration with noise-free data)**
Assume the kernel $k \in C^1$ in (1) is such that $k(t, s) \geq 0$ and $\frac{\partial}{\partial t}k(t, s) \geq 0$ for $0 \leq s, t \leq T$. Let u_n denote n^{th} Richardson iterate defined by (22) for noise-free data f. Then, for sufficiently smooth initial guess u_0 and for $\beta_n \geq \beta > 0$ sufficiently small, the iterates u_n converge to $\bar{u}$ as $n \to \infty$.

The convergence of the Richardson iteration in the case of noisy data is a more difficult problem as this particular iteration seems quite sensitive to noise in the data, even for ν-smoothing problems with ν small. Of course, when perturbed data f^δ is used in place of f in (22), we cannot expect convergence as $n \to \infty$. Instead (as with all iterative methods for ill-posed problems), we expect to see "semiconvergence" (see, e.g., [4] or [11]), meaning that the error $\|u_n^\delta - \bar{u}\|$ decreases for some initial iteration steps, but eventually begins to increase. Therefore, in the presence of noisy data one must determine a stopping criterion, or method of selecting a stopping point $n = n(\delta)$ in the iteration process, for which we have regularized convergence of the iterates as $\delta \to 0$. That is, we require that $n(\delta) \to \infty$ and

$$\|u_{n(\delta)}^\delta - \bar{u}\| \to 0,$$

as $\delta \to 0$.

Vasin discussed a stopping criteria in [70], however simple numerical examples suggest that the Richardson method may be most effective for only moderately ill-posed problems. For example, Plato and Hämarik [69] provided a thorough analysis of a stopping criteria for the Richardson iteration based on classical discrepancy principles, but the theory they provided was limited to a restricted class of operators (strictly sectorial operators) which include the classical Abel operator. The Abel problem is only moderately ill-posed and, indeed, it is an open question as to whether the Richardson iteration is better suited for such problems than for general ν-smoothing (or infinitely-smoothing) problems. In addition, in [64], Plato noted that generally Lavrent'ev's m-times iterated method is superior to Richardson iteration (and also superior to an implicit scheme given in (23) below) for strictly sectorial operators.

Other Iterative Methods There are additional iterative methods, which when applied to equation (1), take full advantage of the Volterra nature of the problem. Among these we mention an implicit iteration,

$$(I + \beta A)u_n = u_{n-1} + \beta f^\delta, \quad n = 1, 2 \ldots, \tag{23}$$

and an alternating directions iteration,

$$u_{n-1/2} = u_{n-1} - \frac{\beta}{2}(Au_{n-1} - f^\delta),$$

$$\left(I + \frac{\beta}{2}A\right)u_n = u_{n-1/2} + \frac{\beta}{2}f^\delta, \quad n = 1, 2, \ldots,$$

both for fixed $\beta > 0$ and for suitable initial $u_0 \in U$. Rigorous analysis of these iterations (and their connections to Cauchy's method), along with theoretically-sound stopping criteria, may be found in Plato [63, 65, 66] and Plato and Hämerik [69]. However, the convergence theory for these iterations as applied to Volterra problems is apparently limited at present to only moderately ill-posed equations such as the classical Abel equation (and other Volterra equations with strictly sectorial operator A), so an open question concerns their applicability to the more general problems under consideration in this paper.

3.5   Differentiation and Mollification

For ν-smoothing problems, we know that the original first-kind equation (1) is equivalent (via ν differentiations) to a second-kind Volterra equation (5). However, since we typically only have available a non-smooth perturbation f^δ of f, the differentiation of f^δ is not a well-posed process. In the usual case of perturbed data, one approach is to replace (5) by

$$u(t) + \int_0^t \frac{\partial^\nu}{\partial t^\nu} k(t,s) u(s)\, ds = L_\alpha f^\delta(t), \quad t \in [0,T], \tag{24}$$

where the operator L_α is constructed to satisfy $L_\alpha f^\delta \to f^{(\nu)}$ (in an appropriate sense) as $\delta \to 0$, provided the selection of α is coordinated with that of δ.

There are many ways of selecting L_α, with different theoretical arguments required for each, so we will only indicate some of the possibilities here. For example, Murio [81] considered the one-smoothing differentiation problem (6) and took as $L_\alpha f^\delta$ the approximate derivative of a particular mollification (a convolution with a Gaussian kernel) of f^δ. Magnicki [80] and Srazhidinov [82] considered the general ν-smoothing problem and viewed L_α as a type of mollified differentiation operator, while Kabanikhin [74] employed a difference operator approach. (In both [82] and [74], theoretical results were given for nonlinear Volterra equations.) A further variation on these ideas may be found in Sergeev [53]. We note that although the above approaches are simple, they suffer if k is not known precisely (since derivatives of k must also be taken); further, the method cannot be extended to infinitely-smoothing problems.

Murio (cf. [81] and the references therein) considered specific mollification approaches for solving a number of applied problems. For example, he used mollification to solve a Volterra equation of Abel type, in this case applying the inverse Abel transform (which requires the differentiation of data, under an integral) to a construction of $L_\alpha f^\delta$ similar to that described above for the differentiation problem. He also applied mollification techniques to the IHCP.

Louis and Maaß [79] developed an abstract formulation of the mollification problem for bounded linear operators A on Hilbert spaces, and Louis extended the overall analysis further in [75–78]. The ideas in [77] will be discussed more fully in the next paragraph. Háo took a general approach through the use of a one-parameter class of mollification operators defined on Banach spaces. His treatment included applications to numerical differentiation and the ICHP, generalizing a number of existing theories (see [72] and the references therein). Hegland and Anderssen [73] provided a general Hilbert space analysis, defining "range mollifications" and "domain mollifications" for an operator A. In applications to numerical differentiation and the Abel equation, they made use of translation operator representations to give estimates of the mollification error.

In what follows we sketch the approach taken by Louis in [77] as it pertains to a first-kind Volterra problem of the form (1); for more complete details, see [76, 77]. The idea is to define a one-parameter family $\{e_\alpha\}$ of mollifiers, $e_\alpha = e_\alpha(t,s)$, for which $E_\alpha u \in U$ given by

$$E_\alpha u(t) := \int_0^T e_\alpha(t,s)u(s)\,ds, \quad t \in [0,T],$$

is a suitable approximation of $u \in U$ for $\alpha > 0$ small. The initial task is to find $\psi_\alpha = \psi_\alpha(t,s)$ satisfying $\psi_\alpha(t,\cdot) \in U$ and

$$A^*\psi_\alpha(t,\cdot) \simeq e_\alpha(t,\cdot), \quad t \in [0,T], \tag{25}$$

a computation which can be made prior to receiving any data. Using the perturbed data f^δ of f, the mollified approximation u_α^δ to $\bar{u}$ is then given by

$$u_\alpha^\delta(t) = \int_0^T \psi_\alpha(t,s)f^\delta(s)\,ds, \quad t \in [0,T], \tag{26}$$

and, under suitable assumptions, it can be shown that $\alpha = \alpha(\delta)$ may be selected guaranteeing $u_\alpha^\delta \to \bar{u}$ as $\delta \to 0$, with order-optimal rate of convergence, under additional conditions (cf. [76, 77]).

Although the adjoint operator A^* plays a role in this regularization method, in the case of certain specific Volterra operators A (in particular, those for which $(A^*)^{-1}$ is easily evaluated), it is possible to pick e_α so that ψ_α determined by (25) has support which is optimal with regard to the Volterra problem, i.e., so that this particular ψ_α leads to a reconstruction of u_α^δ in (26) which essentially reduces to

$$u_\alpha^\delta(t) = \int_0^{t+\varepsilon(t)} \psi_\alpha(t,s)f^\delta(s)\,ds, \quad t \in [0,T], \tag{27}$$

for some $\varepsilon = \varepsilon(t,\alpha) \geq 0$ small. Louis gave an example of such an e_α for the derivative problem (6) in [77] and for a Volterra problem of Abel type in [78], which, for both examples, leads to the "Volterra-type" construction (27). (We note that the constructions of u_α^δ for these two particular cases are also related to the "local regularization" ideas discussed in Section 3.2.) The goal of using these mollification ideas to obtain a Volterra-type regularization method for general equations of the form (1) is considerably more difficult because $(A^*)^{-1}e_\alpha$ is not so easily evaluated in the general case.

4   Discrete Regularization Methods of "Volterra Type"

Discrete approximation methods provide another way to regularize the original problem (1). In this case the regularization parameter is the discretization parameter (or stepsize), and a coordination between this parameter and the amount of noise δ in the problem is required in order to obtain good approximations in the presence of noise. This is sometimes known as the "self-regularization" property of discretizations. An important issue is whether self-regularization is sufficient for the kinds of problems we consider here. In the general case it can be said that further regularization will be required in order to obtain acceptable approximations in the presence of perturbed data.

In this section we describe numerical methods that have been developed over the past few decades for solution of the first-kind Volterra problem (1), focusing in particular on those methods that are of "Volterra type" and for which something can be said about the "self-regularization" properties of the algorithms. We note that most often the theory is limited to one-smoothing problems. We begin the discussion with the case of exact data.

4.1 Standard Numerical Discretizations with Exact Data f

As described in survey papers by Brunner [85, 88], the idea of approximating Volterra equations (admittedly, of second kind) in a finite-dimensional setting actually originated with Volterra himself in the late 1800's. Evidently the first application of similar ideas to first-kind equations was given by Huber [101] in 1939 [88], and it was in the 1960's that researchers (notably Jones [103], Kobayasi [105], and Linz [108, 109]) began to look closely at the theoretical issues associated with adapting traditional numerical integration techniques to the approximate solution of first-kind Volterra equations [88, 114].

Discretizations Based on Numerical Integration Rules. The idea behind numerical methods based on integration rules is quite straightforward. The integral in (1) is replaced by a numerical quadrature, and then it is required that the resulting equation be satisfied exactly at a finite number of points in $[0, T]$.

For example, consider the *rectangular* integration rule, given for integer $N \geq 1$ and $h = T/N$ by

$$\int_0^{t_i} \varphi(t)\, dt \approx h \sum_{j=0}^{i-1} \varphi(t_j),$$

$i = 1, \ldots N$, for continuous φ. Throughout we will make the definition $t_j = jh$, for $j = 0, 1, \ldots N$. Replacing the integral in equation (1) by a sum such as given above, one obtains the *Euler method* for approximate solution of (1),

$$h \sum_{j=0}^{i-1} k(t_i, t_j) u_j = f(t_i), \quad i = 1, 2, \ldots, N, \tag{28}$$

where u_i is an approximation for $u(t_i)$. Equivalently, making the definitions $\mathbf{f}^N = (f(t_1), \ldots, f(t_N))^\top$, $\mathbf{u}^N = (u_0, \ldots, u_{N-1})^\top \in \mathbb{R}^N$, we obtain the matrix equation

$$\mathbf{A}^N \mathbf{u}^N = \mathbf{f}^N \tag{29}$$

where $\mathbf{A}^N$ is a lower-triangular $N \times N$ matrix with entries depending on h and the kernel k. It is easy to see that, under one-smoothing assumptions, the diagonal entries of $\mathbf{A}^N$ are nonzero for all $h > 0$ sufficiently small; thus the Euler approximation algorithm may be solved sequentially for a unique approximation vector $\mathbf{u}^N$.

Theorem 6. [107]   (**Euler Method**) *Let $\bar{u}$ be the solution of equation (1), where it is assumed that (1) is a one-smoothing problem, with k, f, $\bar{u}$ sufficiently smooth. Let $\mathbf{u}^N = (u_0, \ldots, u_{N-1})^\top \in \mathbb{R}^N$ be determined by the Euler method (28) using exact data f.*

Then

$$\max_{0 \leq i \leq N-1} |\bar{u}(t_i) - u_i| \to 0 \quad \text{as} \quad h \to 0,$$

with order of discrete convergence equal to 1; that is, $p = 1$ is the largest such number for which

$$\max_{0 \leq i \leq N-1} |\bar{u}(t_i) - u_i| \leq M h^p$$

for some M independent of h.

Similarly, the composite *midpoint* integration rule,

$$\int_0^{t_i} \varphi(t) \, dt \approx h \sum_{j=0}^{i-1} \varphi(t_{j+1/2}),$$

$i = 1, \ldots, N$, where $t_{j+1/2} = t_j + h/2$, leads to the *midpoint method* for the approximate solution of (1),

$$h \sum_{j=0}^{i-1} k(t_i, t_{j+1/2}) \, u_{j+1/2} = f(t_i), \quad i = 1, 2, \ldots, N, \tag{30}$$

also generating a triangular matrix system of the form (29), uniquely solvable for all $h > 0$ sufficiently small.

Theorem 7. [107] (**Midpoint Method**) *Let $\bar{u}$ be the solution of equation (1), where (1) is assumed to be a one-smoothing problem and k, f, and $\bar{u}$ are sufficiently smooth. Let $\mathbf{u}^N = (u_{1/2}, \ldots, u_{N-1/2})^\top \in \mathbb{R}^N$ be determined by the midpoint method (30) using exact data f.*

Then

$$\max_{0 \leq i \leq N-1} |\bar{u}(t_i + h/2) - u_{i+1/2}| \to 0 \quad \text{as} \quad h \to 0,$$

with order of discrete convergence equal to 2.

. Obviously, one could continue applying higher order integration rules to the integral in (1), but the fact is that the midpoint and Euler methods are the only reasonable approaches that result from the use of standard quadrature rules. For example, the trapezoidal rule results in a method that is convergent of order 2, but which is also considered to be numerically unstable [106, 107]. (We note that the notion of numerical instability is separate from that of ill-conditioning. Ill-conditioning is an inherent feature of the original problem, while numerical instability refers to undesirable aspects of the numerical difference scheme. The idea of numerical stability for Volterra equations is similar to that used in analyzing numerical methods to solve ordinary differential equations.)

It was first observed by Linz [111] (for some representative quadrature rules) and later confirmed by Gladwin and Jeltsch [100] that all of the standard higher order Gregory methods and Newton-Cotes integration rules (such as Simpson's rule and the three-eights rule) lead to unstable numerical methods for solving equation (1) [106]. This is in sharp contrast to numerical approximation of second-kind Volterra equations (where all "reasonable" integration rules lead to "reasonable" numerical approximation methods for (1), with complications due mainly to "stiffness" associated with the particular case of singularly perturbed problems), and is an example of the way in which the ill-posedness of (1) influences numerical implementation even when one is using perfect data. [107]

In summary, of the standard integration-type methods the midpoint method is generally the best choice for the approximate solution of (1) in the presence of noise-free data; we note also that the accuracy of the midpoint method may be improved further using Richardson extrapolation [107].

Collocation-Based Discretization Methods. In spite of the undesirable situation for discretizations based on standard integration rules, in fact arbitrarily high order numerical methods for the solution of equation (1) do exist (under conditions of additional smoothness on the kernel k and true solution $\bar{u}$). There are several different types of these methods (see the end of this section), and we describe here the class of collocation-based methods. As in the case of methods based on integration rules, the theoretical analysis for collocation methods is typically restricted to one-smoothing problems.

One commonly seen implementation of collocation is as follows. Given intervals $\sigma_i = (t_{i-1}, t_i]$, $i = 2, \ldots, N$, $\sigma_1 = [t_0, t_1]$, we seek an approximate solution u^N in the space of piecewise polynomials $S_{m-1}^{(-1)}(N)$,

$$S_{m-1}^{(-1)}(N) = \{u : u|_{\sigma_i} \in \Pi_{m-1}, \ i = 1, \ldots N\},$$

where Π_k denotes the space of real polynomials of degree not exceeding k. Since $\dim S_{m-1}^{(-1)}(N) = Nm$, one must impose the same number of conditions on the approximate solution in order to obtain the solution uniquely. To this end we define Nm collocation points,

$$X_m(N) = \{t_{i,j} = t_{i-1} + c_j h, \ j = 1, \ldots m, \ i = 1, \ldots N\},$$

where the $\{c_j\}$ are "collocation parameters" selected satisfying

$$0 < c_1 < \ldots < c_m \leq 1,$$

and we impose the condition that $u^N \in S_{m-1}^{(-1)}(N)$ must satisfy equation (1) precisely at each collocation point in $X_m(N)$, i.e.,

$$Au^N(t_{i,j}) = g(t_{i,j}), \quad t_{i,j} \in X_m(N). \tag{31}$$

This leads to a system of equations of the form (29) where now the Nm-square matrix $\mathbf{A}^N$ is block-triangular (with N diagonal $m \times m$ blocks, each nonsingular for $h > 0$ sufficiently small provided the problem is one-smoothing), leading to a block-sequential solution method. The following convergence result is due to Brunner [87].

Theorem 8. [87–90] **(Method of Collocation)** *Let the original problem (1) be one-smoothing and assume that k and f are sufficiently smooth to guarantee that $\bar{u} \in C^m$. Let u^N be given by the collocation procedure described above using exact data f.*

Then u^N converges uniformly to $\bar{u}$ (with discrete convergence of order m) if and only if the quantity $\xi_m = \prod_{j=1}^{m} \frac{1-c_j}{c_j}$ satisfies

$$\xi_m \leq 1, \text{ if } m \text{ is odd,} \quad \text{and } \xi_m < 1, \text{ if } m \text{ is even.}$$

In addition, "local superconvergence" at certain discrete interior points of σ_i is achievable under additional conditions.

Thus, even though convergence does not occur for all choices of collocation parameters, the particular choice of $c_m = 1$ (i.e., collocation occurs at each of the original gridpoints t_j, $j = 1, \ldots N$) assures uniform convergence of order m.

In addition, Eggermont showed how the superconvergence (at selected points) indicated in the above theorem may be improved via a postprocessing of the collocation solution. See, for example, [97, 99]. Finally, Kauthen and Brunner considered analogous results for the case of *continuous* collocation approximation spaces $S_{m-1}^{(0)}(N)$ in [104]; methods based on splines with full continuity and of degree exceeding one are known to be divergent [102].

To illustrate the simplicity of the collocation algorithm, we briefly consider collocation in the space $S_0^{(-1)}(N)$ of piecewise constants. In this case, for $i = 1, \ldots, N$, let χ_i be the usual characteristic function on σ_i (i.e., $\chi_i(t) = 1$, for $t \in \sigma_i$, and $\chi_i(t) = 0$, otherwise). Then the collocation solution u^N is given by

$$u^N(t) = \sum_{i=1}^{N} u_i \chi_i(t), \quad t \in [0, T],$$

where the coefficients $u_1, \ldots, u_N$ satisfy

$$\sum_{j=1}^{i} u_j \int_{t_{j-1}}^{t_j} k(t_i, s) \, ds = f(t_i), \quad i = 1, \ldots, N \tag{32}$$

(again, a system of the form (29) with $\mathbf{A}^N$ lower-triangular). From the last theorem we know that equation (32) generates a piecewise constant u^N that converges uniformly to $\bar{u}$ at a rate of $\mathcal{O}(h)$.

For simple kernels k, the moment integrals $\int_{t_{j-1}}^{t_j} k(t_i, s) \, ds$ appearing in (32) (and in the more general problem (31)) may be evaluated analytically, however, more commonly they will need to be evaluated using a discrete quadrature rule. In [87], Brunner provided an analysis of the coordination of the particular integration rule with m so that the overall order of discrete convergence is not lost.

In addition, Brunner showed that a number of other high order methods developed independently for use in solving first-kind Volterra equations (methods such as the block-by-block approach considered in the early 1970's by de Hoog and Weiss [92, 93] and some product integration methods studied, for example, by Young [124], Weiss and Anderssen [121], Linz [110], and McAlevey [113]), can be obtained via collocation if a particular numerical quadrature scheme is used to approximate the moment integrals [86]. Further, collocation appears to offer some flexibility in the approximation of (1) in that specially-selected quadrature rules can be used to construct the moment integrals in the case of a singular, nonsmooth, or rapidly-varying kernel.

Other Higher Order Methods. A comprehensive discussion of higher order methods for first-kind Volterra equations would take us beyond the scope of this paper. Instead we will only mention a few of the many methods available, including linear multistep methods (see, for example, McKee [115] and the references in his survey [114]), generalized linear multistep methods (van der Houwen and te Riele [120]), inverted differentiation formulas (Taylor [119]), reducible quadrature (Wolkenfelt [123]), product integration rules (e.g., see the references given above), and multilag methods [122]. A comprehensive list of references is not possible here. A "unified" convergence theory was given for many of these methods (including collocation) in the case of both linear and nonlinear Volterra problems of the first kind in papers by Scott (née Dixon), McKee, and Jeltsch (see, e.g., [94, 95, 117, 118]).

Discretizations of ν-Smoothing Problems for $\nu > 1$. Those numerical methods described above which rely heavily on the use of discrete evaluations of the kernel along the "diagonal" (i.e., $k(t, t)$) will no longer be of use for ν-smoothing problems for $\nu > 1$ since, in this case, $k(t, t) = 0$. Further, as mentioned above, the vast amount of theoretical analysis for standard methods is restricted to one-smoothing Volterra problems. Among the exceptions are a convergence analysis for collocation methods (Eggermont, [98]) and for linear multistep methods (Andrade, Franco, and McKee [83]), in the case of *two-smoothing* Volterra problems. Of interest in this regard is a comment in [83] that the original goal of the authors was to consider ν-smoothing problems for general $\nu \geq 2$. In fact, they were successful only in the case of $\nu = 2$, and found that even in this case, the stability properties of the method were quite different from the stability of the same methods applied to one-smoothing problems. Thus, from a theoretical point of view, the analysis of standard discretization methods when applied to general ν-smoothing Volterra problems (in the case of ν bigger than one or two) remains an open problem.

For a more comprehensive view of numerical methods for first-kind Volterra equations in the presence of noise-free data, see, for example, [84, 88, 89, 91, 106, 107, 114], survey papers written at different times over the last 25 years of study on this subject. In recent years, many of these numerical methods have been extended to nonlinear first-kind Volterra problems.

4.2 Standard Numerical Discretizations with Perturbed Data f^δ: "Self-Regularization"

We now consider the use of one of the above numerical methods for the solution of equation (1) in the case where f has been replaced by f^δ. In this situation we are interested in the convergence of numerical approximations (constructed using f^δ) to the true solution $\bar{u}$ as both the amount δ of noise in the data and the discretization parameter h go to zero. As $\delta \to 0$ we may let $h \to 0$, but, in general, we cannot let h get as small as we might like for a given value of $\delta > 0$.

Natterer [10] gave a general mathematical context in which to evaluate the regularizing properties of numerical methods such as those given the last section, in particular, those methods which can be formulated as projection methods. In keeping with the ideas of [10], we will say that a discrete numerical method is "robust" for one-smoothing problems if the magnification of data error by the method is of order δ/h as $h \to 0$. (Compare, for example, with Richter [116].)

Linz [107, 112] investigated the handling of error for the midpoint method and found that the method was robust for one-smoothing problems. Eggermont extended the analysis to higher-order numerical methods (such as cyclic linear multistep and reducible quadrature methods) and, using a projection-type analysis, showed that under reasonable circumstances these methods were also robust for one-smoothing problems [96]. In addition, the same results hold for one-smoothing problems when collocation is coupled with certain quadrature methods for moment integrals [96, 99].

Thus, for any of the above methods, we are assured of optimal error handling of the method provided that the original problem is one-smoothing. A very important question is whether this error handling is sufficient when the same method is applied to a ν-smoothing problem for $\nu > 1$, or to an infinitely-smoothing problem such as the IHCP. Practical experience indicates that standard numerical methods alone (of the form described in the last section) are *not* sufficient to handle the more ill-conditioned problems (and, in fact, often give meaningless results) without the use of additional regularization techniques.

4.3 Combined "Volterra-Type" Discretization and Regularization

Effective discretized regularization techniques for Volterra problems involve the pairing of a discretization method of "Volterra type" (such as one of the methods described in Section 4.1) with a continuous regularization method which also preserves the Volterra structure of the original problem (cf. Section 3).

Discretization and Mollification. An approach taken by Linz [112] was to smooth or filter the data first before applying a standard numerical method to (1). In [73], Hegland and Anderssen combined mollification with projection methods. Their analysis provided for theoretical error estimates for mollified finite difference techniques as applied to the differentiation problem and the Volterra equation of Abel type.

Discretization of Lavrent'ev's m-times iterated method. In [67,68], Plato combined a Galerkin method for approximating (1) with Lavrent'ev's m-times iterated method in the case of an accretive Volterra operator A (see Section 3.3). Plato showed that the numerical realization $\mathbf{A}^N$ is lower Hessenberg; efficient solution methods exist for equations governed by such matrices. Order-optimal convergence rates for the combined discretization-regularization method were obtained.

Discretization of the Local Regularization Method. The analysis of a collocation method paired with the ideas of "local regularization" (cf. Section 3.2) was first considered in [56] where it was shown that collocation of the second-kind local regularization equation (13) (with a specific choice of discrete measure η_r) over the space of piecewise constant functions leads to a particularly simple discrete regularization procedure (known as "Beck's method" when applied to the IHCP). See [16] and the references therein for practical application of this procedure to the IHCP. We briefly describe this discrete process in what follows.

Let $h = T/N$ and let $r = \gamma h$ denote the discrete regularization parameter (indicating the length of the local regularization interval) for a fixed integer $\gamma \geq 1$. Then, given $u_1, \ldots, u_{i-1}$ (corresponding to the regularized solution u at $t_1, \ldots, t_{i-1}$), we determine u_i by first "predicting" an optimal constant-valued (i.e., *over*-regularized) solution $\hat{u}$ on the interval $(t_{i-1}, t_{i-1} + \gamma h]$, where $\hat{u}$ is constructed via a least squares fitting to the data at points $t_i, t_{i+1}, \ldots, t_i + \gamma h$. We next "correct" for over-regularization by retaining only the value of $\hat{u}$ at the position t_i; i.e., we set $u_i := \hat{u}$. Then the procedure is repeated, until all u_i have been determined in this sequential process. For obvious reasons, the discretized local method is often called a *predictor-corrector* regularization method.

In [55], a convergence theory was given for this discrete method for the one-smoothing convolution problem, and convergence was shown to be of optimal order. An analysis of the conditioning of the discretized ν-smoothing problem (and the dependence of condition numbers on ν, γ, and h) was discussed in [57]. More recently these regularization ideas were extended in [62] to variable r (i.e., variable γ) and nonconvolution problems, with the addition of an optional penalty term. The ideas were generalized in a different direction in [54], where there one seeks an optimal degree-d polynomial, for integer $d \geq 0$, in the "prediction" step.

In [60], the discrete local regularization approach was modified further, forming the basis for the method of *sequential Tikhonov regularization* for Volterra convolution problems. In this case, a local Tikhonov regularization is performed at each sequential step and, again, only the first component of the local solution is retained at each step. The cost of each local Tikhonov problem is reduced substantially using an efficient algorithm of Eldén [2], one which employs orthogonal transformations and takes advantage of the Toeplitz structure of discretizations of Volterra convolution problems. Although the convergence theory is again limited to one-smoothing problems (with optimal convergence rates), numerical examples in [60] were used to illustrate that the method works well when applied to the (infinitely-smoothing) IHCP. In addition, a preliminary numerical study of discrepancy principles to pick local Tikhonov regularization parameters was undertaken in [59], where it appears that *variable* regularization of solutions (effectively finding steep/sharp areas of solutions) is possible using these ideas.

5 Conclusion

We have reviewed some representative continuous and discrete regularization methods for first-kind Volterra problems with continuous kernels, paying particular attention to those methods which which tend to retain the Volterra structure of the original problem. As seen in the previous sections, there are many interesting open problems in this research area. In particular, the extension of methods and theoretical results to problems which are infinitely-smoothing, or even ν-smoothing for large ν, remains an important issue.

Acknowledgement
This work was supported in part by the National Science Foundation under contract number NSF DMS 9704899.

References

General Theory:

1. Corduneanu, C.: Integral equations and applications, Cambridge University Press, Cambridge, 1991.
2. Eldén, L.: An algorithm for the regularization of ill-conditioned, banded least squares problems, SIAM J. Sci. Statist. Comput. **5** (1984) no. 1, 237–254.
3. Engl, H. W., Gfrerer, H.: A posteriori parameter choice for general regularization methods for solving linear ill-posed problems, Appl. Numer. Math. **4** (1988) no. 5, 395–417.
4. Engl, H. W., Hanke, M., Neubauer, A.: Regularization of inverse problems, Kluwer Academic Publishers Group, Dordrecht, 1996.
5. Faber, V., Manteuffel, T. A., White, Jr., A. B., Wing, G. M.: Asymptotic behavior of singular values and singular functions of certain convolution operators, Comput. Math. Appl. Ser. A **12** (1986) no. 6, 733–747.
6. Faber, V., Wing, G. M.: Asymptotic behavior of singular values of convolution operators, Rocky Mountain J. Math. **16** (1986) no. 3, 567–574.
7. Gripenberg, G., Londen, S.-O., Staffans, O.: Volterra integral and functional equations, Cambridge University Press, Cambridge-New York, 1990.
8. Groetsch, C. W.: The theory of Tikhonov regularization for Fredholm equations of the first kind, Pitman (Advanced Publishing Program), Boston-London, 1984.
9. Lavrent'ev, M. M. Savel'ev, L. Ya.: Linear operators and ill-posed problems, Consultants Bureau, New York, 1995.

10. Natterer, F.: Regularisierung schlecht gestellter Probleme durch Projektionsverfahren, Numer. Math. **28** (1977) no. 3, 329–341.
11. Natterer, F.: The mathematics of computerized tomography, B. G. Teubner, Stuttgart, 1986.
12. Nohel, J. A., Shea, D. F.: Frequency domain methods for Volterra equations, Advances in Math. **22** (1976) no. 3, 278–304.
13. Raus, T.: Residue principle for ill-posed problems. Acta et comment. Univers. Tartuensis **672** (1984) 16–26.
14. Schmaedeke, W. W.: Approximate solutions for Volterra integral equations of the first kind, J. Math. Anal. Appl. **23** (1968) 604–613.

The Inverse Heat Conduction Problem:

15. Alifanov, O. M., Artyukhin, E. A., Rumyantsev, S. V: Extreme methods for solving ill-posed problems with applications to inverse heat transfer problems, Begell House, Inc., New York, 1995.
16. Beck, J. V., Blackwell, B., St. Clair, Jr., C. R.: Inverse heat conduction, Wiley-Interscience, 1985.
17. Eldén, L.: Numerical solution of the sideways heat equation, Inverse problems in diffusion processes (Lake St. Wolfgang, 1994) SIAM, Philadelphia, PA, 1995, pp. 130–150.
18. Eldén, L.: Numerical solution of the sideways heat equation by difference approximation in time, Inverse Problems **11** (1995) no. 4, 913–923.
19. Engl, H. W., Rundell, W. (eds.): Inverse problems in diffusion processes, Society for Industrial and Applied Mathematics (SIAM) Philadelphia, PA, 1995.
20. Frankel, J. I.: Residual-minimization least-squares method for inverse heat conduction, Comput. Math. Appl. **32** (1996) no. 4, 117–130.
21. Hào, D. N., Reinhardt, H.-J.: On a sideways parabolic equation, Inverse Problems **13** (1997) no. 2, 297–309.
22. Hào, D. N., Reinhardt, H.-J.: Recent contributions to linear inverse heat conduction problems, J. Inverse Ill-Posed Probl. **4** (1996) no. 1, 23–32.
23. Janno, J., Wolfersdorf, L. V.: Identification of memory kernels in general linear heat flow, J. Inverse Ill-Posed Probl. **6** (1998) 141–164.
24. Janno, J., Wolfersdorf, L. V.: Inverse problems for identification of memory kernels in heat flow, J. Inverse Ill-Posed Probl. **4** (1996) 39–66.
25. Kurpisz, K., Nowak, A. J.: Inverse thermal problems, Computational Mechanics Publications, Southampton, 1995.
26. Liu, J.: A stability analysis on Beck's procedure for inverse heat conduction problems, J. Comput. Phys. **123** (1996) no. 1, 65–73.
27. Liu, J., Guerrier, B., Bénard, C.: A sensitivity decomposition for the regularized solution of inverse heat conduction problems by wavelets, Inverse Problems **11** (1995) no. 6, 1177–1187.
28. Mejía, C. E., Murio, D. A.: Numerical solution of generalized IHCP by discrete mollification, Comput. Math. Appl. **32** (1996) no. 2, 33–50.
29. Murio, D. A., Liu, Y., Zheng, H.: Numerical experiments in multidimensional IHCP on bounded domains, Inverse Problems in Diffusion Processes (Lake St. Wolfgang, 1994) SIAM, Philadelphia, PA, 1995, pp. 151–180.
30. Murio, D. A., Zheng, H. C.: A stable algorithm for 3D-IHCP, Comput. Math. Appl. **29** (1995) no. 5, 97–110.

31. Regińska, T.: Sideways heat equation and wavelets, J. Comput. Appl. Math. **63** (1995) no. 1-3, 209–214, International Symposium on Mathematical Modelling and Computational Methods Modelling 94 (Prague, 1994).
32. Regińska, T., Eldén, L.: Solving the sideways heat equation by a wavelet-Galerkin method, Inverse Problems **13** (1997) no. 4, 1093–1106.
33. Tautenhahn, U.: Optimal stable approximations for the sideways heat equation, J. Inverse Ill-Posed Probl. **5** (1997) no. 3, 287–307.
34. Wolfersdorf, L. V.: Inverse problems for memory kernels in heat flow and viscoelasticity, J. Inverse Ill-Posed Probl. **4** (1996) 341-354.
35. Zhan, S., Murio, D. A.: Identification of parameters in one-dimensional IHCP, Comput. Math. Appl. **35** (1998) no. 3, 1–16.
36. Zheng, H., Murio, D. A.: 3D-IHCP on a finite cube, Comput. Math. Appl. **31** (1996) no. 1, 1–14.

Singular Perturbation Theory and Regularization Methods:

37. Angell, J. S., Olmstead, W. E.: Singularly perturbed Volterra integral equations, SIAM J. Appl. Math. **47** (1987) no. 1, 1–14.
38. Angell, J. S., Olmstead, W. E.: Singularly perturbed Volterra integral equations. II, SIAM J. Appl. Math. **47** (1987) no. 6, 1150–1162.
39. Asanov, A.: A class of systems of Volterra integral equations of the first kind, Funktsional. Anal. i Prilozhen. **17** (1983) no. 4, 73–74, English transl.: Functional Analysis and its Applications **17** (1983) 303–4.
40. Baev, A. V.: Solution of an inverse problem for the wave equation using a regularizing algorithm, Zh. Vychisl. Mat. i Mat. Fiz. **25** (1985) 140–146, 160, English transl.: USSR Comput. Math. Math. Phys. **25** (1985) no. 1, 93–97.
41. Denisov, A. M.: The approximate solution of a Volterra equation of the first kind, Ž. Vyčisl. Mat. i Mat. Fiz. **15** (1975) no. 4, 1053–1056, 1091, English transl.: USSR Comput. Math. Math. Phys. **15** (1975) 237–239.
42. Denisov, A. M., Korovin, S. V.: A Volterra-type integral equation of the first kind, Vestnik Moskov. Univ. Ser. XV Vychisl. Mat. Kibernet. **1992** (1992) 22–28, 64, English transl.: Moscow Univ. Comp. Math. Cybernetics (1992) 19–24.
43. Denisov, A. M., Lorenzi, A.: On a special Volterra integral equation of the first kind, Boll. Un. Mat. Ital. B (7) **9** (1995) 443–457.
44. Imanaliev, M. I., Asanov, A.: Solutions of systems of nonlinear Volterra integral equations of the first kind, Dokl. Akad. Nauk SSSR **309** (1989) no. 5, 1052–1055, English transl.: Soviet Math. Dokl. **40** (1990) 610–613.
45. Imanaliev, M. I., Khvedelidze, B. V., Gegeliya, T. G., Babaev, A. A., Botashev, A. I.: Integral equations, Differentsial'nye Uravneniya **18** (1982) no. 12, 2050–2069, 2206, English transl.: Differential Equations **18** (1982) 1442–1458.
46. Imomnazarov, B.: Approximate solution of integro-operator equations of Volterra type of the first kind, Zh. Vychisl. Mat. i Mat. Fiz. **25** (1985) no. 2, 302–306, 319, English transl.: USSR Comput. Math. Math. Phys. **25** (1985) 199–202.
47. Imomnazarov, B.: Regularization of dissipative operator equations of the first kind, Zh. Vychisl. Mat. i Mat. Fiz. **22** (1982) no. 4, 791–800, 1019, English transl.: USSR Comput. Math. Math. Phys. **22** (1982) 22-32.
48. Janno, J., Wolfersdorf, L. V.: Regularization of a class of nonlinear Volterra equations of a convolution type, J. Inverse Ill-Posed Probl. **3** (1995) 249–257.
49. Kauthen, J.-P.: A survey of singularly perturbed Volterra equations, Appl. Numer. Math. **24** (1997) no. 2-3, 95–114, Volterra centennial (Tempe, AZ, 1996).

50. Lavrent'ev, M. M.: Numerical solution of conditionally properly posed problems, Numerical solution of partial differential equations, II (SYNSPADE, 1970) (Proc. Sympos., Univ. Maryland, College Park, Md., 1970) Academic Press, New York, 1971, pp. 417–432.
51. Lavrent'ev, M. M.: O nekotorykh nekorrektnykh zadachakh matematicheskoui fiziki, Izdat. Sibirsk. Otdel. Akad. Nauk SSSR, Novosibirsk, 1962, English transl. by Robert J. Sacker: Some improperly posed problems of mathematical physics, Springer-Verlag, Berlin, 1967.
52. Magnickiĭ, N. A.: The approximate solution of certain Volterra integral equations of the first kind, Vestnik Moskov. Univ. Ser. XV Vyčisl. Mat. Kibernet. **1978** (1978) no. 1, 91–96, English transl.: Moscow Univ. Comput. Math. Cybernetics **1978** (1978) no. 1, 74–78.
53. Sergeev, V. O.: Regularization of Volterra equations of the first kind, Dokl. Akad. Nauk SSSR **197** (1971) 531–534, English transl.: Soviet Math. Dokl . **12** (1971) 501–505.

Local Regularization and Sequential Predictor-Corrector Methods:

54. Cinzori, A. C., Lamm, P. K.: Future polynomial regularization of ill-posed Volterra equations, submitted, 1998.
55. Lamm, P. K.: Approximation of ill-posed Volterra problems via predictor-corrector regularization methods, SIAM J. Appl. Math. **56** (1996) no. 2, 524–541.
56. Lamm, P. K.: Future-sequential regularization methods for ill-posed Volterra equations. Applications to the inverse heat conduction problem, J. Math. Anal. Appl. **195** (1995) no. 2, 469–494.
57. Lamm, P. K.: Regularized inversion of finitely smoothing Volterra operators: predictor-corrector regularization methods, Inverse Problems **13** (1997) no. 2, 375–402.
58. Lamm, P. K.: Solution of ill-posed Volterra equations via variable-smoothing Tikhonov regularization, Inverse problems in geophysical applications (Yosemite, CA, 1995) SIAM, Philadelphia, PA, 1997, pp. 92–108.
59. Lamm, P. K.: Variable-smoothing regularization methods for inverse problems, To appear in Conf. Proceedings of 6th Mediterranean Conference on Control and Systems (Sardinia, 1998).
60. Lamm, P. K., Eldén, L.: Numerical solution of first-kind Volterra equations by sequential Tikhonov regularization, SIAM J. Numer. Anal. **34** (1997) no. 4, 1432–1450.
61. Lamm, P. K., Scofield, T. L.: Local regularization methods for the stabilization of ill-posed Volterra problems, preprint, 1998.
62. Lamm, P. K., Scofield, T. L.: Sequential predictor-corrector methods for the variable regularization of Volterra inverse problems, preprint, 1998.

Iterative Methods and Lavrent'ev's m-times Iterated Method:

63. Plato, R.: Iterative and parametric methods for linear ill-posed problems, Habiliationsschrift Fachbereich Mathematik, TU Berlin, 1995.
64. Plato, R.: Lavrentiev's method for linear Volterra integral equations of the first kind, with applications to the non-destructive testing of optical-fibre preforms, Inverse Problems in Medical Imaging and Nondestructive Testing (Oberwolfach, 1996) Springer, Vienna, 1997, pp. 196–211.

65. Plato, R.: On the discrepancy principle for iterative and parametric methods to solve linear ill-posed equations, Numer. Math. **75** (1996) no. 1, 99–120.

66. Plato, R.: Resolvent estimates for Abel integral operators and the regularization of associated first kind integral equations, J. Integral Equations Appl. **9** (1997) no. 3, 253–278.

67. Plato, R.: The Galerkin scheme for Lavrentiev's m-times iterated method to solve linear accretive Volterra integral equations of the first kind, BIT **37** (1997) no. 2, 404–423.

68. Plato, R.: The Lavrentiev-regularized Galerkin method for linear accretive ill-posed problems, Matimyas Matematika (Journal of the Mathematical Society of the Philippines) Special Issue, August 1998, International Conf. Inverse Problems and Applications, Proc. Manila 1998, pp. 57–66.

69. Plato, R., Hämarik, U.: On pseudo-optimal parameter choices and stopping rules for regularization methods in Banach spaces, Numer. Funct. Anal. Optim. **17** (1996) no. 1-2, 181–195.

70. Vasin, V. V.: Monotone iterative processes for nonlinear operator equations and their applications to Volterra equations, J. Inverse Ill-Posed Probl. **4** (1996) no. 4, 331–340.

71. Vasin, V. V.: Monotone iterative processes for operator equations in partially ordered spaces, Dokl. Akad. Nauk **349** (1996) no. 1, 7–9.

Differentiation and Mollification Methods:

72. Hào, D. N.: A mollification method for ill-posed problems, Numer. Math. **68** (1994) no. 4, 469–506.

73. Hegland, M., Anderssen, R. S.: A mollification framework for improperly posed problems, Numer. Math. **78** (1998) no. 4, 549–575.

74. Kabanikhin, S. I.: Numerical analysis of inverse problems, J. Inverse Ill-Posed Probl. **3** (1995) no. 4, 278–304.

75. Louis, A. K.: A unified approach to regularization methods for linear ill-posed problems, Inverse Problems, to appear, May 1998.

76. Louis, A. K.: Application of the approximate inverse to 3D X-ray CT and ultrasound tomography, Inverse Problems in Medical Imaging and Nondestructive Testing (Oberwolfach, 1996) Springer, Vienna, 1997, pp. 120–133.

77. Louis, A. K.: Approximate inverse for linear and some nonlinear problems, Inverse Problems **12** (1996) no. 2, 175–190.

78. Louis, A. K.: Constructing an approximate inverse for linear and some nonlinear problems in engineering, Inverse problems in engineering, ASME, New York, 1998, pp. 367–374.

79. Louis, A. K., Maaß, P.: A mollifier method for linear operator equations of the first kind, Inverse Problems **6** (1990) no. 3, 427–440.

80. Magnickiĭ, N. A.: A method of regularizing Volterra equations of the first kind, Ž. Vyčisl. Mat. i Mat. Fiz. **15** (1975) no. 5, 1317–1323, 1363, English transl: USSR Comput. Math. Math. Phys. **15** (1975) 221–228.

81. Murio, D. A.: The mollification method and the numerical solution of ill-posed problems, John Wiley & Sons, Inc., New York, 1993.

82. Srazhidinov, A.: Regularization of Volterra integral equations of the first kind, Differentsial'nye Uravneniya **26** (1990) no. 3, 521–530, 551, English transl: Differential Equations **26** (1990) 390–398.

80

Numerical Methods for First-Kind Volterra Problems:

83. Andrade, C., Franco, N. B., McKee, S.: Convergence of linear multistep methods for Volterra first kind equations with $k(t, t) \equiv 0$, Computing **27** (1981) no. 3, 189–204.

84. Baker, C. T. H.: Methods for Volterra equations of first kind, Numerical solution of integral equations (Liverpool-Manchester Summer School, 1973) Clarendon Press, Oxford, 1974, pp. 162–174.

85. Brunner, H.: 1896–1996: One hundred years of Volterra integral equations of the first kind, Appl. Numer. Math. **24** (1997) no. 2-3, 83–93, Volterra centennial (Tempe, AZ, 1996).

86. Brunner, H.: Discretization of Volterra integral equations of the first kind, Math. Comp. **31** (1977) no. 139, 708–716.

87. Brunner, H.: Discretization of Volterra integral equations of the first kind. II, Numer. Math. **30** (1978) no. 2, 117–136.

88. Brunner, H.: On the discretization of Volterra integral equations, Nieuw Arch. Wisk. (4) **2** (1984) no. 2, 189–217.

89. Brunner, H.: Open problems in the discretization of Volterra integral equations, Numer. Funct. Anal. Optim. **17** (1996) no. 7-8, 717–736.

90. Brunner, H.: Superconvergence of collocation methods for Volterra integral equations of the first kind, Computing **21** (1978/79) no. 2, 151–157.

91. Brunner, H., van der Houwen, P. J.: The numerical solution of Volterra equations, North-Holland Publishing Co., Amsterdam-New York, 1986.

92. de Hoog, F., Weiss, R.: High order methods for Volterra integral equations of the first kind, SIAM J. Numer. Anal. **10** (1973) 647–664.

93. de Hoog, F., Weiss, R.: On the solution of Volterra integral equations of the first kind, Numer. Math. **21** (1973) 22–32.

94. Dixon, J., McKee, S.: A unified approach to convergence analysis of discretization methods for Volterra-type equations, IMA J. Numer. Anal. **5** (1985) no. 1, 41–57.

95. Dixon, J., McKee, S., Jeltsch, R.:Convergence analysis of discretization methods for nonlinear first kind Volterra integral equations, Numer. Math. **49** (1986) no. 1, 67–80.

96. Eggermont, P. P. B.: Approximation properties of quadrature methods for Volterra integral equations of the first kind, Math. Comp. **43** (1984) no. 168, 455–471.

97. Eggermont, P. P. B.: Beyond superconvergence of collocation methods for Volterra integral equations of the first kind, Constructive methods for the practical treatment of integral equations (Oberwolfach, 1984) Birkhäuser, Basel, 1985, pp. 110–119.

98. Eggermont, P. P. B.: Collocation for Volterra integral equations of the first kind with iterated kernel, SIAM J. Numer. Anal. **20** (1983) no. 5, 1032–1048.

99. Eggermont, P. P. B.: Improving the accuracy of collocation solutions of Volterra integral equations of the first kind by local interpolation, Numer. Math. **48** (1986) no. 3, 263–279.

100. Gladwin C. J., Jeltsch R.: Stability of quadrature rule methods for first kind Volterra integral equations, Nordisk Tidskr. Informationsbehandling (BIT) **14** (1974) 144–151.

101. Huber, A.: Eine Näherungsmethode zur Auflösung Volterrascher Integralgleichungen, Monatsh. Math. Phys. **47** (1939) 240–246.

102. Hung, H. S.: The numerical solution of differential and integral equations by spline functions, Technical Summary Report 1053, Mathematics Research Center, University of Wisconsin, 1970.

103. Jones, J. G.: On the numerical solution of convolution integral equations and systems of such equations, Math. Comp. **15** (1961) 131–142.

104. Kauthen, J.-P., Brunner, H.: Continuous collocation approximations to solutions of first kind Volterra equations, Math. Comp. **66** (1997) no. 220, 1441–1459.

105. Kobayasi, M.: On numerical solution of the Volterra integral equations of the first kind by trapezoidal rule, Rep. Statist. Appl. Res. Un. Japan. Sci. Engrs. **14** (1967) no. 2, 1–14.

106. Linz, P.: A survey of methods for the solution of Volterra integral equations of the first kind, Application and numerical solution of integral equations (Proc. Sem., Australian Nat. Univ., Canberra, 1978) Nijhoff, The Hague, 1980, pp. 183–194.

107. Linz, P.: Analytical and numerical methods for Volterra equations, Society for Industrial and Applied Mathematics (SIAM) Philadelphia, Pa., 1985.

108. Linz, P.: Numerical methods for Volterra integral equations of the first kind., Comput. J. **12** (1969) 393–397.

109. Linz, P.: Numerical methods of Volterra integral equations with applications to certain boundary value problems, Ph.D. thesis, University of Wisconsin, Madison, 1968.

110. Linz, P.: Product integration methods for Volterra integral equations of the first kind, Nordisk Tidskr. Informationsbehandling (BIT) **11** (1971) 413–421.

111. Linz, P.: The numerical solution of Volterra integral equations by finite difference methods, Technical Summary Report 825, Mathematics Research Center, University of Wisconsin, 1967.

112. Linz, P.: The solution of Volterra equations of the first kind in the presence of large uncertainties, Treatment of integral equations by numerical methods (Durham, 1982) Academic Press, London, 1982, pp. 123–130.

113. McAlevey, L. G.: Product integration rules for Volterra integral equations of the first kind, BIT **27** (1987) no. 2, 235–247.

114. McKee, S.: A review of linear multistep methods and product integration methods and their convergence analysis for first kind Volterra integral equations, Treatment of integral equations by numerical methods (Durham, 1982) Academic Press, London, 1982, pp. 153–161.

115. McKee, S.: Best convergence rates of linear multistep methods for Volterra first kind equations, Computing **21** (1978/79) no. 4, 343–358.

116. Richter, G. R.: Numerical solution of integral equations of the first kind with nonsmooth kernels, SIAM J. Numer. Anal. **15** (1978) no. 3, 511–522.

117. Scott, J. A.: A unified analysis of discretization methods for Volterra-type equations, Constructive methods for the practical treatment of integral equations (Oberwolfach, 1984) Birkhäuser, Basel, 1985, pp. 244–255.

118. Scott, J. A., McKee, S.: On the exact order of convergence of discrete methods for Volterra-type equations, IMA J. Numer. Anal. **8** (1988) no. 4, 511–515.

119. Taylor, P. J.: The solution of Volterra integral equations of the first kind using inverted differentiation formulae, Nordisk Tidskr. Informationsbehandling (BIT) **16** (1976) no. 4, 416–425.

120. van der Houwen, P. J., te Riele, H. J. J.: Linear multistep methods for Volterra integral and integro-differential equations, Math. Comp. **45** (1985) no. 172, 439–461.

121. Weiss, R., Anderssen, R. S.: A product integration method for a class of singular first kind Volterra equations, Numer. Math. **18** (1971/72) 442–456.

122. Wolkenfelt, P. H. M.: Modified multilag methods for Volterra functional equations, Math. Comp. **40** (1983) no. 161, 301–316.
123. Wolkenfelt, P. H. M.: Reducible quadrature methods for Volterra integral equations of the first kind, BIT **21** (1981) no. 2, 232–241.
124. Young, A.: The application of approximate product integration to the numerical solution of integral equations, Proc. Roy. Soc. London Ser. A. **224** (1954) 561–573.

Layer Stripping

John Sylvester

Dept. of Mathematics, Box 354350
University of Washington
Seattle, WA 98195
sylvester@math.washington.edu
http://www.math.washington.edu/~sylvest/

Abstract. We describe a rigorous layer stripping approach to inverse scattering for the Helmholtz equation in one dimension. In section 3, we show how the Ricatti ordinary differential equation, which comes from the invariant embedding approach to forward scattering, becomes an inverse scattering algorithm when combined with the principle of causality. In section 4 we discuss a method of stacking and splitting layers. We first discuss a formula for combining the reflection coefficients of two layers to produce the reflection coefficient for the thicker layer built by stacking the first layer upon the second. We then describe an algorithm for inverting this procedure; that is, for splitting a reflection coefficient into two thinner reflection coefficients. We produce a strictly convex variational problem whose solution accomplishes this splitting. Once we can split an arbitrary layer into two thinner layers, we proceed recursively until each reflection coefficients in the stack is so thin that the Born approximation holds (i.e. the reflection coefficient is approximately the Fourier transform of the derivative of the logarithm of the wave speed). We then invert the Born approximation in each thin layer.

1   Introduction

The layer stripping approach to inverse scattering is, in principle, very simple. It can be summarized as follows:

- Born Approximation A thin layer of a medium is easy to recognize from how it reflects an incoming wave. In many layer stripping methods, the layer is infinitesmally thin and the Born Approximation becomes a *trace formula*.
- Causality Principle The reflections from the thin layer nearest the receiver are sensed before the reflections from deeper within the medium.
- Splitting The initial reflection guaranteed by the causality principle, combined with a specific model of wave propagation, provides enough information to determine the upper thin layer and to compute the response of the medium with that thin layer stripped away.

This approach has been investigated in many papers (e.g.[4], [3], [6], [8],[1], [2], [7]) As with any proposed method, the crucial question is stability. In more than one dimension this question is open, but in one dimension we can give an algorithm and a rigorous proof that it must succeed.

In our point of view, probably the most unexpected lesson here is the role of characterization. Most inverse problems have four fundamental parts: uniqueness, reconstruction, continuous dependence, and characterization. In our study of layer stripping, the characterization of the range of the scattering operator has consistently provided the insight which led to our reconstruction algorithms and the proper formulation of continuous dependence.

The next few subsections contain the results (some from [9], some from [10], and some new) for which we will provide proofs and elaborations in the next two sections.

1.1   The 1-D Helmholtz Equation and Travel Time

The one dimensional Helmholtz equation is:

$$\frac{d^2u}{dz^2} + \frac{\omega^2}{c^2(z)}u = 0 \tag{1}$$

We work on the negative half line or a subset thereof (a layer), $-\infty \leq B < z < T \leq 0$. The reflection and transmission coefficients are defined by the following conditions at the top and bottom of the layer:

$$u(z,\omega) \underset{T}{\sim} \sqrt{c_T}\left(e^{\frac{-i\omega(z-T)}{c_T}} + r(\omega)e^{\frac{i\omega(z-T)}{c_T}}\right) \tag{2}$$

$$u(z,\omega) \underset{B}{\sim} \sqrt{c_B}t(\omega)e^{\frac{-i\omega(z-B)}{c_B}} \tag{3}$$

where the symbol $\underset{B}{\sim}$ means "has the same Cauchy data at $z = B$ as". That is, $u \underset{B}{\sim} v$ means that $u(B,\omega) = v(B,\omega)$ and that $u'(B,\omega) = v'(B,\omega)$. This is equivalent to the hypothesis that (1) holds on the whole line and that $c(z)$ is continuous and constant outside the layer. In the case that $B = -\infty$ we understand (3) as a limit.

Because only variations in the wave speed produce reflections, it is convenient to introduce

$$\alpha = -\frac{1}{2}\frac{dc}{dz}$$

At the detector, we observe the reflected waves parameterized by the time it takes the wave to reach them and return. It is convenient to replace the physical depth, z , by the travel time depth, x.

$$x(z) = \int_0^z \frac{dz}{c}$$

In travel time coordinates, (1) becomes

$$u'' + 2\alpha(x)u' + \omega^2 u = 0 \tag{4}$$

and the definition of reflection and transmission for the layer $B < x < T$ changes slightly (but r and t remain the same):

$$u(x,\omega) \underset{T}{\sim} e^{-i\omega(x-T)} + r(\omega)e^{i\omega(x-T)} \tag{5}$$

$$u(x,\omega) \underset{B}{\sim} e^{\int_B^T \alpha} t(\omega)e^{-i\omega(x-B)} \tag{6}$$

Our scattering theory will study the map S between α and r

$$\alpha(x) \overset{S}{\mapsto} r(\omega)$$

and our inverse scattering algorithm will produce α from r. Before proceeding further, we discuss the recovery of $c(z)$ from $\alpha(x)$. First note that, with $T = 0$,

$$z'(x) = c(z(x)) = e^{-2\int_0^x \alpha}$$

so that

$$z(x) = \int_0^x e^{-2\int_0^{x'} \alpha} dx'$$

is monotonone and therefore invertible on its range. Therefore,

$$c(z) = e^{-2\int_0^{x(z)} \alpha} \tag{7}$$

Our inverse scattering theory works with $\alpha \in L^2$. Thus, for some α's, (7) will produce a $c(z)$ defined only on a finite interval, with $c = 0$ at the bottom of that interval. This is as it should be. For example, if $\alpha \equiv 0.5$ (not exactly L^2, but easy to compute) then

$$c(z) = 1 + z$$

This corresponds to a medium whose wave speed decreases to zero as z approaches -1. In this medium, it takes an infinitely long time for a wave to reach $z = 1$ and no wave penetrates deeper than that. Our inversion can therefore do no better than to return the wave speed at depths above 1.

1.2 Characterization and continuous dependence

The Fourier transform of a function in $L^2(-\infty, 0)$ extends to be analytic in the complex upper half plane. The set of such analytic functions from the linear Hardy space, $H^2(C^+)$. The norm on $H^2(C^+)$ is defined to be

$$\|\rho\|_{H^2} = \sup_{b>0} \|\rho(\cdot + ib)\|_{L^2} \tag{8}$$

The Fourier transforms of a real-valued functions belong to

$$\mathcal{H}^2(C^+) = \{\rho \in H^2(C^+) \mid \rho(-\bar{\omega}) = \overline{\rho(\omega)}\} \tag{9}$$

The range of the (nonlinear) scattering map is also a Hardy space, $\mathcal{H}^E(C^+)$. $\mathcal{H}^E(C^+)$ is not linear, but it is a complete metric space (see section 2). We define

$$E(r) := \int e(r)d\omega := \int (-\log(1 - |r|^2))d\omega \tag{10}$$

and $\mathcal{H}^E(C^+)$ to be the subset of $\mathcal{H}^2(C^+)$ such that

$$\mathcal{H}^E(C^+) = \left\{ \rho \in \mathcal{H}^2(C^+) \mid \sup_{b>0} E(\rho(\cdot + ib)) < \infty \right\} .$$

Our basic results on characterization of the range of the scattering map are stated below. They tell us how to recognize a reflection coefficient and how to recognize a reflection coefficient of a finite width layer (a layer of width W means an $\alpha \in L^2(-W, 0)$; when we say that r has width W, we mean that it is the reflection coefficient of a layer of width W).

Theorem 1 (Characterization of Reflection Coefficients).

- The scattering map is a homeomorphism from $L^2(-\infty, 0)$ onto $\mathcal{H}^E(C^+)$.
- The nonlinear Plancherel equality holds

$$E(r) = \pi\|\alpha\|_{L^2}^2$$

- An $r \in \mathcal{H}^E(C^+)$ has width W if and only if

$$\frac{r}{t} \in e^{-i\omega W}\mathcal{H}^2(C^+) \cap e^{i\omega W}\mathcal{H}^2(C^-) \tag{11}$$

The second condition involves the transmission coefficient, t, which can be computed from r, as long as we know $r(\omega)$ for all real ω. t is the $e^{i\omega W}$ times the unique outer function with modulus $\sqrt{1 - |r^2|}$ (see (114)).

1.3   Stacking and Splitting Layers

Suppose that we stack two layers, one with width W_1 and a second with width W_2, the resulting layer is

$$\alpha_{12} = \begin{cases} \alpha_1(x) & 0 > x > -W_1 \\ \alpha_2(x + W_1) & -W_1 > x > -(W_1 + W_2) \end{cases} \tag{12}$$

and the resulting reflection coefficient is given by the formula

$$r_{12} = r_1 \circ \frac{t_1}{\bar{t}_1} r_2 \tag{13}$$

where $\circ$ represents the formula for composition of conformal s of the unit disk onto itself.

$$a \circ b := \frac{a+b}{1+\bar{a}b} \tag{14}$$

Notice that, according to the Plancherel equality, the E-norm of the layer-composition (13) is the sum of the E-norms of the reflection coefficients of the layers.

$$E(r) = E(r_1) + E(r_2) \tag{15}$$

Our inverse scattering algorithm is based on inverting (13).

Theorem 2 (Layer Splitting Decomposition). *Let* $r \in \mathcal{H}^E(C^+)$, *and let* $W_1 > 0$. *Then the strictly convex variational problem*

$$\min_{\substack{\rho \in \mathcal{H}^E(C^+) \\ r - \rho \in e^{2i\omega W_1} \mathcal{H}^2(C^+)}} E(\rho) \tag{16}$$

has a unique minimizer r_1, *and* r_1 *is the first factor in the unique layer decomposition of* r

$$r = r_1 \circ \frac{t_1}{\bar{t}_1} r_2 \tag{17}$$

such that $r_1, r_2 \in \mathcal{H}^E(C^+)$ *and* r_1 *has width* W_1. *Moreover, the rest of the decomposition, namely* t_1 *and* r_2, *can be computed from formulas (136) and (137).*

1.4 Thin Layers and the Born Approximation

Repeated application of theorem 2 allows us to split a reflection coefficient into a composition of layers of small width. Once the width is small enough, we may resort to the Born approximation or linear inverse scattering, which tells us that the reflection coefficient is approximately the Fourier transform of α at 2ω.

Theorem 3 (Born Approximation). *Let* r *have width* W, *then*

$$\|r(\omega) - \hat{\alpha}(2\omega)\|_{L^\infty} \leq 4\|\alpha\|_{L^2(-W,0)}^3 W^{\frac{3}{2}} \tag{18}$$

$$\|r(\omega) - \hat{\alpha}(2\omega)\|_L^2 \leq \|\alpha\|_{L^2(-W,0)}^2 W^{\frac{1}{2}} \tag{19}$$

$$|\log(t) - i\omega| \leq W^{\frac{1}{2}}\|\alpha\|_{L^2} \tag{20}$$

1.5   Complete Layer Decomposition

Combining theorems 3 and 2, we may compute α from r by solving a sequence of convex variational problems ((16)) and then inverting a sequence of Fourier transforms.The theorem below is a corollary of the last two subsections

Theorem 4. *Let* $r \in \mathcal{H}^E(C^+)$ *and* $\{W_i\}$ *be a sequence of positive real numbers and* $\{S_i\}$ *their partial sums. Then* r *has a unique infinite decomposition:*

$$r = r_1 \circ \frac{t_1}{\bar{t}_1} \left(r_2 \circ \frac{t_2}{\bar{t}_2} \left(r_3 \circ \ldots \right. \right. \tag{21}$$

The individual terms in the sum

$$a = \sum_{i=1}^{\infty} \left(e^{2i\omega S_i} \frac{r_i}{t_i}(2\omega) \right)^{\vee} \tag{22}$$

are supported on disjoint intervals of width W_i *and* a *converges to* $\alpha = S^{-1}r$ *in* L^2 *as the width of the* W_i *approach zero.*

We will elaborate on the previous subsections in the next three sections.

2   The Geometry of $H^E(C^+)$

2.1   The Hardy Spaces $\mathcal{H}^p$

We recall, following [5], that for $1 \leq p \leq \infty$

$$H^p(\mathbb{C}^{\pm}) = \{\rho \mid \rho \text{ holomorphic in } \mathbb{C}^{\pm} \text{ and } \sup_{b>0} \|\rho(\cdot + ib)\|_{L^p} < \infty\}$$

All such functions have, and are uniquely determined by, their boundary values on the real axis. We will always demand an additional symmetry :

$$\mathcal{H}^p(\mathbb{C}^{\pm}) = \{\rho \in H^p(\mathbb{C}^{\pm}) \mid \rho(-\bar{\omega}) = \overline{\rho(\omega)}\} \tag{23}$$

We will make use primarily of $\mathcal{H}^2$. In fact, $\mathcal{H}^2(\mathbb{C}^{\pm})$ are exactly the Fourier transforms of real valued L^2 functions supported on the negative (resp. positive) half line (see [5]). With $\mathcal{L}^2$ denoting L^2 functions with $f(-\omega) = \overline{f(\omega)}$, we have

$$\mathcal{L}^2(\mathbb{R}) = \mathcal{H}^2(\mathbb{C}^+) \oplus \mathcal{H}^2(\mathbb{C}^-) \tag{24}$$

We let $P^{\pm}$ denote the projections onto $\mathcal{H}^2(\mathbb{C}^{\pm})$ along $\mathcal{H}^2(\mathbb{C}^{\mp})$. P^+ is called the *Riesz transform*. We shall often write

$$f = f^+ + f^- \tag{25}$$

denoting $P^{\pm}f$ by $f^{\pm}$. We speak of f^+ as the *causal* part of f because it is the Fourier transform of a function supported in the past, and to f^- as the *a-causal* part, because it depends on the future. A reflection coefficient must be causal because reflections cannot arrive at the detector before they have originated from the source.

2.2   The Hardy Space $\mathcal{H}^E$

We shall define $\mathcal{H}^E(\mathbb{C}^+)$ like any other Hardy space :

$$\mathcal{H}^E(\mathbb{C}^+) = \{r \mid r \text{ holomorphic in } \mathbb{C}^+, \sup_{b>0} E(r) < \infty, \text{ and } r(-\bar{\omega}) = \overline{r(\omega)}\}$$

where the L^p norm is replaced by

$$E(r) = \int (-\log(1 - |r|^2))d\omega \tag{26}$$

$$= \sum_{k=1}^{\infty} \frac{\int |r|^{2k}}{k} \tag{27}$$

An immediate consequence of (27) is:

Lemma 1. *$E(r)$ is strictly convex and positive.*

We can use E to define a metric to measure the distance between two reflection coefficients and hence view $\mathcal{H}^E(C^+)$ as a metric space.

$$D_E^2(r, s) := E(-r \circ s) \tag{28}$$

We will call D_E the E-distance or the distance in the E-metric. A little motivation for the above definition is probably in order. Let

$$e(r) = -log(1 - |r|^2) \tag{29}$$

$$p(r) = log\left(\frac{1 + |r|}{1 - |r|}\right) \tag{30}$$

$$d_e(r, s) = e(-r \circ s) \tag{31}$$

$$d_p(r, s) = p(-r \circ s) \tag{32}$$

For the moment, let r and s denote complex numbers in the unit disk. Then $p(r)$ is the Poincaré distance from r to the origin; $d_p(r, s)$ is the Poincaré distance from r to s. The definition (32) can also be described as follows: Choose a conformal map, F, of the unit disk which maps r to the origin, then measure the distance between $F(s)$ and the origin. This definition makes the Poincaré distance conformally invariant. The analogous definition gives the e-metric (and hence the E-metric) the same property.

Our reflection coefficients will take values in the Poincaré disk. Furthermore, the formula (13) shows that when we add a layer, the new reflection coefficient is formed by applying a conformal map to the old one, so that, in a conformally invariant metric, adding the same top layer to two different layers will not change the E-distance between their reflection coefficients.

Lemma 2. *The metrics, d_e, and therefore D_E, are conformally invariant; i.e. for any conformal F of the unit disk onto itself*

$$d_e(a,b) = d_e(F(a), F(b)).$$

<u>Proof</u>   A conformal of the unit disk, $F(z)$ has the form

$$F(z) = e^{i\theta} \frac{a-z}{1-\bar{a}z} \tag{33}$$

where $\theta \in \mathbb{R}$ and a belongs to the unit disk. We use the notation F_a to refer to the in (33) with $\theta = 0$. Now

$$\begin{aligned} d_e(b,c) &= e(-b \circ c) \\ &= e(F_b(c)) \end{aligned}$$

while

$$d_e(G(b), G(c)) = e(F_{G(b)}(G(c))).$$

Now

$$F_{G(b)}(G(z)) : b \mapsto 0$$

so that, according to (33),

$$F_{G(b)}(G(z)) = e^{i\theta} F_b(z)$$

for some θ, so

$$\begin{aligned} d_e(G(b), G(c)) &= e(e^{i\theta} F_b(c)) \\ &= e(F_b(c)) \end{aligned}$$

∎

Theorem 5 (Cauchy Schwartz and Triangle Inequalities).

$$|E(a,b)| \leq E(a)^{\frac{1}{2}} E(b)^{\frac{1}{2}} \tag{34}$$

$$D_E(r,s) \leq D_E(r,\tau) + D_E(\tau,s). \tag{35}$$

In addition, the Cauchy Schwartz inequality holds for the tails of the series expansion for $E(r)$, i.e.

$$|E_M(a,b)| \leq E_M(a)^{\frac{1}{2}} E_M(b)^{\frac{1}{2}} \tag{36}$$

where

$$E_M(b) := \int \sum_{k=M+1}^{\infty} \frac{|b|^{2k}}{k} d\omega \tag{37}$$

Proof

$$E(a,b) = \int \log(1 - \bar{a}b)d\omega$$

$$= \int \sum_{k=1}^{\infty} \frac{\bar{a}b^k}{k} d\omega$$

$$\leq \int \left(\sum_{k=1}^{\infty} \frac{|a|^{2k}}{k}\right)^{\frac{1}{2}} \left(\sum_{k=1}^{\infty} \frac{|b|^{2k}}{k}\right)^{\frac{1}{2}} d\omega$$

$$\leq \left(\int \sum_{k=1}^{\infty} \frac{|a|^{2k}}{k} d\omega\right)^{\frac{1}{2}} \left(\int \sum_{k=1}^{\infty} \frac{|b|^{2k}}{k} d\omega\right)^{\frac{1}{2}}$$

$$= E(a)^{\frac{1}{2}} E(b)^{\frac{1}{2}}$$

$$D_E^2(a,b) = E(-a \circ b)$$
$$= E(a) + E(b) - 2ReE(a,b)$$
$$\leq E(a) + E(b) + 2E(a)^{\frac{1}{2}}E(b)^{\frac{1}{2}}$$
$$= (E(a)^{\frac{1}{2}} + E(b)^{\frac{1}{2}})^2$$
$$= (D_E(a,0) + D_E(0,b))^2.$$

Now, given any C, choose F conformal and mapping 0 to C. Then

$$D_E(a,b) = D_E(F^{-1}(a), F^{-1}(b))$$
$$\leq D_E(F^{-1}(a), 0) + D_E(0, F^{-1}(b))$$
$$= D_E(a,c) + D_E(c,b).$$

Finally, the last assertion follows from the proof of the first, on simply beginning the summations above at $k = M + 1$ instead of $k + 1$. ∎

Corollary 1. *The unit disk, with the metric d_e, and $\mathcal{H}^E(\mathbb{C}^+)$, with the metric D_E, are complete metric spaces.*

Proof Suppose that a sequence $\{r_n\}$ is E-cauchy. According to (28)

$$D_E^2(r_n, r_m) = E(-r_n \circ r_m)$$
$$= \left\| \frac{r_n - r_m}{1 - \bar{r}_n r_m} \right\|_{L^2}^2$$
$$\geq \frac{1}{4} \|r_n - r_m\|_{L^2}^2$$

so that the sequence is L^2-cauchy, and therefore has an L^2 limit. The triangle inequality guarantees that $E(r_n)$ and hence $E(r)$ are bounded above. Another application of the triangle inequality shows that $D_E^2(r_n, r)$ is bounded above by $(E(r_n) + E(r))^2$ so that we may apply the dominated convergence theorem to conclude that $D_E^2(r_n, r)$ goes to zero. Thus the sequence converges in $\mathcal{H}^E$. ∎

Theorem 6 (Weak and Strong Convergence). *Suppose that, for all* $g \in \mathcal{H}^E(\mathbb{C}^+)$

$$E(r_n, g) \to E(r, g)$$

and that

$$E(r_n) \to E(r)$$

then

$$D_E(r_n, r) \to 0.$$

In other words, weak convergence plus convergence of norms implies strong convergence.

Proof

$$
\begin{aligned}
D_E(r_n, r) &= E(r_n) + E(r) - 2\mathrm{Re}\ E(r_n, r) \\
&\to E(r) + E(r) - 2\mathrm{Re}\ E(r, r) \\
&= 0
\end{aligned}
$$

∎

3   The Ricatti Equation

We begin with a layer with top T and bottom B; for the moment, we assume that α is smooth with compact support and that B lies below the support of α. In [9], we worked with general $\alpha \in L^2$, but here we will work with more restricted α's and use our continuous dependence estimates to extend our results to all $\alpha \in L^2$. We return to the Helmholtz equation

$$u'' + 2\alpha(x)u' + \omega^2 u = 0 \tag{38}$$

and the conditions at the ends of the layer

$$u \underset{T}{\sim} \frac{e^{-\int_B^T \alpha}}{t(T, \omega)} \left(e^{-i\omega(x-T)} + r(\omega)e^{i\omega(x-T)} \right) \tag{39}$$

$$u \underset{B}{\sim} e^{-i\omega(x-B)} \tag{40}$$

The condition (39) is equivalent to the pair of equations

$$u(T,\omega) = \frac{e^{-\int_B^T \alpha}}{t(T,\omega)}(1 + r(T,\omega)) \tag{41}$$

$$u'(T,\omega) = -i\omega\frac{e^{-\int_B^T \alpha}}{t(T,\omega)}(1 - r(T,\omega)) \tag{42}$$

and condition (40) insists that u is the unique solution to (38) with Cauchy data

$$u(B,\omega) = 1 \tag{43}$$
$$u'(B,\omega) = i\omega \tag{44}$$

Note that u is independent of B as long as B stays below the support of α. We shall now differentiate (41) and (42) with respect to T to derive the following differential equations for $r(T,\omega)$ and $t(T,\omega)$.

$$r' = 2i\omega r + \alpha(1 - r^2) \tag{45}$$
$$r(B,\omega) = 0 \tag{46}$$
$$t' = i\omega t - \alpha r t \tag{47}$$
$$t(B,\omega) = 1 \tag{48}$$

Strictly speaking, $'$ in the equations above should be differentiation with respect to T, not x; but as $u(x,\omega)$ does not depend on T,

$$\frac{d}{dT}\left(u\big|_{x=T}\right) = \frac{\partial u}{\partial x}\bigg|_{x=T} + \frac{\partial u}{\partial T}\bigg|_{x=T} \tag{49}$$
$$= u'(T,\omega) + 0 \tag{50}$$

hence derivatives with respect to T and x are equivalent. One way to obtain (45) is to divided (42) by (41), obtaining

$$\frac{u'}{u} = -i\omega\left(\frac{1-r}{1+r}\right) \tag{51}$$

Differentiating (51) gives

$$2i\omega\frac{r'}{(1+r)^2} = q\frac{u''}{u} - \left(\frac{u'}{u}\right)^2$$

$$= -\alpha\frac{u'}{u} + (i\omega)^2\frac{u}{u} - \left(\frac{u'}{u}\right)^2$$

$$= \alpha i\omega\left(\frac{1-r}{1+r}\right) + (i\omega)^2 - \left(-i\omega\left(\frac{1-r}{1+r}\right)\right)^2$$

A little algebra now yields (45). Finally, differentiate (41) to obtain a formula for $u'(T, \omega)$ and set it equal to the formula for $u'(T, \omega)$ in (42). Then use (45) to arrive at (47).

3.1 Forward Scattering

We will establish our forward scattering via the Ricatti equation. We first prove:

Theorem 7. *Let $\alpha \in C_0^\infty$ with $\operatorname{supp}(\alpha) \subset [B, 0]$, then there exists a unique solution $r \in C([B, 0], \mathcal{H}^E(C^+))$ satisfying (45) and (46). In addition,*

$$E(r) = \pi \|\alpha\|_{L^2}^2 \tag{52}$$

Proof

- Fix $\omega \in C^+$ and prove local existence of a solution to the integral equation

$$r(x, \omega) = r(x_0) + \int_{x_0}^{x} e^{2i\omega(x-y)} \alpha(y)\left(1 - r^2(y, \omega)\right) dy \tag{53}$$

by using the estimate

$$\left| \int_{x_0}^{x} e^{2i\omega(x-y)} \alpha(y)\left(r^2 - \tilde{r}^2(y, \omega)\right) dy \right| \le \|\alpha\|_{L^2} |x - x_0|^{\frac{1}{2}} |r + \tilde{r}| \|r - \tilde{r}\| \tag{54}$$

to show that the mapping defined by the right hand side of (53) is a contraction on a suitable ball (say $|r| < 2$) if $|x - x_0|$ is small enough.

- Obtain global existence by noting that the a priori estimate $|r| < 1$ follows from multiplying (45) by $\bar{r}$ and taking real parts to obtain

$$|r|^{2'} = \alpha(r + \bar{r})(1 - |r|^2) \tag{55}$$

- Establish the large ω asymptotics of r by proving some bounds on r and its x-derivatives. We start by differentiating (45) to obtain the differential equation,

$$(r')' = (2i\omega - \alpha r)r' + \alpha'(1 - r^2) \tag{56}$$

and the integral representation,

$$r'(x) = r'(B) + \int_{B}^{x} e^{2i\omega(x-y)} e^{2i \int_x^y \alpha r} \alpha'(y)\left(1 - r^2(y, \omega)\right) dy$$

and, after noting that $r'(B) = \alpha(B) = 0$, the estimate

$$|r'(x)| \le 2\|\alpha\|_{L^2} e^{\|\alpha\|_{L^2} |x-B|^{\frac{1}{2}}} |x - B|^{\frac{1}{2}} \tag{57}$$

The point of (57) is that $|r'(x)|$ is bounded independent of ω (but not x). That the same is true of $|r''(x)|$ and higher derivatives can be established in the same way. Now return to (56) and notice that every term except $2i\omega r'$ is bounded; so it must be bounded also. Hence

$$|r'(x)| \le \frac{C}{\omega}$$

and, via integration, r must also satisfy this estimate. Next examine (45) to see that every term except $2i\omega r + \alpha$ is bounded by constant over ω, so that

$$r(x) = -\frac{\alpha}{2i\omega} + O\left(\frac{1}{\omega^2}\right) \tag{58}$$

-- Establish the Plancherel equality (52) by dividing (55) by $1 - |r|^2$.

$$-\log(1 - |r|^2)' = \alpha(r + \bar{r}) \tag{59}$$

Next integrate both sides with respect to ω along the real axis. Note that since r is holomorphic in $\mathbb{C}^+$ and $\bar{r}$ in $\mathbb{C}^-$ with the asymptotics (58), we may perform a residue calculation to establish, for $b > 0$,

$$\int_{-\infty}^{\infty} r(\omega + ib)d\omega = \pi\alpha \tag{60}$$

$$\int_{-\infty}^{\infty} \overline{r(\omega + ib)}d\omega = 0 \tag{61}$$

Although neither of the integrals (60) nor (61) are continuous in b as it passes through zero, the sum of the two is continuous because (58) guarantees the integrability of $r + \bar{r}$. Thus we arrive at

$$\left(-\int \log(1 - |r|^2)d\omega\right)' = \pi\alpha^2 \tag{62}$$

whence integration in x from B to 0 yields (52). ∎

Theorem 8. *Suppose that* $\alpha_n \longrightarrow \alpha$ *in* L^2. *Then* $r_n \longrightarrow r$ *in* $\mathcal{H}^E$.

<u>Proof</u>

-- r_n^b is Cauchy in L^2. Let $b > 0$ and let ω_b and $r^b(x, \omega)$ denote $\omega + ib$ and $r(x, \omega + ib)$, respectively. Divide the interval $(-\infty, 0)$ into a finite number of intervals such that

$$\|\alpha\|_{L^2(x_k, x_{k+1})} < \frac{1}{4}\sqrt{b} \tag{63}$$

and note that, on each such interval, the same estimate will hold with α replaced by α_n, and $\frac{1}{4}$ replaced by a slightly larger constant, as long as n is large enough. We make use of the integral equation

$$r_n^b(x_{k+1}) = r_n^b(x_k) + \int_{x_k}^{x_{k+1}} e^{2i\omega_b(x_{k+1}-y)}\alpha(y)(1 - (r_n^b)^2)dy \tag{64}$$

and define

$$\rho_{nm}(x) = \sup_{x_k < y < x} ||r_n^b(y) - r_m^b(y)||_{L^2(d\omega)} \tag{65}$$

then

$$\rho_{nm}(x_{k+1}) \le \rho_{nm}(x_k) + ||\hat{a}_n - \hat{a}_m||_{L^2(d\omega)} +$$
$$\int_{x_k}^{x_{k+1}} e^{-2b(x_{k+1}-y)} \left|(\alpha_n - \alpha_m)(r_n^b)^2 + \alpha_m \left((r_n^b)^2 - (r_m^b)^2\right)\right|$$

Applying the Cauchy-Schwartz inequlity a few more times,

$$\rho_{nm}(x_{k+1}) \le \rho_{nm}(x_k) + (1 + \frac{1}{\sqrt{b}})||\alpha_n - \alpha_m||_{L^2(x_{k+1},x_k)}$$
$$+\frac{1}{\sqrt{b}}||\alpha_n||_{L^2(x_{k+1},x_k)}\rho_{nm}(x_{k+1})$$

so that

$$\rho_{nm}(x_{k+1}) \le \frac{1 + \frac{1}{\sqrt{b}}}{1 - \frac{||\alpha_n||_{L^2(x_{k+1},x_k)}}{\sqrt{b}}} \left(\rho_{nm}(x_k) + ||\alpha_n - \alpha_m||_{L^2(x_{k+1},x_k)}\right) \tag{66}$$

Applying (66) consecutively to each interval yields

$$\rho_{nm}(0) \le K(b,\alpha)||\alpha_n - \alpha_m||_{L^2} \tag{67}$$

which implies that the r_n^b are Cauchy in L^2.

— $\underline{r_n^k \to r^k \text{ weakly in } L^2}$ As $|r_n|$ and $|r|$ are bounded by 1,

$$|(r_n^b)^k - (r^b)^k| \le K(k)|r_n^b - r^b|$$

so that

$$\rho_n^b := (r_n^b)^k - (r^b)^k \xrightarrow{L^2} 0 \tag{68}$$

For any $h \in \mathcal{H}^2$

$$(\rho_n, h) = (\rho_n, h^b) + (\rho_n, h - h^b)$$
$$= (\rho_n^b, h) + (\rho_n, h - h^b)$$

so that the second term can be made arbitrarily small by choice of b – remember that $||\rho_n||_{L^2}^2$ is bounded by $E(r_n) + E(r)$ which is bounded independently of n – and the first term on the right goes to zero because of (68).

- Weak E-convergence implies strong E-convergence by theorem 5. Verify the weak convergence as follows; let $g \in \mathcal{H}^E$ and E_M be defined as in (37)

$$E(r_n, g) - E(r, g) = \sum_{k=1}^{M} \frac{(r_n^k - r^k, g^k)}{k} + E_M(r_n, g) - E_M(r, g) \qquad (69)$$

According to the Cauchy-Schwartz inequality, (36), and the (independent of n) bound on $E(r_n)$, a sufficiently large choice of M will make the last two terms arbitrarily small. Now each of the terms in the summation approach zero because of the weak L^2 convergence discussed in the previous paragraph. ∎

An immediate consequence of the proof of theorem 8, which we will use later is below. We use the notation $\alpha_{(-\infty,x]}$ to denote α times the characteristic function of the interval $(-\infty, x]$.

Corollary 2. *If $\alpha \in L^2$, $r(x, \omega) = S\alpha_{(-\infty,x]}$, and $b > 0$, then $r^b(x, \omega)$ is the unique solution to the integral equation*

$$r^b(x, \omega) = \int_{-\infty}^{0} e^{2i\omega_b(x-y)} \alpha(y) \left(1 - (r^b)^2\right) dy \qquad (70)$$

The last thing we do in this subsection is prove theorem 3 from the introduction.

Proof of theorem 3

We once again apply the Cauchy-Schwartz inequality to the integral equation

$$r(x) = \int_{-W}^{x} e^{2i\omega(x-y)} \alpha(y) \left(1 - r^2(y)\right) dy \qquad (71)$$

$$= e^{2i\omega x} \hat{\alpha}(2\omega) - \int_{-W}^{0} e^{2i\omega(-y)} \alpha r^2 dy \qquad (72)$$

to obtain the L^2 estimate

$$\|r - \hat{\alpha}(2\omega)\|_{L^2} \leq \|\alpha\|_{L^2} W^{\frac{1}{2}} \|r\|_{L^2} \qquad (73)$$

$$\leq \|\alpha\|_{L^2} W^{\frac{1}{2}} E(r) \qquad (74)$$

$$\leq \|\alpha\|_{L^2}^2 W^{\frac{1}{2}} \qquad (75)$$

which is (19). For the L^∞ estimate, (18), we start with (71) to estimate

$$|r(y)| \leq 2|y|^{\frac{1}{2}} \|\alpha\|_{L^2}$$

and insert this estimate into (72) to obtain

$$|r - \hat{a}(2\omega)| \leq \int_{-W}^{0} \alpha(y)4|y|\|\alpha\|_{L^2(-w,y)}^2 dy$$

$$\leq \frac{4}{3}W^{\frac{3}{2}}\|\alpha\|_{L^2}^3$$

Finally, we integrate the differential equation (47) to obtain

$$t(x,\omega) = e^{i\omega(x+W)}e^{\int_{-w}^{x}\alpha(y)r(y)dy} \qquad (76)$$

and apply the Cauchy-Schwartz inequality to the integral in the exponent to obtain (20). ∎

3.2 Inverse Scattering

We will produce a solution to the inverse problem by solving the Ricatti equation (45), but this time with the initial data given at the top of the layer, which we take to be $x = 0$.

$$r' = 2i\omega r + \alpha(1 - r^2) \qquad (77)$$
$$r(0,\omega) = r_0(\omega) \qquad (78)$$

It will turn out that this single equation will provide an equation for α as well as r. We pass to the integral equation formulation:

$$r(x,\omega) = e^{2i\omega x}r(0,\omega) - \int_{x}^{0} e^{2i\omega(x-y)}\alpha(y)dy + \int_{x}^{0} e^{2i\omega(x-y)}\alpha(y)r^2(y)dy \qquad (79)$$

If we apply $P^{\pm}$, the orthogonal projectors from L^2 onto $\mathcal{H}^2(\mathbb{C}^{\pm})$ defined in (25), to (79), and use the facts that

$$P^+r = r \qquad (80)$$

$$\int_{x}^{0} e^{2i\omega(x-y)}\alpha(y)dy = \left(\alpha_{[x,0]}\right)^{\wedge} \qquad (81)$$

$$P^+ \left(\alpha_{[x,0]}\right)^{\wedge} = 0 \qquad (82)$$

Here

$$\alpha_{[x,0]}^{\wedge} = \int_{x}^{0} e^{-2i\omega y}\alpha(y)dy$$

$\alpha_{[x,0]}$ denotes α times the characteristic function of the interval $[x,0]$, and $\alpha^{\wedge}_{[x,0]}$ denotes its Fourier transform evaluated at 2ω. We obtain a pair of integral equations

$$r(x) = P^+ \left(e^{2i\omega x} r(0) - \int_x^0 e^{2i\omega(x-y)} \alpha(y) r^2(y) dy \right) \tag{83}$$

$$(\alpha_{[x,0]})^{\wedge} = e^{-2i\omega x} P^- \left(e^{2i\omega x} r(0) - \int_x^0 e^{2i\omega(x-y)} \alpha(y) r^2(y) dy \right) \tag{84}$$

for the pair of unknowns $(r(x,\omega), \alpha^{\wedge}_{[x,0]}(\omega))$.

Theorem 9. *If $r_0 \in \mathcal{H}^E(C^+)$, then there exists a unique solution pair $(r,\alpha) \in C((-\infty,0], \mathcal{H}^E(C^+) \oplus L^2(x,0))$ solving (79) (or, equivalently, (83) and (84)). Moreover,*

$$r(x,\omega) = S\alpha_{(-\infty,x]} \tag{85}$$

and the mapping S^{-1} is continuous.

Proof

- Local existence is proved by exhibiting the right hand side of (83) and (84) as a contraction on a ball in $\mathcal{H}^2 \cap \mathcal{H}^\infty \oplus L^2(x,0)$. To see this define

$$\Phi \begin{pmatrix} r \\ \alpha \end{pmatrix} = \begin{pmatrix} 1 & 0 \\ 0 & F^{-1}e^{-2i\omega x} \end{pmatrix} \begin{pmatrix} P^+ \\ P^- \end{pmatrix} e^{2i\omega x} \Psi \tag{86}$$

with

$$\Psi \begin{pmatrix} r \\ \alpha \end{pmatrix} = \left(r(0) - \int_x^0 e^{-2i\omega y} \alpha(y) r^2(y) dy \right) \tag{87}$$

and F^{-1} denoting the inverse Fourier transform. It is straight forward to see that

$$\Psi : \mathcal{H}^2(C^+) \cap \mathcal{H}^\infty \oplus L^2(x,0) \longrightarrow \mathcal{H}^2(C^+) \cap \mathcal{H}^\infty \oplus L^2(x,0) \tag{88}$$

and is a contraction in L^2 norm on an appropriate ball as long as $|x|$ is small enough.

That Φ is a contraction in L^2 norm is an immediate consequence as both $P^\pm$ are bounded from L^2 to itself. What requires some discussion is the assertion that the first component of Φ, $P^+ e^{2i\omega x} \Psi$, remains in $\mathcal{H}^\infty(C^+)$, even though, in general, $P^\pm$ are not bounded on L^∞. Now

$$P^+ e^{2i\omega x} \Psi = e^{2i\omega x} \Psi - P^- e^{2i\omega x} \Psi \tag{89}$$

so it is enough to bound $G := P^{-}e^{2i\omega x}\Psi$ in the L^{∞} norm. This follows because

$$G \in e^{2i\omega x}\mathcal{H}^2(\mathbb{C}^+) \cap \mathcal{H}^2(\mathbb{C}^-) \tag{90}$$

i.e. G is the Fourier transform a function, g, with support in the interval $(0, -x)$. Therefore

$$G = \int_0^{-x} e^{-2i\omega y} g \, dy \tag{91}$$

$$|G| \le |x|^{\frac{1}{2}} \|g\|_{L^2} \tag{92}$$

$$\le |x|^{\frac{1}{2}} \pi \|G\|_{L^2} \tag{93}$$

$$\tag{94}$$

so its L^{∞} norm is bounded by the a constant times its L^2 norm.

– Global existence follows from the Plancherel equality

$$E(r(x_1)) - E(r(x_0)) = \pi \int_{x_0}^{x_1} \alpha^2 dx \tag{95}$$

which itself follows from integrating (62) between x_0 and x_1. Thus (95) implies that $E(r(x))$ decreases as x decreases toward $-\infty$, providing an a priori estimate which allows us to extend our interval of existence.

As a consequence, we have produced a pair (r, α) satisfying (79), in particular, for $N = 1, 2, 3, \ldots$, and $b > 0$ r^b satisfies the integral equation

$$r^b(x, \omega) = r^b(-N, \omega)e^{2i\omega_b(x+N)} + \int_{-N}^{x} e^{2i\omega_b(x-y)}\alpha(1 - (r^b)^2)dy$$

Now, since $\|r^b(x, \cdot)\|_{L^2(d\omega)} \le E(r^b(x, \cdot)) \le E(r^0)$ and $|r^b(x, \omega)|$ is bounded above by one, we may fix x and let $N \to \infty$. We obtain

$$r^b(x, \omega) = 0 + \int_{-\infty}^{x} e^{2i\omega_b(x-y)}\alpha(1 - (r^b)^2)dy \tag{96}$$

which shows, according to corollary 2 that $r = S\alpha$.

– The continuity of S^{-1}
We start by dividing up the half line into k intervals of length W such that

$$\sqrt{E(r)W} < \frac{1}{2} \tag{97}$$

On the k'th interval, $[x_k, x_{k+1}]$, both r and r_n satisfy an integral equation analogous to (86), namely

$$\Phi\left(\begin{array}{c} r(x) \\ \alpha^{\wedge}_{[x_k, x_{k+1}]} \end{array}\right) = \left(\begin{array}{c} P^+ \\ P^- \end{array}\right)e^{2i\omega x}\Psi(x) \tag{98}$$

with

$$\Psi_k(r) = r(x_k) - \int_{x_k}^{x_{k+1}} e^{-2i\omega y}\alpha(y)r^2(y)dy \qquad (99)$$

Define

$$\rho_{n,k} = \sup_{x_k < y < x_{k+1}} ||r_n(y) - r(y)||_{L^2(d\omega)} \qquad (100)$$

$$A_{n,k} = ||\alpha_n(y) - \alpha(y)||_{L^2([x_k, x_{k+1}])} \qquad (101)$$

and subtract the integral equation (98) for r_n from that for r, then

$$A_{n,k} + \rho_{n,k} \le 2||\Psi_k(r) - \Psi_k(r_n)||_{L^2(d\omega)}$$
$$\le 2\rho_{n,k-1} + 2||\alpha||_{l2([x_k, x_{k+1}])}W^{\frac{1}{2}}\rho_{n,k} + 2||r||_{L^2(d\omega)}W^{\frac{1}{2}}A_{n,k}$$

so that, since $\sqrt{E(r)}$ dominates the L^2 norms of both α and r,

$$A_{n,k} \le \frac{\rho_{n,k-1}}{1 - 2\sqrt{WE(r)}}$$

$$\rho_{n,k} \le \frac{\rho_{n,k-1}}{1 - 2\sqrt{WE(r)}}$$

which yields recursively

$$A_{n,k} \le \left(\frac{1}{1 - 2\sqrt{WE(r)}}\right)^k ||r(0) - r_n(0)||_{L^2(d\omega)} \qquad (102)$$

which shows that α_n converges in L^2 of every finite interval, hence weakly in $L^2(-\infty, 0)$, to α. Now the convergence of norms guaranteed by the Plancherel equality yields strong L^2 convergence and the continuity of S^{-1}. ∎

4 Layer Stacking and Splitting

4.1 Stacking Layers

There are several ways to deduce the formulas (13) for stacking layers directly from the Helmholtz equation. As we have already deduced the Ricatti equations, we shall start with the observation that, as a consequence of (45) and (47), $\frac{r}{t}$ and $\frac{1}{t}$ satisfy the linear system of equations:

$$\left(\frac{r}{t}\right)' = i\omega\left(\frac{r}{t}\right) + \alpha\left(\frac{1}{t}\right) \qquad (103)$$

$$\left(\frac{1}{t}\right)' = \alpha\left(\frac{r}{t}\right) - i\omega\left(\frac{1}{t}\right) \qquad (104)$$

For single layer with reflection and transmission coefficients r_1 and t_1, the solution to (103) with intimal data at $x = B$

$$\begin{pmatrix} \frac{r}{t}(B) \\ \frac{1}{t}(B) \end{pmatrix} = \begin{pmatrix} 0 \\ 1 \end{pmatrix} \tag{105}$$

is

$$\begin{pmatrix} \frac{r}{t}(T) \\ \frac{1}{t}(T) \end{pmatrix} = \begin{pmatrix} \frac{r_1}{t_1} \\ \frac{1}{t_1} \end{pmatrix} \tag{106}$$

Now if $\begin{pmatrix} v_1 \\ v_2 \end{pmatrix}$ solve (103) so does $\begin{pmatrix} \overline{v_2} \\ \overline{v_1} \end{pmatrix}$, hence the fundamental solution matrix to (103), mapping data from the bottom to the top of the layer is

$$M_1 = \begin{pmatrix} \frac{1}{\overline{t_1}} & \frac{r_1}{t_1} \\ \frac{\overline{r_1}}{\overline{t_1}} & \frac{1}{t_1} \end{pmatrix} \tag{107}$$

If we stack layer 1 atop layer 2, then we start with initial data $\begin{pmatrix} 0 \\ 1 \end{pmatrix}$ at the bottom of layer 2, the value of the solution to (103) at the top of layer 2 (which is the same as the bottom of layer 1) is $\begin{pmatrix} \frac{r_2}{t_2} \\ \frac{1}{t_2} \end{pmatrix}$ so that, at the top of layer 1

$$\begin{pmatrix} \frac{r_{12}}{t_{12}} \\ \frac{1}{t_{12}} \end{pmatrix} = \begin{pmatrix} \frac{1}{\overline{t_1}} & \frac{r_1}{t_1} \\ \frac{\overline{r_1}}{\overline{t_1}} & \frac{1}{t_1} \end{pmatrix} \begin{pmatrix} \frac{r_2}{t_2} \\ \frac{1}{t_2} \end{pmatrix} \tag{108}$$

which can then be unwound to produce

$$r_{12} = \frac{r_1 + \frac{t_1}{\overline{t_1}} r_2}{1 + \overline{r_1} \frac{t_1}{\overline{t_1}} r_2} \tag{109}$$

$$t_{12} = \frac{t_1 t_2}{1 + \overline{r_1} \frac{t_1}{\overline{t_1}} r_2} \tag{110}$$

We remark that stacking layers is associative, that is, stacking layer 2 on top of layer 3, and then layer 1 on top of the 2-3 stack, had better yield the same thing as stacking layer 1 atop layer 2 and then putting the 1-2 layer atop layer 3. The corresponding formula for reflection coefficients takes the form

$$r_{123} = r_1 \circ \frac{t_1}{\overline{t_1}} \left(r_2 \circ \frac{t_2}{\overline{t_2}} r_3 \right) \tag{111}$$

$$= \left(r_1 \circ \frac{t_1}{\overline{t_1}} r_2 \right) \circ \frac{t_{12}}{\overline{t_{12}}} r_3 \tag{112}$$

We return our attention to (109). Our goal is to start with any $r \in \mathcal{H}^E(C^+)$, choose a width, W_1, and produce a factorization as in (109), with r_1 the reflection coefficient of a layer of width W_1. Once we accomplish this step, we can repeat it on r_2, eventually representing r as a composition of reflection coefficients, each of width W_1. If W_1 is small enough, the Born approximation will yield a good approximation for α; if not we may subdivide each layer again into thinner layers until the Born approximation applies.

Before stating our main decomposition result, we recall some basic relationships between t and r.

Lemma 3. *For a layer, r_1 of width W_1*

$$|t_1|^2 = 1 - |r_1|^2 \tag{113}$$

$$t_1 = e^{i\omega W_1} e^{P^+ \log(1-|r_1|^2)} \tag{114}$$

$$\frac{r_1}{t_1} \in e^{i\omega W_1} \mathcal{H}^2(C^-) \cap e^{-i\omega W_1} \mathcal{H}^2(C^+) \tag{115}$$

<u>Proof</u> We start with (47), and multiply both sides by $\overline{t_1}$ to obtain

$$|t_1|^{2'} = -\alpha_1(r_1 + \overline{r_1}) = (1 - |r_1|^2)' \tag{116}$$

where the last equality makes use of (55). Now integrate both sides, using (48) and (46) to get (113). On the other hand, we may integrate (47) directly to obtain

$$t_1 = e^{i\omega W_1} e^{\int_{-W_1}^{0} \alpha_1 r_1 dy} \tag{117}$$

If we let

$$\tau_1 = e^{-i\omega W_1} t_1 \tag{118}$$

$$\log(\tau_1) = \int_{-W_1}^{0} \alpha_1 r_1 dy \in \mathcal{H}^2(C^+) \tag{119}$$

$$\log(\tau_1) = 2P^+ \operatorname{Re} \log(\tau_1) \tag{120}$$

$$= P^+ \log(|\tau_1|^2) \tag{121}$$

$$= P^+ \log(1 - |r_1|^2) \tag{122}$$

from which (114) follows. Note also that

$$e^{-\|\alpha_1\|_{L^2} W_1^{\frac{1}{2}}} \leq |\tau_1| \leq e^{\|\alpha_1\|_{L^2} W_1^{\frac{1}{2}}} \tag{123}$$

so that $\frac{r_1}{\tau_1} \in \mathcal{H}^2(C^+)$ which together with (118) gives the second inclusion in (115). To see the first inclusion, let $r_2 \in \mathcal{H}^E(C^+)$ be the reflection coefficient of α_2 and let r be given by (109). Then

$$\pi \left(\|\alpha_1\|^2 + \|\alpha_2\|^2 \right) = E(r_1) + E(r_2) + \int \log(1 + \frac{\overline{r_1}}{t_1} t_1 r_2) d\omega \tag{124}$$

but the Plancherel equality tells us that the first two terms on the right exactly equal the two terms on the left, so that

$$\int \log(1 + \frac{\overline{r_1}}{t_1} t_1 r_2) d\omega = 0 \tag{125}$$

Since we may choose r_2 arbitrarily small, we must have

$$\int \frac{\overline{r_1}}{t_1} t_1 r_2 d\omega = 0 \tag{126}$$

but r_2 can be an arbitrary function in $\mathcal{H}^E(C^+)$, so that $\frac{\overline{r_1}}{t_1} t_1 \in \mathcal{H}^E(C^+)$. Since $\frac{1}{t_1} \in e^{i\omega W_1} \mathcal{H}^\infty(C^+)$, we have $\frac{\overline{r_1}}{t_1} \in e^{-i\omega W_1} \mathcal{H}^2(C^+)$. Taking complex conjugates gives the remainder of (115). ∎

Theorem 10. *Let* $r \in \mathcal{H}^E(C^+)$ *and* $W_1 > 0$. *The following are equivalent*

1. r_1 *is the unique minimizer of the strictly convex variational problem*

$$\begin{array}{c} \min \\ \rho \in \mathcal{H}^E(C^+) \\ r - \rho \in e^{2i\omega W_1} \mathcal{H}^2(C^+) \end{array} \quad E(\rho) \tag{127}$$

2. r_1 *satisfies*

$$r - \rho \in e^{2i\omega W_1} \mathcal{H}^2(C^+) \tag{128}$$

$$\frac{r_1}{t_1} \in e^{i\omega W_1} \mathcal{H}^2(C^-) \cap e^{-i\omega W_1} \mathcal{H}^2(C^+) \tag{129}$$

3. *The unique layer decomposition of* r *into a stack of two layers with top layer of width* W_1 *is*

$$r = r_1 \circ \frac{t_1}{t_1} r_2 \tag{130}$$

Proof

- **3 implies 2** Any reflection coefficient, r_1 of with W satisfies (129) according to (115). If we use the equality (113), we may rewrite (130) as

$$r = r_1 + t_1^2 \left(1 + \frac{\overline{r_1}}{t_1} t_1 r_2\right)^{-1} r_2{}' \tag{131}$$

which makes it apparent that the second term belongs to $e^{2i\omega W} \mathcal{H}^2(C^+)$, establishing (128).

2 implies 1   Let $\sigma \in e^{2i\omega W} \mathcal{H}^2(C^+)$ and compute

$$\frac{d}{d\epsilon}\Big|_{\epsilon=0} E(r_1 + \epsilon\sigma) = \int \frac{\overline{r_1}}{\overline{t_1}} \frac{\sigma}{t_1} \tag{132}$$

$$= 0 \tag{133}$$

because

$$\frac{\sigma}{t_1} \in e^{i\omega W_1} \mathcal{H}^2(C^+) \tag{134}$$

we conclude that

$$\frac{\overline{r_1}}{\overline{t_1}} \in e^{-i\omega W_1} \mathcal{H}^2(C^+) \tag{135}$$

so that r_1 is a critical point for the variational problem. But as the variational problem is strictly convex, that critical point can only be the unique minimizer.

1 implies 3   In theory, we know that the layer decomposition (130) exists and is unique because we have proved existence and uniqueness of the inverse problem in theorem 9. The point here is that we can compute r_1 by performing the convex minimization. Once we have r_1 in hand, according to (113), we can produce

$$t_1 = e^{i\omega W_1} e^{P^+ \log(1-|r_1|^2)} \tag{136}$$

and then

$$r_2 = \frac{\overline{t_1}}{t_1} \left(-r_1 \circ r\right) \tag{137}$$

so that the entire layer decomposition is in hand.

■

Acknowledgement
Partially supported by NSF grant DMS–9801068 and ONR grants N00014–93-1-0295

References

1. A. BRUCKSTEIN AND T. KAILATH, *Inverse scattering for discrete transmission-line models*, SIAM Review, 29(3) (1987), 359–389.
2. CHEN, Y. AND ROKHLIN, V., On the Inverse Scattering Problem for the Helmholtz Equation in One Dimension, *Inverse Problems* **8**, pp. 356–391 (1992)
3. CHENEY, M., ISAACSON, D., SOMERSALO, E., AND ISAACSON, E., A layer stripping approach to impedance imaging, 7th Annual Review of Progress in Applied Computational Electromagnetics, Naval Postgraduate School, Monterrey, (1991).

4. J. P. Corones, R.J. Krueger, and Davison., Direct and inverse scattering in the time domain via invariant embedding equations, J. Acoust. Soc. Am. 74 (1983), 1535–1541.

5. H. Dym and H. P. McKean, *"Fourier Series and Integrals"*, Academic Press, New York (1972).

6. Somersalo, E., Layer stripping for time-harmonic Maxwell's equations with high frequency. *Inverse Problems* **10** (1994), 449–466

7. Sylvester, J., A Convergent Layer Stripping Algorithm for the Radially Symmetric Impedance Tomography Problem, *Communications in PDE* **17**, No.12, pp.1955–1994 (1992)

8. Sylvester, J., Impedance tomography and layer stripping. *Inverse problems: principles and applications in geophysics, technology, and medicine*, 307–321, Math. Res., 74, Akademie-Verlag, Berlin, (Potsdam, 1993) 1993.

9. Sylvester, J., Winebrenner, D. and Gylys-Colwell, F., Layer Stripping for the Helmholtz Equation, *SIAM Journal of Applied Mathematics* **56**(3) pp. 736–754 June 1996

10. Sylvester, J., Winebrenner, D., Nonlinear and Linear Inverse Scattering, *SIAM Journal of Applied Mathematics* **59**(2) pp. 669–699 April 1999

11. Symes, W. W., Impedance profile inversion via the first transport equation, J. Math. Anal. App., 94 (1983), 435–453.

The Linear Sampling Method in Inverse Scattering Theory

David Colton and Andreas Kirsch[2] and Peter Monk[1]

[1] Department of Mathematical Sciences, University of Delaware, Newark, Delaware 19716, USA
[2] Mathematisches Institut II, Universität Karlsruhe, D–76128 Karlsruhe, Germany

Abstract. A survey is given of the linear sampling method for solving the inverse scattering problem of determing the support of an inhomogeneous medium from a knowledge of the far field pattern of the scattered field.An application is given to the problem of detecting leukemia in the human body.

1 Introduction

Until recently there have been two main methods for solving multi–dimensional inverse scattering problems for acoustic and electromagnetic waves. The first, and oldest, of these methods is the use of linearized models such as the Born and physical optics approximations [14]. More recently there has been a tremendous amount of energy invested in nonlinear optimization methods [4], [11]. Although both approaches have achieved success in certain cases, they each have serious limitations. In particular, the linearized models ignore multiple scattering whereas nonlinear optimization methods are numerically extremely expensive for realistic three dimensional problems. Furthermore, both methods require knowledge of the physical properties of the scatterer such as boundary conditions and this information may in fact be unknown.

In 1996 two of us introduced a new method for solving the inverse scattering problem called the *linear sampling method* that avoids the limitations of these older methods [1]. More specifically, without making any a priori assumptions on the scattering process, this method allows one to determine from measured scattering data whether or not a given point is inside a scattering object. Remarkably enough, although the inverse scattering problem is nonlinear, this is accomplished by solving a *linear* integral equation of the first kind and requires essentially no a priori knowledge of the physical properties of the scatterer. Since the publication of [1], a second version of this method has been discovered that provides a particularly elegant formulation for the cases of either non absorbing media or low frequencies [12], [13]. Further extensions have been made to practical problems involving near field data and piecewise homogeneous background media [5], [6], [7], [9]. The purpose of this paper is to describe some of these recent developments.

For pedagogical reasons we will restrict our description of the linear sampling method to the special case of scattering by a two dimensional inhomogeneous medium of compact support. In the next two sections we will further restrict ourselves to the case of plane wave incidence and far field scattering data. In these sections we will describe the original linear sampling method of Colton and Kirsch [1], [4], [8] as well as the modified version recently developed by Kirsch [12], [13]. In both cases the analysis will rely heavily on the spectral theory of the far field operator F as well as a study of the range and null space of either F or $(F^*F)^{1/4}$. In this connection, a crucial role is played by an unusual boundary value problem for elliptic equations called the *interior transmission problem* [4], [11]. Finally, in the last section, we will show how the linear sampling method developed in [1], [4] and [8] can be extended to the physically important case of point sources as incident fields, near field data and piecewise constant background media. More specifically, we will briefly describe recent results of Colton and Monk on the application of the linear sampling method for using microwaves to detect leukemia in the human body [5], [6], [7]. Further results in this direction can be found in the recent Ph.D. thesis of Joe Coyle [9].

2   The Linear Sampling Method

We begin by considering the two dimensional scattering problem of determining u from the equations

$$\Delta_2 u + k^2 n(x)\,u = 0 \quad \text{in} \quad \mathbb{R}^2 \tag{2.1}$$

$$u = u^i + u^s \tag{2.2}$$

$$\lim \ \sqrt{r}\left(\frac{\partial u^s}{\partial r} - iku^s\right) = 0 \tag{2.3}$$

where $u^i(x) = e^{ikx\cdot\hat{\theta}}$, $|\hat{\theta}| = 1$, $n(x)$ is continuously differentiable for $x \in \bar{D}$ where $\bar{D}$ denotes the (bounded) support of $1 - n$, $\text{Im}\,n \geq 0$ and $k > 0$ is the wave number. The unit vector $\hat{\theta}$ denotes the direction of propagation of the incident field u^i, the Sommerfield radiation condition (2.3) for the scattered field u^s is assumed to hold uniformly for $\hat{x} = x/|x| \in \Omega := \{x : |x| = 1\}$ as $r = |x|$ tends to infinity and we will allow the possibility that n has a jump discontinuity across ∂D where ∂D is assumed to be smooth. The problem of finding u from (2.1)–(2.3) is called the *direct scattering problem*. It can be shown that there exists a unique solution $u \in H^2_{\text{loc}}(\mathbb{R}^2)$ to this problem and

$$u^s(x) = \frac{e^{ikr}}{\sqrt{r}}\ u_\infty(\hat{x}, \hat{\theta}) + O(r^{-3/2}) \tag{2.4}$$

as r tends to infinity where u_∞ is the *far field pattern* corresponding to the incident field $e^{ikx\cdot\hat{\theta}}$ [4], [11]. The *inverse scattering problem* is to determine n, or something about n, from $u_\infty(\hat{x}, \hat{\theta})$ for $\hat{x}$, $\hat{\theta} \in \Omega$ and k fixed. In two dimensions and fixed $k > 0$ it is not known if u_∞ uniquely determines n. However, it can be shown that for fixed k, u_∞ uniquely determines the support $\bar{D}$ [15], [17]:

Theorem 1. *For each fixed* $k > 0$, u_∞ *uniquely determines the support* $\bar{D}$ *of* $1 - n$.

We will now proceed to derive a numerical technique for determining the support $\bar{D}$ from the measured far field data u_∞ for fixed $k > 0$. To this end we define the *far field operator* $F : L^2(\Omega) \to L^2(\Omega)$ by

$$(Fg)(\hat{x}) := \int_\Omega u_\infty(\hat{x}, \hat{\theta})\, g(\hat{\theta})\, ds(\hat{\theta}) \qquad (2.5)$$

and note that Fg is the far field pattern corresponding to the direct scattering problem (2.1)–(2.3) with $u^i(x) = e^{ikx \cdot \hat{\theta}}$ replaced by the *Herglotz wave function*

$$v_g^i(x) := \int_\Omega e^{ikx \cdot \hat{\theta}}\, g(\hat{\theta})\, ds(\hat{\theta}) \qquad (2.6)$$

with kernel $g \in L^2(\Omega)$. The basic identity used to study the far field operator is the following theorem [3]:

Theorem 2. *Let* v_g *be the solution of the direct scattering problem (2.1)–(2.3) with* $u^i = v_g^i$. *Then*

$$k^2 \iint_D \operatorname{Im} n(x) |v_g(x)|^2\, dx = \sqrt{8\pi k}\, \operatorname{Im}(e^{-i\pi/4} Fg, g) - k\, \|Fg\|^2$$

where $(\cdot, \cdot)$ *denotes the scalar product in* $L^2(\Omega)$ *and* $\| \cdot \|$ *the norm.*

Using this theorem and Lidski's Theorem [4], [16] one can now easily deduce the following result [4], [11] which will be important in the next section of our paper:

Theorem 3. *If* $\operatorname{Im} n(x) = 0$ *for* $x \in D$, F *is normal and all the eigenvalues lie on the circle* $\sqrt{8\pi k}\, \operatorname{Im}(e^{-i\pi/4}\lambda) - k|\lambda|^2 = 0$. *If* $\operatorname{Im} n(x) > 0$ *for some* $x \in D$ *then* F *has an infinite number of eigenvalues and they all lie inside this circle.*

It can be shown [3] that if n is a known constant and $\operatorname{Im} n > 0$ then a knowledge of how far the eigenvalues move away from the above circle gives a lower bound to the area of the support $\bar{D}$.

In what follows, the injectivity of F will be important, i.e. whether or not $\lambda = 0$ is an eigenvalue of F. In particular, if $\operatorname{Im} n(x) > 0$ for some $x \in D$ and $Fg = 0$ then from the above basic identity and the unique continuation principle for elliptic equations [4] we see that $v_g(x) = 0$ for $x \in D$. From the Lippmann–Schwinger integral equation formulation of the direct scattering problem [4], [11] we now have that the Herglotz wave function $v_g^i(x) = 0$ for $x \in D$ and hence $g = 0$, i.e. F is injective. If $\operatorname{Im} n(x) = 0$ for $x \in D$ then in general F is no longer injective. In particular, if $n = n(r)$ is a function only of $r = |x|$ then there exists a discrete set of values of k, called *transmission eigenvalues*, such that F is not injective [4], [11]. For general $n(x)$ satisfying certain growth conditions as x tends to ∂D it can be shown that the set of transmission eigenvalues is at most a discrete set [2], [4].

We now introduce the far field equation. To this end let

$$\Phi(x,y) := \frac{i}{4} H_0^{(1)}(k|x-y|), \quad y \in D \tag{2.7}$$

be the fundamental solution to the Helmholtz equation where $H_0^{(1)}$ denotes a Hankel function of the first kind of order zero and note that from the asymptotic behavior of the Hankel function we have that the far field pattern of Φ is given by

$$\Phi_\infty(\hat{x},y) = \frac{e^{i\pi/4}}{\sqrt{8\pi k}} e^{-ik\hat{x}\cdot y}. \tag{2.8}$$

The *far field equation* is given by $Fg = \Phi_\infty$, i.e.

$$\int_\Omega u_\infty(\hat{x},\hat{\theta})\, g(\hat{\theta})\, ds(\hat{\theta}) = \Phi_\infty(\hat{x},y) \tag{2.9}$$

for $y \in D$ where $g(\hat{\theta}) = g(\hat{\theta};y)$. It is easy to show using Rellich's lemma that the far field equation has a solution if and only if there exists a function w such that, for v the Herglotz wave function with kernel g, v and w satisfy the *interior transmission problem* [4], [11]

$$\begin{aligned} \Delta_2 w + k^2 n(x)\, w = 0 \\ \Delta_2 v + k^2 v = 0 \end{aligned} \quad \text{in} \quad D \tag{2.10}$$

$$\begin{aligned} w - v = \Phi(\cdot,y) \\ \tfrac{\partial}{\partial \nu}(w-v) = \tfrac{\partial}{\partial \nu}\Phi(\cdot,y) \end{aligned} \quad \text{on} \quad \partial D \tag{2.11}$$

where ν is the unit outward normal to ∂D. In this connection, the following theorem plays a central role [4], [8]:

Theorem 4. *Assume* $\operatorname{Im} n(x) > 0$ *for* $x \in \bar{D}$. *Then there exists a unique (weak) solution to the interior transmission problem and* v *can be approximated in* $L^2(D)$ *by a Herglotz wave function.*

We note that the concept of a weak solution to (2.10), (2.11) is a bit unorthodox in the sense that for such a solution it is only assumed that v and w are in $L^2(D)$ with $w-v$ in the Sobolev space $H^2(D)$; for details we refer the reader to [4] and [8].

The above theorem now allows us to deduce the fact that for every $\epsilon > 0$ there exists a function $g = g(\cdot,y) \in L^2(\Omega)$ such that $\|Fg - \Phi_\infty\| < \epsilon$ and both $\|g(\cdot,y)\|$ and $\|v_g^i(\cdot,y)\|$ become unbounded as y tends to ∂D where v_g^i is the Herglotz wave function with kernel g. This fact now suggests a numerical procedure for determining the support $\bar{D}$ of $1-n$ from noisy far field data u_∞^δ. In particular, let u_∞^δ be the measured far field data, i.e. $\|u_\infty^\delta - u_\infty\| < \delta$, and assume g is such that $\|Fg - \Phi_\infty\| < \epsilon$. If F_δ is the operator F with the kernel u_∞ replaced by u_∞^δ then we want to find an approximation to g by solving $F_\delta \varphi = \Phi_\infty$, i.e. we view both the operator and the right hand side as being inexact. For each fixed y we now determine φ by minimizing the Tikhonov functional

$$\|F_\delta \varphi - \Phi_\infty(\cdot, y)\|^2 + \alpha \, \| \varphi(\cdot, y)\|^2 \tag{2.12}$$

where the regularization parameter $\alpha = \alpha(y)$ is chosen by Morozov's generalized discrepancy principle [18]. In particular, assuming that $\epsilon << \delta$, $\alpha = \alpha(y)$ is chosen such that $\|F_\delta \varphi - \Phi_\infty(\cdot, y)\| \approx \delta\|\varphi(\cdot, y)\|$. Note that $\alpha = \alpha(y)$ tends to zero as y tends to ∂D and hence the level curves of both $\|g(\cdot, y)\|$ and $\alpha(y)$ can be used to advantage in determining the support $\bar{D}$ [8].

In practice we determine φ from (2.12) for y on some partition P of a rectangle known a priori to contain D and determine those points $y \in P$ where $\|g(\cdot, y)\|$ begins to sharply increase. Note that this calculation is very rapid since the operator F_δ does not change as y varies over P. A numerical example using this approach for the more complicated situation of a piecewise homogeneous background medium and near field data will be given in the last section of this paper.

3   A Modified Linear Sampling Method

In this section we will modify the approach of the previous section and prove an explicit characterization of the support of $m = 1 - n$. Again, we restrict ourselves to the two dimensional case. We recall that the far field operator $F : L^2(\Omega) \longrightarrow L^2(\Omega)$ is defined by

$$Fg(\hat{x}) := \int_\Omega u_\infty(\hat{x}, \hat{\theta}) \, g(\hat{\theta}) \, ds(\hat{\theta}), \quad \hat{x} \in \Omega. \tag{3.1}$$

Here again, $u_\infty(\hat{x}, \hat{\theta})$ denotes the far field pattern of the scattered field u^s corresponding to the incident field $u^i(x) = \exp(ik x \cdot \hat{\theta})$, $x \in \mathbb{R}^2$.

We make the following assumption on the index of refraction n which we will later refer to by **Assumption (A)**:

Assume that the exterior of D is connected and $\partial D \in C^2$. Let $\Re n(x) < 1$ in D (or $\Re n(x) > 1$ in D) and assume that there exists $\gamma > 0$ with $|\Im n(x)| \leq \gamma |1 - \Re n(x)|$ for all $x \in D$.

As a first step towards our final result we want to factorize F in the form $F = GTG^\#$ where T is an isomorphism and $G^\#$ is the adjoint of an operator G defined in a moment.

For technical reasons we have to introduce the weighted space $L^2(D, |m|)$ which is defined as the completion of $C(\bar{D})$ with respect to the inner product

$$\langle \varphi, \psi \rangle_m := \iint_D \varphi(x) \, \overline{\psi(x)} \, |m(x)| \, dx. \tag{3.2}$$

The standard space $L^2(D)$ is continuously imbedded in $L^2(D, |m|)$.

Writing the Helmholtz equation for the scattered field u^s in the form

$$\Delta u^s + k^2 n u^s = k^2 m u^i \tag{3.3}$$

we observe that u^s is the solution of the more general problem:

Given $h \in L^2(D, |m|)$, determine the solution v of

$$\Delta v + k^2 n v = k^2 m h \quad \text{in } \mathbb{R}^2, \tag{3.4}$$

which satisfies the radiation condition

$$\lim_{r \to \infty} \sqrt{r} \left(\frac{\partial v}{\partial r} - ikv \right) = 0. \tag{3.5}$$

The solution has to be understood in a weak sense. The radiation condition yields the asymptotic form

$$v(x) = \frac{\exp(ikr)}{\sqrt{r}} v_\infty(\hat{x}) + \mathcal{O}(r^{-3/2}), \quad r = |x| \to \infty. \tag{3.6}$$

We now define the operator $G : L^2(D, |m|) \longrightarrow L^2(\Omega)$ by $Gh = v_\infty$ where v_∞ is the far field pattern (3.6) of the solution v of (3.4). Then the following result is known [13]:

Theorem 5. *The operator F can be factorized in the following form:*

$$F = -\frac{1}{\gamma} G T G^\# \tag{3.7}$$

where $G^\# : L^2(\Omega) \longrightarrow L^2(D, |m|)$ is the adjoint of G and T is defined by

$$T\varphi(x) := \frac{1}{k^2} \operatorname{sign}(m(x)) \varphi(x) + \iint_D |m(y)| \varphi(y) \overline{\Phi(x,y)} \, dy, \quad x \in D. \tag{3.8}$$

The constant $\gamma \in \mathbb{C}$ is given by $\gamma = \frac{\exp(-i\pi/4)}{\sqrt{8\pi k}}$, and $\operatorname{sign}(m) = \frac{m}{|m|}$ denotes the sign of $m \in \mathbb{C}$.

The operator T is closely related to the Lippmann-Schwinger integral equation, and one can show that it is an isomorphism from $L^2(D, |m|)$ onto itself. The interpretation of the factorization (3.7) is that the operator F is smoothing "double as much" as the operator G, a fact which we will make more precise for absorbing and non-absorbing media separately.

We begin with the case where n is real valued and make the further assumption that $m = 1 - n$ has no sign change in D and k^2 is not a transmission eigenvalue. We formulate the results for the case $n(x) < 1$ on D, i.e. $m(x) > 0$. The case $n(x) > 1$ can be treated analogously.

We have seen in the previous section that in this case the far field operator F is compact, one-to-one and normal. From basic functional analysis (see [16]) we conclude that there exists eigenvalues $\lambda_j \in \mathbb{C}$ and a complete orthonormal system of eigenfunctions $\psi_j \in L^2(\Omega)$ of F.

Theorem 6. *In addition to assumption (A) we assume that n is real-valued and k^2 is not a transmission eigenvalue. Then we have for the range $\mathcal{R}(G)$ of G:*

$$\mathcal{R}(G) = \mathcal{R}((F^*F)^{1/4}) = \left\{ \sum_{j=1}^{\infty} \rho_j \, \psi_j : \sum_{j=1}^{\infty} \frac{|\rho_j|^2}{|\lambda_j|} < \infty \right\}. \tag{3.9}$$

As an application of this result one can derive a simple characterization of the support D of m:

Theorem 7. *In addition to assumption (A) we assume that n is real-valued and k^2 is not a transmission eigenvalue. For any $y \in \mathbb{R}^2$ define the function $\Phi_\infty(\cdot, y) \in L^2(\Omega)$ by (2.8), i.e.*

$$\Phi_\infty(\hat{x}, y) := \frac{\exp(i\pi/4)}{\sqrt{8\pi k}} e^{-ik\,\hat{x}\cdot y}, \quad \hat{x} \in \Omega. \tag{3.10}$$

Then $y \in D$ if and only if $\Phi_\infty(\cdot, y) \in \mathcal{R}(G)$, i.e. by the previous theorem

$$y \in D \iff \sum_{j=1}^{\infty} \frac{|\rho_j^{(y)}|^2}{|\lambda_j|} < \infty$$

where $\rho_j^{(y)} \in \mathbb{C}$ are the expansion coefficients of $\Phi_\infty(\cdot, y)$ with respect to ψ_j.

Now we consider absorbing media, i.e. we allow n to have a nonvanishing imaginary part. Then F fails to be normal. As mentioned in the previous section, there still exist eigenvalues but the corresponding eigenfunctions don't form an orthogonal system anymore. The analysis of the preceding case cannot be extended to this case (to the authors knowledge). Therefore, we propose a different approach which is also based on the factorization of F.

We make still the assumption (A). From the factorization

$$\tilde{F} := -\gamma F = GTG^{\#}$$

we conclude that its adjoint $\tilde{F}^* : L^2(\Omega) \longrightarrow L^2(\Omega)$ is given by

$$\tilde{F}^* = GT^{\#}G^{\#}$$

and thus for the self adjoint part

$$\tilde{F}_{sa} := \frac{1}{2}(\tilde{F} + \tilde{F}^*) = GT_{sa}G^{\#}$$

where

$$T_{sa} := \frac{1}{2}(T + T^{\#}) : L^2(D, |m|) \longrightarrow L^2(D, |m|).$$

Then it can be shown that the operator T_{sa} is positive definite for sufficiently small k which guarantees the existence of the positive definite square root $\sqrt{T_{sa}} : L^2(D, |m|) \to L^2(D, |m|)$ of T_{sa}. Therefore,

$$\tilde{F}_{sa} = (G\sqrt{T_{sa}})(G\sqrt{T_{sa}})^{\#}.$$

Using a singular system $\{\sigma_j, \varphi_j, \psi_j\}$ of $G\sqrt{T_{sa}}$, it is easy to see that $\{\psi_j : j \in \mathbb{N}\}$ is a complete orthonormal system in $L^2(\Omega)$. Analogously to Theorem 3.2 one can show [13]:

Theorem 8. *Under the assumption (A) there exists $k_0 > 0$ such that for all $0 < k \le k_0$ the ranges of G and $\sqrt{\tilde{F}_{sa}}$ coincide, i.e.*

$$\mathcal{R}(G) = \mathcal{R}\left(\sqrt{\tilde{F}_{sa}}\right) = \left\{ \sum_{j=1}^{\infty} \rho_j \psi_j : \sum_{j=1}^{\infty} \frac{|\rho_j|^2}{\lambda_j} < \infty \right\}$$

where λ_j and ψ_j are the eigenvalues and eigenfunctions, respectively, of the operator $\tilde{F}_{sa} = -\frac{1}{2}(\gamma F + \bar{\gamma} F^*)$.

The next theorem is the exact analogue of Theorem 3.3.

Theorem 9. *Again we make the assumption (A) and define the function $\Phi_\infty(\cdot, y)$ by (2.8) for any $y \in \mathbb{R}^2$. Then there exists $k_0 > 0$ such that for all $0 < k \le k_0$ the function $\Phi_\infty(\cdot, y)$ belongs to the range of G if and only if $y \in D$, i.e.*

$$y \in D \iff \sum_{j=1}^{\infty} \frac{|\rho_j^{(y)}|^2}{\lambda_j} < \infty$$

where $\rho_j^{(y)} \in \mathbb{C}$ are the expansion coefficients of $\Phi_\infty(\cdot, y)$ with respect to the eigenfunctions ψ_j of $\tilde{F}_{sa}$.

4   The Detection of Leukemia

In a recent series of papers [5], [6], [7] two of us have examined the possibility of applying the linear sampling method to detect leukemia in the leg by microwave interrogation. The viability of such an approach rests on the fact that the increased capacitance of diseased cells should cause the permittivity of the bone marrow to increase and the conductivity to decrease significantly. In the above papers a simple mathematical model was formulated for this problem based on the solution of an inverse scattering problem for a two dimensional reduced wave equation. This inverse scattering problem was then solved by modifying the linear sampling method described in section two of this paper in order for it to be applicable to anomalies in a piecewise homogeneous background medium, point sources as incident fields and near field data. In realistic applied problems such a situation is typical and hence modifying the linear sampling method in this manner is of basic importance for practical applications. The purpose of this section of our paper is to describe what these modifications are, referring the reader to the above mentioned references for further details. For another application of the linear sampling method to the determination of anomalies in a piecewise constant background medium see [9].

The model we are about to consider is based on the following assumptions:

1. Line sources parallel to the leg are placed in a sheath surrounding the leg. The sheath is made of collodion (which has roughly the same electrical properties as skin) and the line sources inside the sheath are immersed in siliconized oil (which has roughly the same electrical properties as body fat).
2. The portion of the leg imaged is viewed as a cross section of a cylinder with the permittivity and conductivity varying only along a plane perpendicular to the cylinder.
3. We know the location and (constant) index of refraction of the fat, muscle, bone and bone marrow in the leg and ignore the presence of arteries and veins. The thin collodion/skin layer is modeled by an impedance sheet approximation due to Harrington and Mautz [10].

The scattered field due to the line sources is now measured at points on the same curve through which the line sources pass and from this information it is desired to determine the presence of (inhomogeneous) tumors in the bone marrow.

Under the above assumptions the total electric field is $E(x)\, e^{-i\omega t}\, e_z$ where ω is the frequency, $x \in \mathbb{R}^2$, t is time and e_z is a unit vector parallel to the axis of the leg. Then $u = E$ satisfies the *resistive boundary value problem*

$$\Delta_2 u + k^2 n^2(x)\, u = 0 \quad \text{in} \quad D_0 \tag{4.1a}$$

$$\Delta_2 u_0 + k^2 n_0^2 u_0 = \delta(x - y) \quad \text{in} \quad \mathbb{R}^2 \backslash \bar{D}_0 \tag{4.1b}$$

$$u_0(x) = \frac{i}{4} H_0^{(1)}(k n_0 |x - y|) + u^s(x) \tag{4.2a}$$

$$u^s \quad \text{is bounded in} \quad \mathbb{R}^2 \backslash \bar{D}_0 \tag{4.2b}$$

$$u_0 = u \quad \text{on} \quad \partial D_0 \tag{4.3a}$$

$$\lambda u_0 + \left(\frac{\partial u_0}{\partial \nu} - \frac{\partial u}{\partial \nu} \right) = 0 \quad \text{on} \quad \partial D_0 \tag{4.3b}$$

where D_0 is the cross section of the leg, $x = y \in \mathbb{R}^2 \backslash \bar{D}_0$ is the location of the source, $k > 0$ is the wave number, n^2 is the index of refraction in the leg and n_0^2 is the index of refraction of the siliconized oil, ν is the unit outward normal to ∂D_0 and λ, $\operatorname{Im} \lambda > 0$, is the resistivity parameter. We denote by D the region of the bone marrow that is proliferated by cancer cells and assume that outside D the index of refraction is piecewise constant.

The inverse scattering problem we are interested in is to determine the support $\bar{D}$ of the cancer cells from a knowledge of $u^s(x) = u^s(x, y)$ for x, y on a curve C lying in $\mathbb{R}^2 \backslash \bar{D}_0$. Let u_b^s denote the scattered field when cancer is not present, i.e. the scattering due to the background medium alone. The linear sampling method for determining the support $\bar{D}$ from a knowledge of $u^s(x, y)$ for

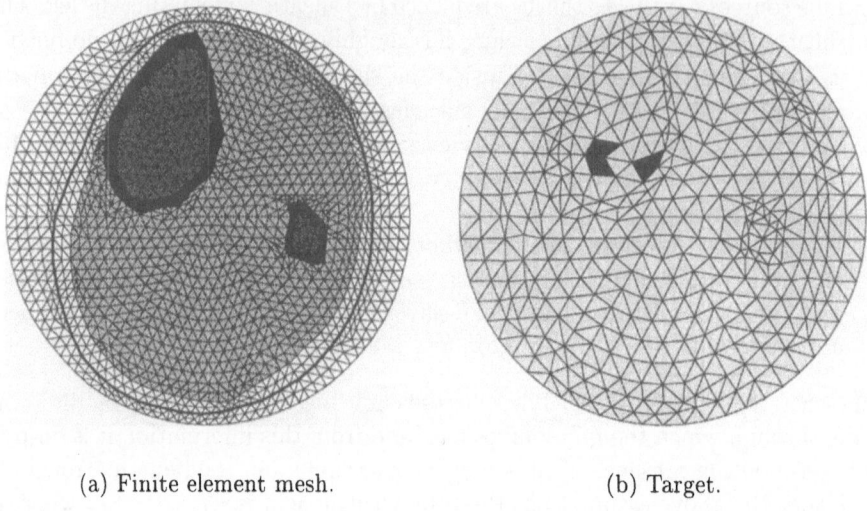

(a) Finite element mesh.　　　　　　　　　　(b) Target.

Fig. 1. The mesh and target used in this paper. In panel a) the black region is bone, the dark gray region is marrow, the light gray region is muscle and the white region is the fat or oil layers. The thick curve marks the skin/plastic layer. This mesh is used for all forward calculations and for inverse calculations. Panel b) shows the region in the marrow of proliferated cells.

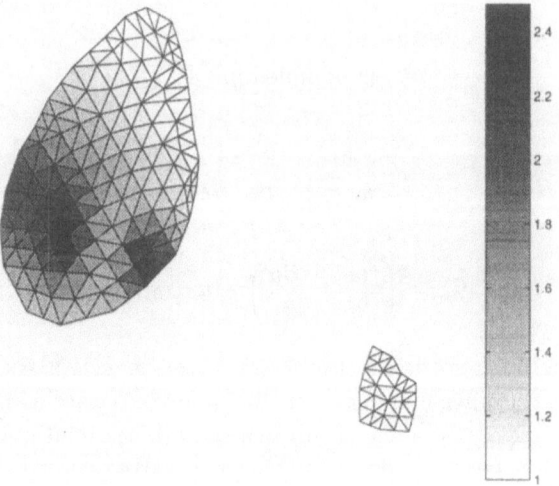

Fig. 2. Here we plot $1/\|g\|$ as a function of position in the bone marrow. The proliferated tissue shows as a darkened area. Here we use $\lambda = 1.5 + 0.8i$ and frequency 1.6 GHz. The noise level δ is 1%.

$x, y \in C$ is, given $y_0 \in D$, to find a (regularized) solution $g \in L^2(C)$ of the *near field equation*

$$(Fg)(x) := \int_C [u_b^s(x,y) - u^s(x,y)] \, g(y;y_0) \, ds(y) = \Gamma(x,y_0), \ x \in C \qquad (4.4)$$

where Γ is the Green's function for (4.1)–(4.3) when leukemia is not present. It can then be shown [7] that the L^2 norm of this regularized solution becomes unbounded as y_0 tends to ∂D, i.e. ∂D is characterized by points y_0 where the L^2 norm of a function sharply increases. More specifically, it was shown in [7] that F is injective in $L^2(C)$ and for every $\epsilon > 0$ there exists $g = g(\cdot; y_0) \in L^2(C)$ such that

$$\|Fg - \Gamma(\cdot, y_0)\|_{L^2(C)} < \epsilon$$
$$\lim_{y_0 \to \partial D} \|g(\cdot; y_0)\|_{L^2(C)} = \infty \qquad (4.5)$$
$$\lim_{y_0 \to \partial D} \|V_g\|_{L^2(D)} = \infty$$

where

$$V_g(x) := \int_C \Gamma(x,y) \, g(y) \, ds(y), \quad x \in D. \qquad (4.6)$$

In order to solve our inverse scattering problem we can now proceed as in section two of this paper, i.e. if Tikhonov regularization with the generalized Morozov principle is applied to (4.4) with a given noise level δ on the kernel of F one can expect to obtain an approximation to the function g satisfying (4.5) for ϵ and δ sufficiently small. This expectation is indeed confirmed by numerical experiments (figures 1 and 2). In particular, the linear integral equation (4.4) is solved for y_0 on a grid containing the bone marrow and ∂D is determined by those points on the grid where the L^2 norm of the regularized solution begins to sharply increase. Note that, as in the analysis of section two of this paper, we are making explicit use of the improperly posed nature of the inverse scattering problem, i.e. we are looking for an unbounded solution of the first kind integral equation (3.4).

Acknowledgement

The research of the authors was supported in part by a grant from the United States Air Force Office of Scientific Research.

References

1. Colton, D. and Kirsch, A., A simple method for solving inverse scattering problems in the resonance region, *Inverse Problems* **12** (1966), 383–393.
2. Colton, D., Kirsch, A. and Päivärinta, A., Far field patterns for acoustic waves in an inhomogeneous medium, *SIAM J. Math. Anal.* **20** (1989), 1472–1483.
3. Colton, D. and Kress, R., Eigenvalues of the far field operator for the Helmholtz equation in an absorbing medium, *SIAM J. Appl. Math.* **55** (1995), 1724–1735.
4. Colton, D. and Kress, R., *Inverse Acoustic and Electromagnetic Scattering Theory*, Second Ed., Springer–Verlag, New York, 1998.
5. Colton, D. and Monk, P., A linear sampling method for the detection of leukemia using microwaves, *SIAM J. Appl. Math.* **58** (1998), 926–941.
6. Colton, D. and Monk, P., Mathematical problems in microwave medical imaging, in *Computational Radiology and Imaging: Therapy and Diagnostics*, C. Börgers and F. Natterer, eds., Springer–Verlag, New York, 1999, 137–156.
7. Colton, D. and Monk, P., A linear sampling method for the detection of leukemia using microwaves II, submitted for publication.
8. Colton, D., Piana, M. and Potthast, R., A simple method using Morozov's discrepancy principle for solving inverse scattering problems, *Inverse Problems* **13** (1997), 1477–1493.
9. Coyle, J., *Direct and Inverse Problems in Electromagnetic Scattering from Anisotropic Objects*, Ph.D. Thesis, University of Delaware, Newark, 1998.
10. Harrington, R. and Mautz, J., An impedance sheet approximation for thin dielectric shells, *IEEE Trans. Antennas and Propagation* **23** (1975), 531–534.
11. Kirsch, A., *An Introduction to the Mathematical Theory of Inverse Problems*, Springer–Verlag, New York, 1996.
12. Kirsch, A., Characterization of the shape of the scattering obstacle by the spectral data of the far field operator, *Inverse Problems* **14** (1998), 1489–1512.
13. Kirsch, A., Factorization of the far field operator for the inhomogeneous medium case and an application in inverse scattering theory, *Inverse Problems*, to appear.
14. Langenberg, K. J., Applied inverse problems for acoustic, electromagnetic and elastic wave scattering, in *Basic Methods of Tomography and Inverse Problems*, P.C. Sabatier, ed., Adam Hilger, Bristol, 1987, 125–467.
15. Potthast, R., On a concept of uniqueness in inverse scattering for a finite number of incident waves, *SIAM J. Appl. Math.* **58** (1998), 666–682.
16. Ringrose, J. R., *Compact, Non–Self–Adjoint Operators*, Van Nostrand Reinhold Co., London, 1971.
17. Sun, Z. and Uhlmann, G., Recovery of singularities for formally determined inverse problems, *Comm. Math. Physics* **153** (1993), 431–445.
18. Tikhonov, A. N., Goncharsky, A. V., Stepanov, V. V. and Yagola, A. G., *Numerical Methods for the Solution of Ill–Posed Problems*, Kluwer, Dordrecht, 1995.

Carleman Estimates and Inverse Problems in the Last Two Decades

Michael V. Klibanov

University of North Carolina at Charlotte, Charlotte, NC 28223, U.S.A.

Abstract. Carleman estimates are a powerful tool which was originally proposed by T. Carleman in 1939 for proofs of uniqueness results for ill-posed Cauchy problems. Since 1981 this tool has been applied to inverse problems for PDEs. The goal of this paper is to provide a tutorial-like short review of the role which Carleman estimates play in three fundamental issues of inverse problems: uniqueness, stability, and numerical methods.

1 Introduction

Given an inverse problem for a PDE, three most important questions should be addressed: (1) uniqueness result, (2) stability estimate, and (3) numerical solution. Initially, Carleman estimates were introduced in the theory of inverse problems simultaneously and independently by A. L. Bukgeim and M. V. Klibanov in 1981 [2,3,12,13] as a powerful tool for proofs of global uniqueness results for multidimensional inverse problems with a single measurement: no such results existed before 1981. It was also immediately clear to these authors that a little modification of this idea would allow one to prove Hölder stability estimates for these problems (cf. comment on p. 577 of [17]), because such estimates were obtained earlier in [24] for some ill-posed Cauchy problems. Gradually, however, different authors started to apply successfully Carleman estimates to the proofs of the Lipshitz, rather than weaker Hölder stability of hyperbolic [11,20] and most recently parabolic [9] ill-posed Cauchy problems as well as inverse problems. Finally and quite surprisingly, Carleman estimates found their application in the third topic: numerics, which is important for applications. First, they were applied to construct numerical schemes and establish convergence rate for the elegant Quasi-Reversibility method [28] for numerical solutions of ill-posed hyperbolic and elliptic Cauchy problems [18-20]. Carleman estimates have been also applied both for the theoretical construction of numerical methods [7,21-23] and to the real numerics for inverse problems [6,24-27]. Thus, it has been proven in the last two decades that the tool of Carleman estimates can be successfully used in all three main topics of inverse problems.

The goal of this paper is rather tutorial. Specifically, we outline here the basic ideas of applications of Carleman estimates to all three above topics of inverse problems. Hence, throughout the paper the author will intentionally present rather simple examples, referring the reader to other publications for the additional technical details. Therefore such things as the minimal smoothness of the coefficients of PDE and their solutions, the minimal requirements on domains, boundaries, etc. are of the least concern of the author. Basically, it will be assumed throughout the paper that the coefficients of all the operators involved are smooth enough, so that the solutions of the corresponding PDEs of the second order are at least in C^3.

2 TWO MAIN CLASSES OF INVERSE PROBLEMS UNDER CONSIDERATION

While the method of Carleman estimates can be applied to a broad class of inverse problems, we focus in this paper only on inverse problems for linear hyperbolic and parabolic PDEs. More general cases can be found in [3,4,13,17]. In particular, this method can be applied to inverse problems for non-linear elliptic and parabolic PDEs [14-16].

Let $\Omega \subset R^n$ be a domain with piecewise boundary $\partial\Omega$ and $\Gamma \subseteq \partial\Omega$ be a part of $\partial\Omega$. We will use Γ to assign the data for the inverse problems on this part of $\partial\Omega$. For a constant $T > 0$ denote $\Omega_T^+ = \Omega \times (0,T)$, $\Omega_T = \Omega \times (-T,T)$, $S_T^+ = \partial\Omega \times (0,T)$, $\Gamma_T^+ = \Gamma \times (0,T)$, $S_T = \partial\Omega \times (-T,T)$ and $\Gamma_T = \Gamma \times (-T,T)$. Let $L(x,D)$ be an elliptic operator of the second order in Ω with sufficiently smooth coefficients produced its principal part $L_0(x,D)$

$$L(x,D) = \sum_{|\alpha|\leq 2} a_\alpha(x, D_x^\alpha),$$

$$L_0(x,D) = \sum_{|\alpha|=2} a_\alpha(x, D_x^\alpha),$$

$$\beta_2|\xi|^2 = \sum_{|\alpha|=1} a_\alpha(x)\,\xi^\alpha \leq \beta_1|\xi|^2, \text{ for all } \xi \in R^n \setminus \{0\},$$

where $\beta_1, \beta_2 = \text{const.} > 0$ and $\beta_2 \leq \beta_1$. Consider the hyperbolic boundary value problem

$$u_{tt} = L(x,D)u \text{ in } \Omega_T^+ \tag{2.1a}$$

$$u(x,0) = f_0(x), \; u(x,0) = f_1(x) \tag{2.1b}$$

$$u|_{S_T^+} = \varphi(x,t) \tag{2.1c}$$

Inverse Problem 1 (IP1) Let a_{α_0} be one of the coefficients of the operator L in (2.1). Suppose that the functions $a_{\alpha_0}(x)$ is unknown, but all the coefficients of L are given. Determine the function $a_{\alpha_0}(x)$ given the normal derivative of the function u on Γ,

$$\frac{\partial u}{\partial n}\bigg|_{\Gamma_T^+} = \psi(x,t) \tag{2.1d}$$

Now, consider a similar parabolic boundary value problem

$$v_t = L(x, D) v \text{ in } \Omega_T^+ \tag{2.2a}$$

$$v(x, 0) = f_0(x) \tag{2.2b}$$

$$v|_{S_T^+} = g(x, t) \tag{2.2c}$$

Inverse Problem 2 (IP2) Let the coefficient $a_{a_0}(x)$ of the operator L in (2.2) be unknown, whereas all other coefficients of L are given. Determine the function $a_{a_0}(x)$ given the function $h(x, t)$

$$\left. \frac{\partial v}{\partial n} \right|_{\Gamma_T^+} = h(x, t) \tag{2.2d}$$

It is easier to prove the uniqueness and stability results for the hyperbolic, rather than for the parabolic case. In the latter case however, such results can be proven (at this point) only if relying on an analog of the Laplace transform connecting functions u and v.

Another question to ask is: In a "real life" problem, how to obtain the Neumann boundary conditions (2.1d) and (2.2d), given the Dirichlet conditions (2.1c) and (2.2c)?

The point here is that in many applications the PDEs (2.1a) and (2.2a) are satisfied in domains Ω_{1T}^+, which are larger than Ω_T^+, i.e. $\Omega_T^+ \subset \Omega_{1T}^+$, where $\Omega \subset \Omega_1$, and the target function $a_{a_0}(x)$ is usually given outside of Ω, i.e., in $\Omega_1 \setminus \Omega$, cf. section 5. In such a case the Dirichlet boundary data on $\partial \Omega_1 \times (0, T)$ are usually given (often they are equal to zero). Whereas the data (2.1c) and (2.2c) are results of measurements of the output signals. Although such measurements can be performed only on discrete sets of points, they can be interpolated on the whole boundary $\partial \Omega$. Finally, solving the corresponding forward boundary problems in $(\Omega_1 \setminus \Omega) \times (0, T)$, one can obtain the normal derivative on $\partial \Omega \times (0, T)$.

Now we return to the connection between IP1 and IP2. Let, for example, both functions $u(x, t)$ and $v(x, t)$ be solutions of the hyperbolic and parabolic Cauchy problems respectively, rather than the boundary value problems (thus, the function $a_{a_0}(x)$ is given outside of a bounded domain Ω in this case). Suppose that in (2.1a) $T = \infty$ and $f_1(x) \equiv 0$ in (2.1b). Then these functions are connected through the following analog of the Laplace transform (cf. formula (9.2.1) in [10])

$$v(x, t) = \frac{1}{2\sqrt{2\pi t}} \int_0^\infty \exp\left(-\frac{\tau^2}{4t}\right) u(x, \tau) \, d\tau \tag{2.3}$$

In particular, (2.3) implies that the function $v(x, t)$ is analytic as a function of the real variable $t > 0$, and the same connection is valid for the pairs (φ, g) and

(ψ, h). Therefore, since the transform (2.3) is one-to-one, then, in this case, a uniqueness result for IP1 implies a uniqueness result for IP2.

We mention here that this Carleman estimate can also be applied to the following two cases: (1) when several coefficients of the operator L rather than a single one are unknown, and (2) the non-linear parabolic and elliptic PDEs, when the unknown coefficient(s) depends on the solution of this PDE. For example, $u_t = \Delta u + q(u, x_2, \ldots, x_n)$. The first case was discussed in [4]. The results for the second case can be found in [14-16].

3 CARLEMAN ESTIMATES AND ILL-POSED CAUCHY PROBLEMS

Beginning from the remarkable work of T. Carleman [5], "his" estimates have been traditionally applied to the proofs of uniqueness and stability results for the ill-posed Cauchy problems. Therefore, it is natural to outline here this traditional topic. First, we introduce a simple definition of the Carleman estimate. Keeping the stability issue in mind, we intentionally introduce a somewhat non-standard form of this estimate.

3.1 Definition of the Carleman estimate

Let $G \subset R^n$ be a domain with a piecewise smooth boundary ∂G (in terms of section 1, one can regard here G as $G = \Omega_T \subset R^{n+1}$ and $x = (x_1, \ldots, x_n, t)$). Let $A(x, D)$ be a differential operator of the second order,

$$A(x, D) = \sum_{|\alpha| \leq 2} b_\alpha(x) D^\alpha$$

and $s \in C^\infty(\bar{G})$ be a positive function with $\nabla s(x) \neq 0$ in $\bar{G}$. For a $k > 0$ we define $G_k = G \cap \{\varphi > k\}$, $\Phi_k = \{s(x) = k\} \cap G$, and $\partial_1 G_k = \partial G \cap G_k \neq \emptyset$.

Definition. The operator $A(x, D)$ satisfies the Carleman estimate in G_k with the weight function $\exp(2\lambda s(x))$ if

$$\int_{G_k} [A(x, D) u]^2 \exp(2\lambda s) \, dx$$

$$\geq \frac{C_0}{\lambda} \sum_{|\alpha|=2} \int_{G_k} (D^\alpha u)^2 \exp(2\lambda s) \, dx$$

$$+ C_1 \lambda \sum_{|\alpha| \leq 1} \int_{G_k} (D^\alpha u)^2 \exp(2\lambda s) \, dx \tag{3.1}$$

$$- B_0 \lambda^{m_0} \sum_{|\alpha|=2} \int_{\partial_1 G_k} (D^\alpha u)^2 \exp(2\lambda s)\, d\sigma$$

$$- B_1 \lambda^{m_1} \sum_{|\alpha|\leq 1} \int_{\partial_1 G_k} (D^\alpha u)^2 \exp(2\lambda s)\, d\sigma,$$

for all functions $u \in H^2(G_k)$ such that $D^\alpha u = 0$ on Φ_k for $|\alpha| \leq 2$, and for all $\lambda \geq \lambda_0$, where the constants C_0, C_1, B_0, B_1, λ_0, m_0 and m_1 do not depend on the function u. All these constants are positive, except for possible C_0 and B_0, which are non-negative; $B_0 = 0$ if and only if $C_0 = 0$.

Both constants C_0 and B_0 are positive in the elliptic and parabolic cases and they are equal to zero in the hyperbolic case. The latter fact allows one to apply Carleman estimates to the inverse problem for non-linear elliptic and parabolic PDEs [14-16]. We note that the last two terms in the right hand side of (3.1) are integrals over a part ∂G_k of the boundary ∂G, where the data are given. So, if for example, the data $D^\alpha u = 0$ on ∂G_k and $A(x, D)u = 0$ in G, then (3.1) implies immediately a certain uniqueness result.

Remark 3.1. The estimate (3.1) implies a similar estimate for the principal part of $A_0(x, D)$ of the operator $A(x, D)$ and vice versa. This can be easily derived from the fact that the lower order derivatives are involved with a large parameter λ in (3.1).

Now, formally at least, the above definition can be applied only to a differential operator $A(x, D)$ whose principal part $A_0(x, \xi)$, $\xi \in R^n$, is a uniform polynomial of the second order, with respect to ξ. However, a similar definition is valid for the case of operators with non-uniform principal parts, such as parabolic operators, for example [10]. We will use such an extended definition below without actually formulating it.

3.2 Uniqueness and Hölder stability of the Cauchy problem

Let $A_0(x, D)$ be the principal part of the operator $A(x, D)$

$$A_0(x, D) = \sum_{|\alpha|=2} a_\alpha(x, D)^\alpha$$

Consider a differential inequality

$$|A_0(x, D)u| = M(|\nabla u| + |u| + |f|), \quad \text{in } G \tag{3.2a}$$

$$u|_\Gamma = u_0(x), \qquad \left.\frac{\partial u}{\partial n}\right|_\Gamma = u_1(x), \tag{3.2b}$$

where $u \in H^2(G)$, $f \in L_2(G)$, $u_0 \in H^1(\Gamma)$, $u_1 \in L_2(\Gamma)$, $M = \text{const.} > 0$, $\Gamma \subseteq \partial G$, $\Gamma \in C^1$ is a part of the boundary ∂G and n is the outward unit normal vector of Γ. We can assume that

$$\| u_0 \|_{H^1(\Gamma)} + \| u_2 \|_{L_1(\Gamma)} \leq \varepsilon, \tag{3.2c}$$

where $\varepsilon > 0$ is sufficiently small: otherwise one can consider the function $\tilde{u} = u/d$ with a sufficiently large positive constant d. If in (3.1) $C_0 \neq 0$ and $B_0 \neq 0$, then one should replace in (3.2c) $H^1(\Gamma)$ with $H^2(\Gamma)$ and $L_2(\Gamma)$ with $H^1(\Gamma)$. However, we will not consider this case here, for the sake of simplicity.

We want to prove the Hölder stability for a simpler case when $B_0 = C_0 = 0$ in (3.1). The case of B_0, $C_0 > 0$ can be handled along the same lines.

Theorem 3.2 *(Hölder stability) Assume that the Carleman estimate (3.1) is valid for the operator $A_0(x, D)$, where $k > 0$ is sufficiently small and $B_0 = C_0 = 0$. Also, assume that $\partial G_k \neq \emptyset$, $\partial G_k \subseteq \Gamma$, and $G_{3k} \neq \emptyset$ (the latter implies $G_{2k} \neq \emptyset$ and $G_{3k} \subset G_{2k} \subset G_k$), where Γ is a part of ∂G as per (3.2b). Then there exists a small positive constant $\varepsilon_0 > 0$ such that for any $\varepsilon \in (0, \varepsilon_0)$ the following Hölder stability estimate is valid for every function $u \in H^2(G)$ satisfying (3.2)*

$$\| u \|_{H^1(G_{3k})} \in C\varepsilon^\gamma \left[1 + \| u \|_{H^1(G_k)} + \| f \|_{L_2(G_k)}\right] \tag{3.3}$$

In particular, if in (3.2c) $\varepsilon = 0$, then $u = 0$ in G_{3k}, which is a uniqueness result.

Proof. Consider a non-negative function $\chi(x) \in C^\infty(\bar{G})$ such that $\chi(x) = 1$ in G_{2k} and $\chi(x) = 0$ in $G_k \setminus G_{3/2k}$. Let

$$v(x) = \chi(x) u(x) \tag{3.4}$$

Then

$$v(x) = u(x) \text{ in } G_{2k} \tag{3.5}$$

and the Carleman estimate (3.1) is valid for $v(x)$, since $D^\alpha v(x) = 0$ $(|\alpha| \leq 2)$ for $x \in \Phi_k = \{s(x) = k\} \cap G$. Likewise, (3.2a) and (3.4) imply

$$|A_0 v| \leq M_1 (|\nabla u| + |u| + |f|) \text{ in } G, \tag{3.6}$$

where the positive constant M_1 depends on the constant M in (3.2a), as well as on the $C^2(\bar{G})$-norm of the function χ and on $C^1(\bar{G})$-norm of the coefficients of the operator A_0 Below in this proof M_1 will denote different positive constants depending on these parameters.

Hence, (3.1) and (3.6) lead to

$$M_1 \int_{G_k} \left(|\nabla u|^2 + |u|^2 + |f|^2\right) \exp(2\lambda s) \, dx$$

$$\geq \int_{G_k} (A_0 v)^2 \exp(2\lambda s) \, dx$$

$$\geq C_1 \lambda \int_{G_k} \left(|\nabla v|^2 + |v|^2\right) \exp(2\lambda s) \, dx$$

$$- B_1 \lambda^{m_1} \int_{\partial_1 G_k} \left(|\nabla v|^2 + |v|^2\right) \exp(2\lambda s) \, d\sigma$$

Now, since $G_{2k} \subset G_k$ and $\partial G_k \subseteq \Gamma$, then this inequality, (3.2b), (3.4), and (3.5) lead to

$$M_1 \int_{G_k} \left(|\nabla u|^2 + |u|^2 + |f|^2 \right) \exp\left(2\lambda s\right) dx \qquad (3.7)$$

$$\geq C_1 \lambda \int_{G_{2k}} \left(|\nabla u|^2 + |u|^2 \right) \exp\left(2\lambda s\right) dx$$

$$- M_1 \lambda^{m_1} \int_{\Gamma} \left(|\nabla u_0|^2 + |u_0|^2 + |u_1|^2 \right) \exp\left(2\lambda s\right) d\sigma$$

Choose a sufficiently large positive constant λ_0 such that $\frac{1}{2} C_1 \lambda_0 > M_1$. Then by (3.7) for all $\lambda \geq \lambda_0$

$$M_1 \int_{G_k \backslash G_{2k}} \left(|\nabla u|^2 + |u|^2 + |f|^2 \right) \exp\left(2\lambda s\right) dx \qquad (3.8)$$

$$\geq \frac{1}{2} C_1 \lambda \int_{G_{2k}} \left(|\nabla u|^2 + |u|^2 \right) \exp\left(2\lambda s\right) dx$$

$$- \dot{M}_1 \lambda^{m_1} \int_{\Gamma} \left(|\nabla u_0|^2 + |u_0|^2 + |u_1|^2 \right) \exp\left(2\lambda s\right) d\sigma$$

$$\geq \frac{1}{2} C_1 \lambda \int_{G_{3k}} \left(|\nabla u|^2 + |u|^2 \right) \exp\left(2\lambda s\right) dx$$

$$- M_1 \lambda^{m_1} \int_{\Gamma} \left(|\nabla u_0|^2 + |u_0|^2 + |u_1|^2 \right) \exp\left(2\lambda s\right) dS.$$

Clearly,

$$\exp\left(2\lambda s\right) \leq \exp\left(4\lambda k\right) \text{ in } G_k \backslash G_{2k} \qquad (3.9a)$$

and

$$\exp\left(2\lambda s\right) \geq \exp\left(6\lambda k\right) \text{ in } G_{3k} \qquad (3.9b)$$

Denote $c = \max_{G_k} \left(s\left(x\right) \right)$. Then (3.2c), (3.8), and (3.9) lead to

$$M_1 \exp\left(4\lambda k\right) \left[\| u \|_{H^1(G_k)}^2 + \| f \|_{L_2(G_k)}^2 \right] + M_1 \lambda^{m_1} \exp\left(2\lambda c\right) \varepsilon^2$$

$$\geq \frac{1}{2} C_1 \lambda \exp\left(6\lambda k\right) \| u \|_{H^1(G_{3k})}^2$$

Or

$$\| u \|_{H^1(G_{3k})}^2 \leq M_1 \exp\left[2\lambda \left(c - 3k\right)\right] \varepsilon^2 \qquad (3.10)$$

$$+ M_1 \exp\left(-2\lambda k\right) \left[\| u \|_{H^1(G_k)}^2 + \| f \|_{L_2(G_k)}^2 \right]$$

126

The first term in the right hand side of (3.10) approaches infinity as $\lambda \to \infty$, and it approaches zero, if $\varepsilon \to 0$. The second term, however does not depend on ε, and it approaches zero as $\lambda \to \infty$. Thus, to balance these two terms, we assume that $\varepsilon \in (0, \varepsilon_0)$, where $\varepsilon_0 = \varepsilon_0(c, k, M_1)$ is sufficiently small and choose $\lambda = \lambda(\varepsilon)$ such that

$$\exp\left[2\lambda(c - 3k)\right]\varepsilon^2 = \exp\left(-2\lambda k\right) \tag{3.11}$$

This means that

$$\lambda(\varepsilon) = -\frac{\ln(\varepsilon)}{c - 2k}$$

Hence, $\lim\limits_{\varepsilon \to 0} \lambda(\varepsilon) = \infty$. Let

$$\gamma = \frac{1}{c - 2k}$$

By (3.11) $\gamma \in (0, 1)$. Thus (3.10) implies

$$\| u \|^2_{H^1(G_{3k})} \leq M_1 \varepsilon^{2\gamma} \left[1 + \| u \|^2_{H^1(G_k)} + \| f \|^2_{L_2(G_k)}\right] \quad \blacksquare$$

3.3 Examples of Carleman estimates

In this section we give examples of Carleman estimates for three main types of the operators of the second order. In doing so we modify, in a certain way, the results of [29, Chapter 4].

1. The hyperbolic operator.

Let $Au = u_{tt} - \Delta u$, $\Omega = \{(x) < R\}$, and $T =$const. $> R$. Denote $G = \Omega \times (-T, T) = \Omega_T$. Let $s(x, t) = |x|^2 - \alpha t^2$, where $\alpha = $ const. $\in (0, 1)$. For $k > 0$ denote $G_k = \left\{|x|^2 - \alpha t^2 > k\right\} \cap \{(x) < R\}$. Then $|\nabla s| \neq 0$ in $\bar{G}_k$. Choose a $k > 0$ and $\alpha = \alpha(k) \in (0, 1)$ such that the domain G_k does not intercept the top and the bottom sides of the time cylinder G, i.e., $G_k \cap \{t = \pm T\} = \emptyset$.

Consider the hyperbolic inequality

$$|u_{tt} - \Delta u| \leq M\left(|\nabla_{x,t} u| + |u| + |f(x, t)|\right), \text{ in } G \tag{3.12a}$$

$$u|_{S_T} = u_0, \quad \left.\frac{\partial u}{\partial u}\right|_{S_T} = u_1, \tag{3.12b}$$

where $u_0 \in H^1(S_T)$, $u_1 \in L_1(S_T)$, and $f \in L_2(G)$.

The Carleman estimate (3.1) holds with this function s for the operator $A = \partial^2/\partial t^2 - \Delta$ and with $B_0 = C_0 = 0$. Hence, Theorem 3.2 implies Hölder stability of the problem (3.12)

Moreover, taking $k \to 0$, $\alpha \to 1^-$, and $T \to R^+$, we obtain uniqueness result in the cone $\{|x| < t < R\}$.

2. The elliptic operator.

Let

$$A_0 u = \sum_{|\alpha|=2} a_\alpha(x) D^\alpha u, \qquad x \in G \subset R^n$$

where

$$\beta_2 |\xi|^2 \le \sum_{|\alpha|=2} a_\alpha(x) \xi^\alpha \le \beta_1 |\xi|^2, \quad \text{for } \xi \in R^{n_1} \{0\}$$

and $\beta_1, \beta_2 = $ const. > 0. Without loss of generality we assume that the boundary ∂G contains a part Γ of a hyperplane. That is, let $y = (x_2, \ldots x_n)$. Then

$$\Gamma = \{x_1 = 0, \ |y| < c\} \subset \partial G,$$

where $c = $ const. > 0. We can also assume that $\{0 < x_1 < \frac{1}{4}, \ |y| < c\} \subset G$. Let

$$s(x) = \left(x_1 + \frac{1}{c^2} |y|^2 + \frac{1}{4} \right)^{-\nu},$$

where $\nu = $ const. > 0. For $k \in \left(\frac{1}{4}, \frac{1}{2} \right)$ let

$$G_k = \{s(x) > k^{-\nu}\} = \left\{ x_1 + \frac{1}{c^2} |y|^2 < k \right\} \tag{3.14}$$

Hence, G_k is a paraboloid, $G_k \subset G$, and $\partial_1 G_k = \{x_1 = 0, \ |y| \le c\left(k - \frac{1}{4}\right)\} \subset \Gamma \subset \partial G$.

Theorem 3.3 [29] *There exist sufficiently large positive constants ν_0 and λ_0 depending only on the C^1-norms of the coefficients of the operator A_0 and the number c, such that for all $\nu \ge \nu_0, \lambda \ge \lambda_0$ the Carleman estimate (3.14) is valid with $C_0 > 0$ and $B_0 > 0$.*

Remark 3.4 Hence, one can obtain Hölder stability and uniqueness in G_{3k} for the problem (3.2) with the elliptic operator (3.13). It is also clear that any bounded domain G can be covered by a finite number of "G_k-like" domains (3.14). Therefore, a slight modification of the proof of Theorem 3.1 leads to a more general result. Specifically: (1) for any subdomain $G^1 \subset G$ one can estimate $\| u \|_{H^1(G^1)}$ through ε^γ as well as $\| u \|_{H^1(G)}$ and $\|f\|_{H^1(G)}$, and (2) one can prove that if in (3.2c) $\varepsilon = 0$, then $u(x) \equiv 0$ in G.

3. The parabolic operator.

Let $\Omega \subset R^n$ be a bounded domain such that

$$\Gamma = \{x_1 = 0, |y| < c\} \subset \partial\Omega,$$

where $y = (x_2, \ldots, x_n)$ and $c = $ const. > 0. Choose a constant $T > 0$, and let $G = \Omega \times (-T, T) = \Omega_T$. Similarly to the above, let

$$s(x, t) = \left(x_1 + \frac{1}{c^2} |y|^2 + \frac{1}{T^2} t^2 + \frac{1}{4} \right)^{-\nu},$$

where $\nu = $ const. > 0. For $k \in \left(\frac{1}{4}, \frac{1}{2}\right)$ let

$$G_k = \left\{ s\left(x, t\right) > k^{-\nu} \right\} = \left\{ x_1 + \frac{1}{c^2} |y|^2 + \frac{1}{T^2} t^2 + \frac{1}{4} < k \right\}$$

Further, let $B_0\left(x, t, D\right)$ be the elliptic operator in G,

$$B_0\left(x, t, D\right) u = \sum_{|\alpha|=2} a_\alpha\left(x, t\right) D_x^\alpha u,$$

where

$$\beta_2\left(\xi\right)^2 \leq \sum_{|\alpha|=2} a_\alpha\left(x, t\right) \xi^2 \leq \beta_1\left(\xi\right)^2, \; \xi + R_1\left\{0\right\},$$

and $\beta_1, \beta_2 = $ const. > 0. Consider this parabolic operator

$$A_0\left(x, t, D\right) u = u_t - B_0\left(x, t, D\right) u$$

Theorem 3.5 *There exist sufficiently large positive constants ν_0 and λ_0 depending only on C^1-norm of the operator B_0 and on the constants c and T such that for all $\nu \geq \nu_0$ and $\lambda \geq \lambda_0$ the Carleman estimate (3.1) holds with $C_0 > 0$ and $B_0 > 0$.*

Remark 3.6 In Theorems 3.3 and 3.5 one can either count or discount both constants C_0 and B_0 simultaneously. These constants are not included in the Carleman estimates of [29]. However, it was observed in [15] that they can be included indeed in the Carleman bounds of [29].

3.4 Lipshitz stability

In this section we show how to obtain a Lipshitz stability rather than a weaker Hölder stability for the hyperbolic case using Carleman estimates. We follow here [11,20]. As we saw in section 3.3, the level surfaces of the function s for the hyperbolic and parabolic cases do not coincide with the top and bottom sides $t = \pm T$ of the time cylinder $G = \Omega \times (-T, T)$. For this reason, the domains G_k were only the sub-domains of the time cylinder. Thus, we were unable to estimate the function $u\left(x, t\right)$ in the whole domain G, which led to Hölder rather than stronger Lipshitz stability.

In this section we show, however, that the Lipshitz stability can be still obtained in the hyperbolic case mainly due to the fact that the hyperbolic equation can be solved both upwards and downwards in t-direction. Recently O. Imanuvilov and M. Yamamoto [9] obtained the Lipshitz stability for the parabolic inverse problem. The key point of [9] is a novel Carleman estimate for the parabolic operator derived in [8]. Note that, unlike its hyperbolic counterpart, the parabolic equation cannot be solved both upwards and downwards in t-direction. Thus, the result [9] is an interesting one, because its analog cannot be obtained using the "traditional" function s as in section 3.3.

Let $\Omega = \{(x) < R\}$, $T = \text{const.} > R$, $G = (-T,T)$ and $S_T = \partial\Omega \times (-T,T)$. Consider the hyperbolic inequality

$$|u_{tt} - \Delta u| \leq M\left(|\nabla_{x,t}u| + |u| + |f(x,t)|\right), \text{ in } G \qquad (3.15a)$$

$$u|_{S_T} = u_0(x,t), \qquad \frac{\partial u}{\partial n}\bigg|_{S_T} = u_1(x,t), \qquad (3.15b)$$

$$M = \text{const.} > 0, \qquad u_0 \in H^1(S_T), \qquad u_1 \in L_2(S_T), \qquad f \in L_2(G) \qquad (3.15c)$$

Theorem 3.7. *Let in (3.15) $T > R$. Then there exists a positive constant C depending only on M, R, and T such that the following Lipshitz estimate is valid for any function $u \in H^2(G)$ satisfying (3.15)*

$$\| u \|_{H^1(G)} \leq C \left[\| u_0 \|_{H^1(S_T)} + \| u_1 \|_{L_2(S_T)} + \| f \|_{L_2(G)}\right] \qquad (3.16)$$

Proof. In this proof C will denote different positive constants depending only on M, R, and T. The resulting constant C will be the maximal among them. For an $\alpha \in (0,1)$, $i = 1,2$ and $x_0^{(i)} \in \Omega$ let $s_i(x,t) = \left|x - x_0^{(i)}\right|^2 - \alpha t^2$. Let

$$G_k^{(i)} = \{s_i(x,t) > k\} \cap \{(x) < R\}$$

Then $|\nabla\varphi_i| \neq 0$ in $\bar{G}_k^{(i)}$. We take $x_0^{(1)} = 0$ and $x_0^{(2)} \neq x_0^{(1)}$. Choose an $\alpha \in (0,1)$ and sufficiently small positive constants ξ, k, and δ such that $\left|x_0^{(2)}\right| < \xi$, $G_k^{(i)} \subset \{|t| < T\}$, $G_{3k}^{(i)} \neq 0$, and

$$\{|x| < R, |t| < \delta\} \subset \left(G_{3k}^{(1)} \cup G_{3k}^{(2)}\right) \qquad (3.17)$$

This can be done, since $T > R$.

The Carleman estimate (3.1) is valid for the operator $A(x,D) = \partial^2/\partial t^2 - \Delta$ with either of the functions s_1 or s_2. Let $F(x,t) = (u_0(x,t), u_1(x,t))$ be a vector valued function with the norm

$$\| F \| = \left[\| u_0 \|_{H^1(S_T)}^2 + \| u_1 \|_{L_2(S_T)}^2\right]^{1/2}$$

Then by Theorem 3.2 there exists a sufficiently small constant

$$0 < \varepsilon_0 \ll 1 \qquad (3.18)$$

depending only on M, T, and R, such that if

$$\| F \| < \varepsilon_0, \qquad (3.19)$$

then

$$\| u \|_{H^1\left(G_{3k}^{(i)}\right)}^2 \leq C \left[1 + \| u \|_{H^2(G)}^2 + \| f \|_{L_2(G)}^2\right] \| F \|^{2\gamma}, \qquad (3.20)$$

where $i = 1, 2$ and $\gamma = \gamma(M, R, T) = \text{const.} \in (0, 1)$.

Here (3.17) and (3.20) imply that there exists a constant $t_0 \in (-\delta, \delta)$ such that

$$\int_\Omega \left(|\nabla_{x,t} u|^2 + |u|^2 \right) (x, t_0) \, dx \tag{3.21}$$

$$\leq \frac{C}{2\delta} \left[1 + \| u \|_{H^1(G)}^2 + \| f \|_{L_2(G)}^2 \right] \| F \|^{2\gamma}$$

Now we can apply the standard energy estimates to the operator $\partial^2 / \partial t^2 - \Delta$ in both time cylinders $\Omega \times (t_0, T)$ and $\Omega \times (-T, t_0)$ using (3.15) and (3.21). Hence, these estimates imply

$$\| u \|_{H^1(G)}^2 \leq C \left[1 + \| u \|_{H^1(G)}^2 + \| f \|_{L_2(G)}^2 \right] \| F \|^{2\gamma} + C \| f \|_{L_2(G)}^2 \tag{3.22}$$

Let

$$\sigma = \min \left[\left(\frac{1}{4c} \right)^{1/2\gamma}, \ \frac{1}{\sqrt{8c}} \right]$$

Let in (3.19) $\varepsilon_0 \leq \sigma$, as well as

$$\| f \|_{L_2(G)} \leq \sigma \tag{3.23}$$

Then (3.19), (3.22), and (3.23) lead to

$$\| u \|_{H^1(G)}^2 \leq \frac{2}{3} < 1 \tag{3.24}$$

Introduce functions $\tilde{F}(x, t)$, $\tilde{f}(x, t)$ and $\tilde{u}(x, t)$ as

$$\tilde{F} = (\tilde{u}_0, \tilde{u}_1) = \left[\frac{F}{\| F \| + \| f \|_{L_2(G)}} \right] \cdot \sigma, \quad \tilde{f} = \left[\frac{f}{\| F \| + \| f \|_{L_2(G)}} \right] \cdot \sigma,$$

$$\tilde{u} = \left[\frac{u}{\| F \| + \| f \|_{L_2(G)}} \right] \cdot \sigma$$

Then $\| \tilde{F} \| \leq \sigma$, $\| \tilde{f} \|_{L_2(G)} \leq \sigma$, and by (3.15)

$$|\tilde{u}_{tt} - \Delta \tilde{u}| \leq M \left(|D_{x,t} \tilde{u}| + |\tilde{u}| + |\tilde{f}| \right)$$

$$\tilde{u}|_{S_T} = \tilde{u}_0, \quad \left. \frac{\partial \tilde{u}}{\partial u} \right|_{S_T} = \tilde{u}_1$$

Hence, by (3.24) $\| \tilde{u} \|_{H^1(G)} \leq 1$. Therefore,

$$\| u \|_{H^1(G)} \leq \frac{1}{\sigma} \left(\| F \| + \| f \|_{L_2(G)} \right) \ \blacksquare$$

In fact, this result can be generalized on the case when the Dirichlet data are given on the whole boundary $\partial \Omega$, whereas the Neumann data are given only on a part Γ of $\partial \Omega$, see [11]. A similar result is described in [10], where a different technique of the so-called multipliers is used.

4 UNIQUENESS OF INVERSE PROBLEMS BY CARLEMAN ESTIMATES

4.1 Historically: Why Carleman estimates?

By the end of the seventies only the so-called *local* uniqueness results were proven for the multidimensional inverse problems with a single measurement (such as those introduced in section 2), cf. [1,29]. That is, in these results an unknown coefficient was assumed to be either sufficiently small, or piecewise analytic, or being represented through a finite Fourier-like series, etc. So, many experts in inverse problems (including the author) dreamed to prove *global* uniqueness results. That is, in such a result those restrictions on the unknown coefficient would be lifted.

The first result of this sort was established by R. G. Muhometov in 1975 [30,31] for the so-called inverse kinematic problem, which is similar to the problem of Radon transform. The fundamental difference between [30,31] and the latter problem, however, is that in the case of [30,31] the integrations are carried out along the geolesic lines in a Riemannian metric rather than along straight lines as in the case of Radon transform. In retrospect, the author remembers that the first presentation of this result by Muhometov on the seminar of Dr. M. M. Lavrentiev (Novosibirsk, Russia) in 1975 was quite a dramatic one (the member of Russian National Academy of Sciences Mikhail Mikhailovich Lavrentiev is one of the founders of the Russian scientific school of inverse problems; he was the author's "teacher" in this field in the seventies). However, the inverse kinematic problem is very essentially different from posed in section 2.

In the August of 1980 Carleman estimates came to the author's attention when he was reading about this tool in chapter 4 of the book [29], which was originally published in Russian in 1980. Initially the author was amazed by a striking similarity between the hyperbolic/parabolic inverse problems and the ill-posed Cauchy problems. The latter were successfully "cured" by Carleman estimates. An obvious similiarity is that in both cases the data are given on the lateral side of the time cylinder $\Omega \times (-T, T)$. However, an inverse problem has an additional "trouble making" element: the unknown coefficient $a_{\alpha_0}(x)$. Thus, one basically should deal with one PDE which, however, has two unknown functions: the solutions $u(x, t)$ of this PDE and the unknown coefficient $a_{\alpha_0}(x)$. The only thing to keep the balance is the presence of a given initial condition at $t = 0$, the factor, which is not presented in a similar ill-posed Cauchy problem.

Thus, a very first and vague idea of the author was that it might be possible to use somehow this "balancing" factor in order to reduce the original inverse problem to a sort of an ill-posed Cauchy problem, for which the powerful tool of Carleman estimates would be applicable. Finally, in October of 1980, after many exhausting days and sleepless nights, that initially vague idea was crystallized in the author's mind. Specifically, he realized that: (1) first the unknown coefficient $a_{\alpha_0}(x)$, should be eliminated from the original PDE by the differentiation with respect to t since $D_t[a_{\alpha_0}(x)] \equiv 0$, (2) next, the use of the initial condition at $t = 0$ will lead to an analog of an ill-posed Cauchy problem for an integro-differential, rather than differential PDE with the Volterra-like integrals $\int_0^t (\cdots)\, d\tau$, and, finally (3) the Carleman estimates can be applied to this PDE, *regardless* of the presence of these integrals.

In the end of January '81 the author has presented this novel approach publicly, for the first time, on the seminar of M. M. Lavrentiev. This talk was quite an emotional one for both the audience and the author. All the events of this seminar are still unforgettable to the author, even 19 years later, in 2000. On the same seminar A. L. Bukhgeim has also presented, for the first time, very similar results, which he had obtained independently from the author. First, both Bukhgeim and the author have published their papers separately in the same volume of proceedings [2,12]. Next, they published a joint paper [3], in which they, in particular, acknowledged their independence in introducing Carleman estimates in the field of inverse problems.

4.2   Uniqueness

Because of the tutorial nature of this publication, we consider only the hyperbolic IP1 (section 2) for the case when the operator L in (2.1a) is $Lu = \Delta u + a(x)u$, where the coefficient $a(x)$ is unknown. We refer to [3,4,10,13-17] for more general cases, including the non-linear parabolic and elliptic PDEs. The uniqueness result of the parabolic inverse problem (2.2) can be obtained using its connections with the hyperbolic problem via (2.3). Alternatively, one should assume first the conditions (2.2) are satisfied in $\Omega \times (-T,T)$, rather in $\Omega \times (0,T)$, and use a modification of the method of this section.

Since a uniqueness result usually deals with the difference of two possible solutions, we follow this general philosophy by considering two pairs $(u_1(x,t), a_1(x))$ and $(u_2(x,t), a_2(x))$ satisfying the same PDE (2.1a) with the same initial and boundary conditions (2.1b-d). Let $v = u_1 - u_2$ and $\tilde{a}(x) = a_1 - a_2$. Then (2.1) implies

$$u_{tt} = \Delta v + a_1 v + \tilde{a}(x) u_2, \text{ in } \Omega_T^+ \tag{4.1a}$$

$$v(x,0) = 0, \qquad v_t(x,0) = 0 \tag{4.1b}$$

$$v\big|_{S_T^+} = 0, \qquad \frac{\partial v}{\partial n}\bigg|_{S_T^+} = 0 \tag{4.1c}$$

Again, for the sake of simplicity we assume that $\Omega = \{|x| < R\}$. Then, by results of section 3, we should take $T > R$.

Since the Carleman estimate for the operator $\partial^2/\partial t^2 - \Delta$ holds in the domain $G = \Omega \times (-T,T)$, rather than in $\Omega_T^+ = \Omega \times (0,T)$, we extend the functions v and u_2 into $\Omega_T^- = \Omega \times (-T,0)$ as even functions of the variable t. We assume that $v, u_2 \in C^3\left(\overline{\Omega_T^+}\right)$. Hence, (4.1) implies

$$v \in H^3(G) \cap C^3\left(\overline{\Omega_T^+}\right) \cap C^3\left(\overline{\Omega_T^-}\right), \tag{4.2a}$$

$$u_2 \in C^3\left(\overline{\Omega_T^+}\right) \cap C^3\left(\overline{\Omega_T^-}\right). \tag{4.2b}$$

An obvious inconvenience of the equation (4.1a) is that it has two unknown functions $v(x,t)$ and $\tilde{a}(x)$. Hence, one of the key steps is to eliminate the function $\tilde{a}(x)$ from (4.1).

Assume that $u_2(x,0) \neq 0$ in $\bar{\Omega}$. Then $u_2(x,t) \neq 0$ for sufficiently small t with $|t| < \delta$. Dividing (4.1a) by $u_2(x,t)$, we obtain for these t

$$\tilde{a}(x) = \frac{1}{u_2(x,t)}(u_{tt} - \Delta v - a_1 v), \text{ in } \Omega \times (-\delta,\delta). \tag{4.3}$$

Since $\frac{\partial}{\partial t}[a(x)] \equiv 0$, then the differentiation of (4.3) with respect to t leads to

$$[u_{tt} - \Delta v - a_1 v]_t = \frac{u_{2t}}{u_2}[u_{tt} - \Delta v - a_1 v], \text{ in } \Omega \times (-\delta,\delta). \tag{4.4}$$

Now, the use of the initial condition (4.1b) allows one to obtain an integro-differential PDE. Indeed, let

$$w = v_t - \frac{u_{2t}}{u_2} \cdot v \tag{4.5}$$

Since $v(x,0) = v_t(x,0) = 0$, then (4.2) implies $w \in H^2(\Omega \times (-\delta,\delta))$. Further, considering (4.5) as an ordinary or differential equation with respect to v with $v(x,0) = 0$, we obtain

$$v(x,t) = \int_0^t K(x,t,\tau) w(x,\tau) d\tau, \tag{4.6}$$

where

$$K(x,t,\tau) = \frac{u_2(x,t)}{u_2(x,\tau)} \tag{4.7}$$

Substituting (4.6) into (4.4), we obtain

$$w_{tt} - \Delta w - a_1 w - \sum_{i=1}^n \int_0^t K_i(x,t,\tau) w_{x_i}(x,\tau) d\tau \tag{4.8a}$$

$$+ \int_0^t K_0(x,t,\tau) w(x,\tau) d\tau, \text{ in } \Omega \times (-\delta,\delta).$$

where the kernels K_i are generated by the function K in (4.7). Also, (4.1c) implies

$$w|_{S_\delta} = \left.\frac{\partial w}{\partial n}\right|_{S_\delta} = 0, \tag{4.8b}$$

where $S_\delta = \Omega \times (-\delta, \delta)$. Thus, we obtained a lateral problem for the hyperbolic integro-differential equation (4.8a) with the Volterra-like integrals. Therefore, it seems to be reasonable to try to apply Carleman estimate here. The *key* remaining question would be whether or not the integrals would still enable one to do this. Fortunately, the answer on this question is positive.

Theorem 4.1. (Uniqueness) *Let $\Omega = \{(x) < R\}$ and $T > R$. Assume that $u_2(x, 0) \neq 0$ for all $x \in \Omega$ and $v, u_2 \in C^3(\bar{\Omega}_T^+)$. Then $\tilde{a}(x) \equiv 0$ in Ω and $v(x, t) \equiv 0$ in Ω_T^+.*

Proof. We choose the function s as in section 3.3, $s(x, t) = |x|^2 - \alpha t^2$, where $\alpha = \text{const.} \in (0, 1)$. Choose a sufficiently small constant $k > 0$ such that $G_{3k} \neq 0$ and the domain $G_k = \{s(x, t) > k\} \cap \Omega$ is a part of the strip $\Omega \times (-\delta, \delta)$, i.e.,

$$[\{s(x, t) > k\} \cap \Omega] \subset [\Omega \times (-\delta, \delta)], \tag{4.9a}$$

and $G_{3k} \neq 0$. That is, we take k as

$$0 < k < \frac{R^2 - \alpha\delta^2}{3}, \tag{4.10}$$

Then equation (4.8a) holds in G_k. In order to apply Theorem 3.2, it is sufficient to show that

$$\int_{G_k} \left(\int_0^t |f(x, t)| \, dt \right)^2 \exp(2\lambda s) \, dx dt \leq M \int_{G_k} f^2 \exp(2\lambda s) \, dx dt, \tag{4.11}$$

for every function $f \in L_2(G_k)$, where the positive constant M depends only on G_k. Let $G_k = G_k^+ \cup G_k^-$, where $G_k^+ = G_k \cap \{t \geq 0\}$ and $G_k^- = G_k \cap \{t \leq 0\}$. Also, let $G_k^0 = G_k \cap \{t = 0\}$. For every point $x \in G_k^0$ let $(x, t^+(x))$ and $(x, t^-(x))$, be the points on the surface ∂G_k, where $t^+(x) > 0$ and $t^-(x) < 0$. Note that the function $s(x, t)$ is decreasing and increasing with respect to t in G_k^+ and in G_k^-, respectively. Hence,

$$\int\limits_{G_k^+} \left(\int\limits_0^t |f(x,\tau)|\, d\tau \right)^2 \exp\left(2\lambda s\right) dx dt$$

$$\leq M \int\limits_{G_k^+} \left(\int\limits_0^t f^2(x,\tau)\, d\tau \right) \exp\left(2\lambda s\right) dx dt$$

$$= M \int\limits_{G_k^0} \left(\int\limits_0^{t^+(x)} \left(\int\limits_0^t f^2(x,t)\, d\tau \right) \exp\left(2\lambda s\right) dt \right) dx$$

$$= M \int\limits_{G_k^0} \left[\int\limits_0^{t^+(x)} f^2(x,t)\, d\tau \int\limits_\tau^{t^+(x)} \exp\left[2\lambda s(x,t)\right] dt \right] dx$$

$$\leq M \int\limits_{G_k^0} \left\{ \int\limits_0^{t^+(x)} f^2(x,t) \exp\left[2\lambda s(x,t)\right] dt \int\limits_\tau^{t(x)} dt \right\} dx$$

$$\leq M \int\limits_{G_k^0} \left[\int\limits_0^{t^+(x)} f^2(x,t) \exp\left[2\lambda s(x,t)\right] dt \right] dx$$

$$= M \int\limits_{G_k^+} f^2 \exp\left(2\lambda s\right) dx dt$$

This proves that

$$\int\limits_{G_k^+} \left(\int\limits_0^t f(x,\tau)\, d\tau \right)^2 \exp\left(2\lambda s\right) dx dt \leq M \int\limits_{G_k^+} f^2 \exp\left(2\lambda s\right) dx dt \qquad (4.12)$$

Similarly,

$$\int\limits_{G_k^-} \left(\int\limits_0^t f^2 \exp\left(x,\tau\right) d\tau \right)^2 \exp\left(2\lambda s\right) dx dt \leq M \int\limits_{G_k^-} f^2 \exp\left(2\lambda s\right) dx dt \qquad (4.13)$$

Obviously, (4.11) follows from (4.12) and (4.13).

Therefore, by (4.11) and Theorem 3.2 $w(x,t) = 0$ in G_{3k}. Hence, (4.6) implies $v(x,t) = 0$ in G_{3k}. Further, (4.3) leads to $\tilde{a}(x) = 0$ in G_{3k}^0. Taking in (4.10) $\alpha \to 1$, we obtain

$$a(x) = 0 \text{ for } R^2 - \delta^2 < |x|^2 < R^2 \qquad (4.14)$$

We have taken $|t| < \delta$ only because we divided by $u_2(x,t)$ in (4.3), and there is no guarantee that $u_2(x,t) \neq 0$ for $|t| > \delta$. However, as soon as we established (4.14), there is no need to divide by u_2 as long as $x \in \Omega^\delta = \{\sqrt{R^2 - \delta^2} < |x| < R\}$. Hence consider now (4.1) for $x \in \Omega^\delta$, $t \in (-T,T)$. By (4.14), we obtain

$$v_{tt} = \Delta v + a_1(x)\, v_1 \text{ in } \Omega^\delta \times (-T,T) \qquad (4.15a)$$

$$v|_{S_T} = 0, \qquad \frac{\partial v}{\partial n}\bigg|_{S_T} = 0 \qquad (4.15b)$$

Consider now the Carleman estimate for the function v with the operator $\partial^2/\partial t^2 - \Delta$ with the function $\tilde{s}(x,t,t_0) = |x|^2 - \alpha(t-t_0)^2$, where the constant t_0 is such that $\partial/v_2 - T < t_0 < T - \delta/v_2$. Then Theorem 3.2 and (4.1) imply that $v(x,t) = 0$ for $(x,t) \in \{|x|^2 - \alpha(t-t_0)^2 > R^2 - \delta^2\}$. Taking $\alpha \to 1$, we obtain that $v(x,t) = 0$ for $(x,t) \in \{|x|^2 - (t-t_0)^2 > R^2 - \delta^2\}$ for $t_0 \in (\delta - T, T - \delta)$. Now we replace Ω with Ω^δ and T with $T - \delta$ and repeat the above process. Since δ is sufficiently small, we can always assume that $R = m\delta$, where m is an integer. Thus, repeating this process m times, we often $\tilde{a}(x) = 0$ in Ω. ■

Imposing a more stringent condition $u_2(x,t) \neq 0$ for $(x,t) \in \Omega \times (0,T)$, rather than $u_2(x,0) \neq 0$ and using the method of Theorem 3.7 and using the method of Theorem 3.7 in combination with the estimate (4.11), one can obtain a Lipshitz stability estimate for the function w, and therefore, for the function v. However, this estimate does not automatically imply an estimate for the unknown coefficient $\tilde{a}(x)$, because one would need to estimate the L_2-norm of Δw in such a case. Nevertheless, the Lipshitz stability for the function $\tilde{a}(x)$ can be obtained by the differentiation of equation (4.8a) with respect to x_i and, thus, obtaining the estimates for the H^1-norms of derivatives w_{x_i} [33].

The condition $u_2(x,0) \neq 0$ in $\bar{\Omega}$ is not quite suitable for many practical applications. It would be better, of course, to assume that $u_2(x,0) = 0$ and $u_{2t}(x,0) = \delta(x - x_0)$ for an $x_0 \in \Omega$. However, the above result (as well as its analogs for the elliptic and parabolic cases) is the only type of global uniqueness results which is known for non-overdetermined multidimensional inverse problems with a single measurement. The only second and third types of global uniqueness results for some non-overdetermined multidimensional inverse problems, which are quite different from our problems, are due to A. Nachman [32] and R. G. Muhometov [30,31].

5   NUMERICAL METHODS

5.1   General comments

In the studies of numerical methods one can relax some conditions which are necessary to prove a valuable uniqueness result. This thought enabled the author (with collaborators) to develop some numerical methods for the multidimensional inverse problems using the above ideas. In these studies the basic inconvenient restriction of Theorem 4.1-like results, which is $u(x,0) \neq 0$ in $\bar{\Omega}$, is eliminated at the price of imposing more stringent but numerically valuable conditions on the functions involved. Specifically, a practically valuable initial condition $u(x,0) = \delta(x - x_0)$ is used in these studies. This can be done because these numerical schemes rely on the assumption that a function associated with the function $u(x,t)$ can be represented through a finite generalized Fourier series with respect to t. Clearly, such an assumption cannot be used in a uniqueness result. Numerical results show, however, that this assumption is quite reasonable in the practical numerics.

The above mentioned numerical schemes fall into two categories.

1. The globally convergent numerical methods [7,21-23]. In these references globally convex cost functionals are constructed by the use of the weight functions $\exp(2\lambda s)$, which are involved in Carleman estimates. Therefore, the absence of the local minima of such a functional is guaranteed.

2. The locally convergent algorithms [6,24-27] i.e., those which are based on the perturbation approach. The key element of this approach is the solution of a certain linearized problem. It turns out that this linearized problem amounts to the solution of a well-posed boundary value problem for a coupled system of elliptic PDEs. That is why we call this approach the elliptic system method (ESM).

Out of these two approaches only the ESM is tested numerically at this point. These tests have been performed for a parabolic inverse problem arising in optical tomography (OT), both for the computationally simulated [6,24] and for the experimental data [26,27]. We outline the ESM in section 5.2.

In addition, the Carleman estimates can also be used for the proof of the convergence rate of the elegant quasi-reversibility method of R. Lattes and J.-L. Lions [28]. V. Isakov calls the algorithm of [28] "variational regularization" [10]. This method can be applied very effectively to the numerical solutions of ill-posed Cauchy problems. We omit here the description of this algorithm referring to [18,19] for details, which also include results of numerical tests.

5.2   Elliptic System Method

The inverse problem described here has straightforward applications in optical imaging of localized abnormalities hidden in turbid media, such as, for example, tumors in female breasts. In this case light is generated by an ultrashort laser pulse of about 1 picosecond (ps) durations, i.e., $\approx 10^{-12}$ second. Then $u(x,t)$ is the time resolved light intensity, $D(x)$ is the diffusion coefficient, and $a(x)$ is the absorption term. Since this is our target application, we are interested in imaging of *small* inclusions rather than the background. We also note that tumors absorb light from the surrounding tissues [27]. Hence the values of the absorption term $a(x)$ within tumors are greater than in the surrounding tissues.

Let $\Omega \subset G \subset R^n$ $(n = 2, 3)$ be two convex bounded domains with the boundaries ∂G, $\partial \Omega \in C^\infty$. For the purpose of theoretical derivation we assume that the source position $x_0 \in G \backslash \bar{\Omega}$. In practice, however, $x_0 \in \partial \Omega$ will also assume that the diffusion coefficient $D(x)$ and the absorption coefficient $a(x)$ are sufficiently smooth in $\tilde{G}$. Let $T = \text{const.} > 0$. Consider the parabolic boundary value problem

$$u_t = dv\left(D(x)\nabla u\right) - a(x)u, \text{ in } G_T^+, \tag{5.1a}$$

$$u(x, 0) = \delta(x - x_0), \tag{5.1b}$$

$$u|_{\partial G} = 0. \tag{5.1c}$$

In the case of OT c is the speed of light in the medium, $\mu_s'(x)$ is the reduced scattering coefficient and $\mu_a(x)$ is the absorption coefficient. We assume that the medium is low absorbing with $\mu_s' \gg \mu_a$. In biological tissues, for example, this condition is satisfied. In these tissues the ranges of optical parameters are [26,27]

$$c \simeq 0.225\frac{\text{mm}}{\text{ps}}, \mu_s'(0.4, 1.25)\frac{1}{\text{mm}}$$

$$\mu_a \in (0.004, 0.03)\frac{1}{\text{mm}}$$

The formulas for the diffusion coefficient $D(x)$ and for the absorption term $a(x)$ are [26,27]

$$D(x) = \frac{c}{3\mu_s'(x)}, \qquad a(x) = c\mu_a(x)$$

Further, usually in OT Ω is the domain of interest, and air is outside of Ω. In this case ∂G is the so-called "extrapolated boundary" of $\partial\Omega$, which is used to impose "correct" boundary conditions (5.1c) [27]. Specifically, let $x \in \partial\Omega$ and $n(x)$ be the outward unit normal vector to $\partial\Omega$ at the point x. Then $\partial G = \{x'(x) : x \in \partial\Omega$ and $x'(x) = x + zn(x)\}$, where $z = \text{const.} > 0$ is the distance between ∂G and $\partial\Omega$. Usually, $z\mu_s' \in (1.44, 5)$.

Let $\{x_i\}_{i=1}^n \subset \partial\Omega$ be a set of detectors or $\partial\Omega$ and $T_0 = \text{const.} \in (0, T)$. We consider the following:

Inverse Problem. *Given detectors readings* $\varphi_i(t_1) = u(x_i, t)$ *for* $t \in (T_0, T)$, *determine either of the functions* $D(x)$ *or* $a(x)$ *for* $x \in \Omega$, *assuming that both of these are given for* $x \in G\backslash\Omega$.

Interpolating detectors readings over the entire boundary $\partial\Omega$, we obtain the function $\varphi(x, t)$,

$$u(x, t) = \varphi(x, t), \text{ for } (x, t) \subset \partial\Omega \times (T_0, T) \tag{5.2}$$

We observe the importance of the condition $T_0 > 0$, because in a practical scenario of OT light arrives at a detector x_i at a moment of time $T(x_i) > 0$:

this is a well known discrepancy between the reality and its description by the parabolic equation.

First, we consider a simpler case where the coefficient $a(x)$ is unknown. We assume that $a(x) = a_0(x) + h(x)$, where the given function $a_0(x)$ is a prior guess about the background absorption, and $h(x)$ is its small perturbation due to the presence of localized abnormalities, as well as an inaccuracy of the estimate of this background, $\| h \|_{L_2(G)} \ll \| a_0 \|_{L_2(G)}$. Let $u_0(x_i)$ be the solution of the forward problem (5.1) with $a = a_0$. Consider the normalized solution

$$H(x,t) = \left(\frac{u}{u_0}\right)(x,t) - 1$$

Note that by the maximum principle $u_0 \geq$ const. > 0 in $\bar{\Omega} \times [\varepsilon, T]$, for every $\varepsilon \in (0, T)$. One can prove that if the distance between $\partial\Omega$ and its extrapolated boundary ∂G is sufficiently small, then [27]

$$\frac{\partial H}{\partial n}(x,t) \approx 0, \text{ for } (x,t) \in \partial\Omega \times (T_0, T). \tag{5.3}$$

In addition, for a broad class of coefficients $D(x)$ (including the case $D \equiv$ const.) the following limit is valid

$$\lim_{t \to 0} H(x,t) = 0 \tag{5.4}$$

This limit was established in [24] for the case when both functions u and u_0 are solutions of the Cauchy problem. We have also conjectured in [27] that (5.4) still holds for a boundary value problem.

Substituting $u = (H+1)u_0$ into (5.1a) and using the equation for u_0, we obtain

$$H_t = \text{div}(D\nabla H) + 2\frac{\nabla u_0}{u_0} \cdot \nabla H + h + hH, \text{ in } \Omega_T^+ \tag{5.5a}$$

Linearization of this equation with respect to $h(x)$ leads to

$$H_t = \text{div}(D\nabla H) + 2\frac{\nabla u_0}{u_0} \cdot \nabla H + h(x), \text{ in } \Omega_T^+ \tag{5.5b}$$

$$H(x,0) = 0 \tag{5.5c}$$

Now we follow the above idea of the elimination of the perturbation term $h(x)$ from (5.5b) and obtaining an integro-differential equation this way. Denote $p(x,t) = H_t$. Then (5.5c) implies

$$H(x,t) = \int_0^t p(x,\tau) \, d\tau \tag{5.6}$$

Differentiating both sides of (5.5b) with respect to t and using (5.6), we obtain

$$p_t = \text{div}\,(D\nabla p) + 2\frac{\partial}{\partial t}\left[\frac{\nabla u_0}{u_0}\int_0^t \nabla p\,(x,\tau)\,d\tau\right], \quad \text{in } \Omega_T^+ \qquad (5.7a)$$

In addition, by (5.2) and (5.3)

$$p\,(x,t) = \frac{\partial}{\partial t}\left(\frac{\varphi}{u_0}\right), \quad \text{for } (x,t) \in \partial\Omega \times (T_0,T) \qquad (5.7b)$$

$$\frac{\partial p}{\partial n} = 0, \text{ for } (x,t) \in \partial\Omega \times (T_0,T), \qquad (5.7c)$$

The function $\varphi\,(x,t)$ is given with noise. Hence, special care should be taken in the differentiation of this function in (5.7b). Two effective approaches to this are described and tested in [24,27]. Suppose, the function $p\,(x,t)$ is found. Then one should take $H\,(x,t)$ as in (5.6) and by (5.5b)

$$h\,(x) = \frac{1}{T-T_0}\int_{T_0}^T \left(\text{div}\,(D\nabla H) + 2\frac{\nabla u_0}{u_0}\nabla H - H_t\right) dt$$

We take an average value over (T_0,T) here, because the integrand might depend on t in practical computations.

Hence, we should focus now on the determination of the function $p\,(x,t)$ from (5.7). The major inconvenience of (5.7) is that the data (5.7b,c) are given on the interval (T_0,T), whereas the integration in (5.7c) is carried out from $T = 0$. Thus, there is a "gap" between $t = 0$ and $t = T_0$. In addition, the vector valued function $\nabla u_0/u_0$ has a singularity at $t \to 0$. Thus, to tackle these, we assume that the function $p\,(x,t)$ can be represented through a finite generalized Fourier series with respect to t. Note that because of (2.3), $p\,(x,t)$ is an analytic function of the real variable $t \in (0,T)$, at least in the case of the Cauchy problem in (5.1). Hence, it is reasonable to assume that the same holds for the case of the boundary value problem.

Thus, let $\{\tilde{a}_k\,(t)\}_{k=1}^{\infty} \subset C^1\,[0,T]$ be an orthonormal basis in $L_2\,(0,T)$, where all functions $\tilde{a}_k\,(t)$ are real valued and analytic in $(0,T)$. Then

$$p\,(x,t) = \sum_{k=1}^{\infty} \tilde{a}_k\,(t)\,\tilde{Q}_k\,(x)\,, \text{ in } \Omega_T^+,$$

$$\text{where } \tilde{Q}_k\,(x) = \int_0^T p\,(x,t)\,\tilde{a}_k\,(t)\,dt$$

Re-orthogonalizing $\{\tilde{a}_k\,(t)\}$ on (T_0,T), we obtain an orthonormal basis $\{\tilde{a}_k\,(t)\}_{k=1}^{\infty}$ in $L_2\,(T_0,T)$. Let $N \geq 1$ be an integer. We assume that

$$p\,(x,t) \approx \sum_{k=1}^{N} a_k\,(t)\,Q_k\,(x)\,, \text{ in } \Omega_T^+, \qquad (5.8)$$

where the generalized Fourier coefficients $Q_k(x)$ are unknown,

$$Q_k(x) = \int_{T_0}^{T} p(x,t)\, a_k(t)\, dt$$

For $a_k(t)$ we use Legendre polynomials orthonormal on (T_0, T). Other sets of functions can be also tried. The number N is the regularization parameter of this problem. It was established in the numerical experiments [6,24,27] that usually $N = 3$ or 4. The data $(\varphi/u_0)(x_i, t)$ for $t \in (T_0, T)$ at the detectors are approximated well for these values of N. However, an increase of N usually leads to growing artifacts in images, likely because the original problem is ill-posed.

Substitute (5.8) into (5.7) and use the fact that

$$\int_{T_0}^{T} a_k(t)\, a_s(t)\, dt = \begin{cases} 1, \text{ for } k = s \\ 0, \text{ for } k \neq s. \end{cases}$$

Also, denote $Q(x) = (Q_1(x), \ldots, Q_n(x))$, and $\alpha(x) = (\alpha_1(x), \ldots, \alpha_n(x))$, where

$$\alpha_k(x) = \int_{T_0}^{T} \frac{\partial}{\partial t}\left(\frac{\varphi}{u_0}\right)(x,t)\, a_k(t)\, dt$$

Then we obtain a boundary value problem for a coupled system of elliptic PDEs

$$A(Q) := \operatorname{div}(D\nabla Q) - \sum_{i=1}^{n} B_j(x)\, Q_{x_i} - C(x)\, Q = 0, \text{ in } \Omega \qquad (5.9a)$$

$$Q|_{\partial\Omega} = \alpha(x), \qquad \left.\frac{\partial Q}{\partial u}\right|_{\partial\Omega} = 0. \qquad (5.9b)$$

Where $B_i(x)$ and $C(x)$ are $N \times N$ matrices depending on the functions $u_0(x,t)$ and $a_k(t)$ $(k = 1, \ldots N)$.

The resulting system (5.9) is the *core* of the ESM. A very attractive feature of (5.9) is the differential rather than more conventional integral form of this system. Hence, one should anticipate that if applying either the finite element or the finite difference method the problem (5.9) can be solved by the factorization of a sparse matrix, which can be done rapidly.

Still, however, this is an overdetermined problem, because of two boundary conditions (5.9b), rather than one. Our attempt [6] (jointly with T. R. Lucas and Yu. A. Gryazin) to solve the system (5.9a) using only the first boundary condition (5.9b) led to an image whose quality was inferior to those obtained in the early publication [24]. Hence, we use both boundary conditions by replacing (5.9) with the following minimization problem:

142

Find

$$\min_{P} \| A(P) \|^2_{L_2(\Omega)}, \text{ for } P \in H^2(\Omega)$$

subject to the boundary conditions

$$P|_{\partial\Omega} = \alpha(x), \qquad \frac{\partial P}{\partial n}\bigg|_{\partial\Omega} = 0$$

By the variational principle this problem is equivalent to

$$(A^* A)(P) = 0, \tag{5.10a}$$

$$P|_{\partial\Omega} = \alpha(x), \qquad \frac{\partial P}{\partial n}\bigg|_{\partial\Omega} = 0, \tag{5.10b}$$

where A^* is the operator formally adjoint to the operator A. This is a resulting boundary value problem for the elliptic system of the 4th order, which we solve numerically by the finite element method. One can also prove that the problem (5.9) is well-posed [24] and that $P \approx Q$ [27]. Likewise, an iterative Newton/Kantorovich-like method can be arranged to solve the original non-linear equation (5.5a) [27].

Further, in the case when the diffusion coefficient $D(x)$ is unknown, we replace first this function $u(x,t)$ with $v(x,t) = u\sqrt{D}$ [6]. Suppose, $a(x) \equiv 0$. Then

$$v_t = D(x)\Delta v - \Delta\left(\sqrt{D}\right)v \tag{5.11a}$$

$$v|_{t=0} = \delta(x - x_0) \tag{5.11b}$$

$$v|_{\partial G} = 0 \tag{5.11c}$$

As the first step (the only one which is tested at the time being), we simulate the data (5.2) by the solution of the forward problem (5.1). However, in the inverse algorithm, we drop the term $\Delta\left(\sqrt{D}\right)v$ in (5.11a), thus coming up with

$$v_t = D\Delta v \tag{5.12}$$

In the future we plan to study the case of the unknown coefficient $D(x)$ with more details. Let $D(x) = D_0(x) + h(x)$. And v_0 be the solution of (5.12), (5.11b,c) with $D = D_0$. Denote $w = v - v_0$. Then the equation for w linearized with respect to $h(x)$ is

$$w_t = D_0\Delta w + h(x)\Delta v_0 \tag{5.13}$$

Following the ESM, one should now divide (5.12) by Δv_0 and differentiate it then with respect to t. However, there is no guarantee that $\Delta v_0 \neq 0$ in Ω_T^+, unlike the previous case, when we divided by u_0. Thus we introduce a new function

$$\tilde{w} = \int_0^t w(x,\tau)\,d\tau$$

Note that $\Delta v_0 = v_{0t}/D_0$. Hence

$$\int_0^t \Delta v(x,\tau)\,d\tau = \frac{1}{D_0}\int_0^t v_\tau(x,\tau)\,d\tau = \frac{v_0(x,t)}{D_0}$$

By the maximum principle $v_0(x,t) > 0$ in $\bar{\Omega} \times [\varepsilon, T]$ for every $\varepsilon \in (0,T)$. Hence by (5.13)

$$\tilde{w}_t = D_0 \Delta \tilde{w} + h(x)\,v_0$$

Next, we consider the function $H(x,t) = \tilde{w}/v_0$,

$$H(x,t) = \frac{\tilde{w}(x,t)}{v_0(x,t)} = \frac{\int_0^t (v - v_0)(x,\tau)\,d\tau}{v_0(x,t)} \tag{5.14}$$

and follow the process described above. To obtain the data on the boundary, we replace in (5.14)

$$\int_0^t (\cdots)\,d\tau \text{ with } \int_{T_1}^t (\cdots)\,d\tau,$$

where $T_1 = \text{const.} > 0$ is an appropriate number.

Numerical testing of the ESM in the 2-dimensional case was conducted in [24,26,27] for the case when the coefficient $a(x)$ is unknown, and in [6] for the case of the coefficient $D(x)$. These tests have shown that the ESM images quite well small tumor-like inclusions, including the case of experimental data for an optical model of the female breast. It was also demonstrated in [27] that the ESM has a rather low sensitivity to the prior estimate of the background medium. The latter is quite a useful observation, because there are always certain errors in the prior estimates of the background. Likewise, the ESM produces images in a matter of several minutes on a Silicon Graphics Indigo (SGI), because the resulting matrix system is sparse. Another important factor of such a high speed of the ESM is that there is no need to choose a regularization parameter, since it is pre-determined for all media: $N = 3$ or 4. We note that the procedure of the choice of the regularization parameter is time consuming in a number of numerical algorithms for inverse problems, and this parameter is usually quite sensitive both to the noise and to the background medium.

6 DISCUSSION

In this tutorial-like paper we have described the basic ideas of applications of Carleman estimates to inverse problems. Emerging initially in 1981 exclusively as a tool for proofs of global uniqueness results for inverse problems, this approach casts now all three basic topics of this field: uniqueness, stability, and numerics. Interestingly enough, numerics works exactly for the case $u(x,0) = \delta(x - x_0)$, which is not covered by the original idea. The major unsolved problem, which has been crossing the author's mind for almost two decades is a proof of a global uniqueness result for this case. Hopefully, a method of such a proof would also lead to more powerful numerical algorithms.

7 Acknowledgement

This work was partially supported by the National Science Foundation grant DMS-9704923.

8 References

1. Yu. E. Anikonov, *Multidimensional Inverse and Ill-Posed Problems for Differential Equations*, VSP, Netherlands, 1995

2. A. L. Bukhgeim, Carleman estimates for Volterra operators and uniqueness of inverse problems, in *Non-classical Problems of Mathematical Physics*, published by Computing Center of Siberian Branch of Soviet Academy of Science, Novosibirsk, 1981, 56-69 (in Russian).

3. A. L. Bukhgeim and M. V. Klibanov, Uniqueness in the large of a class of multidimensional inverse problems, *Soviet Math. Dokl.* 24 (1981), 244-247.

4. A. L. Bukhgeim, *Introduction to the Theory of Inverse Problems*, Nauka, Norosibirsk, 1988 (in Russian).

5. T. Carleman, Sur un probléme d'unicité pour les systémes d'équations aux dérivées partielles á deux variables indépendantes, *Ark. Mat. Astr. Fys.*, 26B, No. 17 (1939), 1-9.

6. Yu. A. Gryazin, M. V. Klibanov, and T. R. Lucas, Imaging the diffusion coefficient in a parabolic inverse problem in optical tomography, *Inverse Problems*, 15 (1999), 373-397.

7. S. Gutman, M. V. Klibanov, and A. V. Tikhonrarov, Global convexity in a single source 3-D inverse scattering problem, *IMA J. Appl. Math.*, 55 (1995), 281-302.

8. O. Yu. Imanuvilov, Boundary controllability of parabolic equations, *Russian Math. Surveys*, 48 (1993), 192-194.

9. O. Yu. Imanuvilov and M. Yamamoto, Lipshitz stability in inverse parabolic problems by the Carleman estimate, *Inverse Problems*, 14 (1998), 1229-1245.

10. V. Isakov, *Inverse Problems for Partial Differential Equations*, Springer-Verlag, New York, 1998.

11. M. A. Kazemi and M. V. Klibanov, Stability estimates for ill-posed Cauchy problems involving hyperbolic equations and inequalities, *Applicable Analysis*, 50 (1993), 93-102.

12. M. V. Klibanov, Uniqueness in the large of some multidimensional inverse problems, in *Non-classical Problems of Mathematical Physics*, published by Computing Center of Siberian Branch of Soviet Academy of Science, Novosibirsk, 1981, 101-114 (in Russian).

13. M. V. Klibanov, Inverse problems in the "large" and Carleman bounds, *Differential Equations*, 20 (1984), 755-760.

14. M. V. Klibanov, On a class of inverse problems, *Soviet Math. Dokl.*, 26 (1982), 248-250.

15. M. V. Klibanov, Uniqueness in the large of solutions of inverse problems for a class of differential equations, *Differential Equations*, 21 (1985), 1390-1395.

16. M. V. Klibanov, A class of inverse problems for nonlinear parabolic equations, *Siberian Math. J.*, 27 (1987), 698-707.

17. M. V. Klibanov, Inverse problems and Carleman estimates, *Inverse Problems*, 8 (1992), 575-596.

18. M. V. Klibanov and F. Santosa, A computational quasi-reversibility method for Cauchy problems for Laplace's equation, *SIAM J. Appl. Math.*, 51 (1991), 1653-1675.

19. M. V. Klibanov and Rakesh, Numerical solution of a timelike Cauchy problem for the wave equation, *Math. Methods, Appl. Sci.* 15 (1992), 554-570.

20. M. V. Klibanov and J. Malinsky, Newton-Kantorovich method for 3-dimensional potential inverse scattering problem and stability of the hyperbolic Cauchy problem with time-dependent data, *Inverse Problems*, 7 (1991), 577-596.

21. M. V. Klibanov and O. V. Ioussoupova, Uniform strict convexity of a cost functional for 3-D inverse scattering problem, *SIAM J. Math. Anal.*, 26 (1995), 147-179.

22. M. V. Klibanov, Global convexity in diffusion tomography, *Nonlinear-World*, 4 (1997), 247-265.

23. M. V. Klibanov, Global convexity in a three-dimensional inverse acoustic problem, *SIAM J. Math. Anal.*, 28 (1997), 1371-1388.

24. M. V. Klibanov, T. R. Lucas, and R. M. Frank, A fast and accurate imaging algorithm in optical/diffusion tomography, *Inverse Problems*, 13 (1997), 1341-1361.

25. M. V. Klibanov and T. R. Lucas, Method and apparatus for detecting an abnormality within a host medium, United States Patent No. 5,963,658; issue date October 5, 1999.

26. M. V. Klibanov, T. R. Lucas, and R. M. Frank, Image reconstruction from experimental data in diffusion tomography, in Computational Radiology Imaging, IMA Proceedings, 110, 157-181, Springer-Verlag, New York, 1999.

27. M. V. Klibanov and T. R. Lucas, Numerical solution of a parabolic inverse problem in optical tomography using experimental data. *SIAM J. Appl. Math.*, 59 (1999), 1763-1789.

28. R. Lattes and J.-L. Lions, *The Method of Quasi-Reversibility: Applications to Partial Differential Equations*, Elsevier, New York, 1969.

29. M. M. Lavrentiev, V. G. Romanov, and S. P. Shishatskii, *Ill-Posed Problems of Mathematical Physics and Analysis*, AMS, Providence, R.I., 1986.

30. R. G. Muhometov, Inverse seismic kinematic problem on the plane, in *Mathematical Problems of Geophysics,* published by Computing Center of Siberian Branch of Acad. of Sci., Novosibirsk, 1975, 243-252 (in Russian).

31. R. G. Muhometov, The reconstructions problem of a two-dimensional Riemanian metric and integral geometry, *Soviet Math. Dokl.*, 18 (1977), 32-35.

32. A. Nachman, Global uniqueness for a two-dimensional inverse boundary value problem, *Ann. Math.*, 142 (1995), 71-96.

33. J. P. Puel and M. Yamamoto, On a global estimate in a linear inverse hyperbolic problem, *Inverse Problems*, 12 (1996), 995-1002.

Local Tomographic Methods in Sonar

Alfred K. Louis[1] and Eric Todd Quinto[2]

[1] Fachbereich Mathematik
Universität des Saarlandes
D-66121 Saarbrücken, GERMANY
louis@num.uni-sb.de

[2] Department of Mathematics
Tufts University
Medford, MA 02155 USA
equinto@math.tufts.edu

Abstract. Tomographic methods are described that will reconstruct object boundaries in shallow water using sonar data. The basic ideas involve microlocal analysis, and they are valid under weak assumptions even if the data do not correspond exactly to our model.

1   Introduction

Integrals over spheres are important in pure mathematics [12], [20], [22] and in applications in partial differential equations [15] and for physical problems including sonar [10] [21], seismic testing [21], and radar [4]. In this article, we will describe the application to sonar and geophysical testing and prove a general uniqueness theorem for local data. We will give a singularity detection method for the linear problem that requires only local data. We will explain why this method is valid for data that do not fit our model as long as certain fairly weak assumptions hold. Our results are all valid in any dimension, in particular, $n = 2$ and $n = 3$.

In each of these applied problems, after a linearization, the original inverse problem is reduced to an inverse problem for spherical integrals over spheres with restricted centers. Let A be a hypersurface in $\mathbb{R}^n$ and let $a \in A$. Let $r > 0$. Then, the sphere centered at a and of radius r is defined

$$S(a, r) = \{x \in \mathbb{R}^n \,|\, |x - a| = r\}. \tag{1}$$

Now, let f be a continuous function, $f \in C(\mathbb{R}^n)$. We define the spherical average of f over $S(a, r)$ to be

$$Rf(a, r) = \int_{x \in S(a, r)} f(x) dA(x) \tag{2}$$

where dA is the area measure on this sphere.

In seismology or sonar the acoustic wave equation is

$$n^2(x)u_{tt} = \Delta u + \delta(t)\delta(x - a_0) \quad \text{where} \quad a_0 \in A$$

148

and A is a small section of the surface of the earth. After linearization, the determination of $n^2(x)$ from back-scattered data is equivalent to inversion of $R(n^2)(a,r)$, with centers on A [16], [21]. Knowing n^2 or at least the discontinuities of n^2 tells boundaries of objects in the water.

This linearized model is reasonable from a practical standpoint when the speed of sound in the ambient water is fairly constant. This would occur in water of depth less than one hundred feet with fairly constant temperature (private communication, R. Barakat). Since the speed of sound is constant in shallow water with constant temperature, a pulse travels from a point source, a, making a spherical wavefront. The sound that is reflected back to the source at time t gives the amount reflected back from the sphere centered at a and radius $t/2$ times the speed of sound (assuming no multiple reflections). See also [14] for practical information about sonar.

Another inverse scattering problem is to find the scatterer, $q(x)$

$$u_{tt} = \Delta u + q(x)u + \delta(t)\delta(x - a_0) \text{ where } a_0 \in A$$

and A is a small section of the surface of the earth. After linearization, the determination of $q(x)$ from the response at a_0 is equivalent to inversion of the spherical transform R.

A two-dimensional linearized travel-time problem which reduces to integrals over circles with centers on a curve is discussed in [10], [16].

In each of these problems, one wants to find a function or distribution f from integrals over spheres (or circles) with centers on a given surface A (or curve in the plane). In the case of sonar or geophysical testing, A is some part of the surface of the earth. In these practical problems, the distribution f is assumed to be zero on one side of the surface (its support, supp f, is on the other side of this surface).

Much is known in the case when the surface $A = \mathcal{P}$ is a hyperplane in $\mathbb{R}^n$ and $Rf(a,r)$ is known for all $a \in \mathcal{P}$ and all $r > 0$. If f is odd about the hyperplane $\mathcal{P}$, then all spherical integrals over spheres centered on $\mathcal{P}$ are zero by symmetry. Courant and Hilbert [8] proved that any continuous *even* function is uniquely determined by its spherical integrals for spheres with centers on a hyperplane. Thus, the null space of this transform is the set of all odd functions. Therefore, any function supported on one side of $\mathcal{P}$ is uniquely determined by spherical integrals.

Inversion formulas are given for the spherical transform over spheres centered on A when A is a circle in the plane [17], when A is a plane in $\mathbb{R}^3$ [10], and when A is a hyperplane in $\mathbb{R}^n$ [4]. The formulas in [10] and [4] involve back projection, a dual operator to R, composed with a non-local Fourier integral operator. Palamodov [18] and Denisjuk [9] developed mappings which reduce this problem to inversion of the classical Radon transform. Their inversion method is local for odd dimensions (as would be expected from a dimension count). We will discuss this approach a little more in §4. These inversion methods require data $Rf(a,r)$ for spheres of arbitrary large radius to recover the value of $f(x)$ because the back projection requires this.

Very little is known if the set of centers, A, is not a hyperplane or circle. For the problem of integration over circles in the plane, the main theorem of [2] shows that, if f is compactly supported, then f is determined by integrals over circles with centers on all curves A except Coxeter systems of lines (lines intersecting at one point with equally spaced angles). This says that if A is any curve in the plane that is not a line segment, then inversion of $Rf(a, r)$ with centers $a \in A$ is possible. Partial results exist in $\mathbb{R}^n$ (e.g., [3]). It is shown in [1] that if A is the boundary of a compact smooth set in $\mathbb{R}^n$, then f is determined by spherical integrals over spheres with centers on A if f decreases sufficiently rapidly at infinity.

Much work has been done on other inverse scattering problems including models using double integrals over spheres [5] and inversion methods with error estimates for scattering with one direction of incidence and all directions of scatter [7].

This article is organized as follows. In §2, we will develop the basic ideas for understanding singularity detection. In §3, we will describe how R detects singularities, and we will also prove new uniqueness and support theorems for this transform with local data. Finally in §4, we will discuss practical aspects of the problem including numerical implementations and limitations of the model as well as cases in which the model is not satisfied, but the method will still find singularities.

2   The Mathematical Preliminaries

In this section, we talk about singularities using the ideas of Fourier transforms, Sobolev spaces, and wavefront sets. For $f \in L^1(\mathbb{R}^n)$ the Fourier transform and its inverse evaluated on f are

$$\mathcal{F}f(y) = \int_{x \in \mathbb{R}^n} f(x)e^{-ix \cdot y} dx$$

$$\mathcal{F}^{-1}f(x) = \frac{1}{(2\pi)^n} \int_{y \in \mathbb{R}^n} f(y)e^{ix \cdot y} dy \tag{3}$$

Sobolev spaces are generalizations of L^2 spaces that categorize which derivatives of a function are in L^2. The Sobolev space $H^s(\mathbb{R}^n)$ is defined for $s \in \mathbb{R}$ as the set of all distributions f for which the Fourier transform $\mathcal{F}f$ is a function that satisfies

$$\|f\|_s^2 = \int_{y \in \mathbb{R}^n} |\mathcal{F}f(y)|^2 (1 + |y|^2)^s dy < \infty. \tag{4}$$

We can use these ideas and localize in the Fourier domain to get more precise information about singularities, the wavefront set.

Definition 1. *Let $f \in \mathcal{D}'(\mathbb{R}^n)$ and let $x_0 \in \mathbb{R}^n$ and $\xi_0 \in \mathbb{R}^n \setminus 0$. Then, f is smooth microlocally near (x_0, ξ_0) if and only if there is a cut-off function*

$\varphi \in C_c^\infty(\mathbb{R}^n)$ with $\varphi(x_0) \neq 0$ and there is an open cone V containing ξ_0 such that $\mathcal{F}(\varphi f)(y)$ is rapidly decreasing in V. If f is **not** smooth microlocally near (x_0, ξ_0), then we say $(x_0, \xi_0) \in \mathrm{WF}(f)$.

One can define Sobolev wavefront set, which captures more precise information about singularities: singularities that are not in H^s microlocally [19].

Definition 2. Let f be a distribution and let $x_0 \in \mathbb{R}^n$ and $\xi_0 \in \mathbb{R}^n \setminus 0$. Let $s \in \mathbb{R}$. Then, f is microlocally in H^s near (x_0, ξ_0) if and only if there is a cut-off function $\varphi \in C_c^\infty(\mathbb{R}^n)$ with $\varphi(x_0) \neq 0$, and there is an open cone V containing ξ_0 such that $\int_{\xi \in V} |\mathcal{F}(\varphi f)(\xi)|^2 (1 + |\xi|^2)^s d\xi < \infty$. If f is **not** microlocally H^s near (x_0, ξ_0), then we say $(x_0, \xi_0) \in \mathrm{WF}^s(f)$.

If $(x_0, \xi_0) \notin \mathrm{WF}(f)$, then for any s, f is microlocally H^s near (x_0, ξ_0). It can be shown using this definition (and a compactness argument on S^{n-1}) that if f is in H^s in every direction at every point in $\mathbb{R}^n$, then f is in $H^s(\mathbb{R}^n)$.

These definitions generalize to manifolds by having $(x; \xi)$ live on the cotangent space of the manifold. We will consider only the manifolds $\mathbb{R}^n$ and $A \times (0, \infty)$, so we will use the standard basis of $T^*\mathbb{R}^n$: $\{\mathbf{dx}_j \mid j = 1, \ldots, n\}$ where $\mathbf{dx}_j$ is the dual covector to $\partial/\partial x_j$. For $x \in \mathbb{R}^n$, this gives global coordinates on $T_x^*\mathbb{R}^n$. Let $w = (w_1, \ldots, w_n) \in \mathbb{R}^n$, then we define

$$w \cdot \mathbf{dx} = \sum_{j=1}^n w_j \mathbf{dx}_j.$$

So, if $\xi_0 \in \mathbb{R}^n$ is the vector in Definitions 1 and 2, then $(x_0, \xi_0 \cdot \mathbf{dx})$ is the corresponding covector in the wavefront set.

Let A be a hypersurface. We get covectors on T^*A as follows. Let $a \in A$ and let T_a be the hyperplane in $\mathbb{R}^n$ tangent to A at a. Then, for $w \in T_a - a$, the translate of T_a to the origin,

$$w \cdot \mathbf{dx} \in T_a^*A.$$

So, a covector in $T^*(A \times (0, \infty))$ is of the form $(a, r; w \cdot \mathbf{dx} + s \mathbf{dr})$ where $w \in T_a - a$ and $s \in \mathbb{R}$.

3   Mathematical Results

First, we give a precise description of how the spherical transform detects singularities, then we prove local uniqueness theorems.

Theorem 3 (Microlocal regularity of R). Let $f \in \mathcal{D}'(\mathbb{R}^n)$ and let A be a smooth hypersurface. Let $a_0 \in A$ and let T_{a_0} be the hyperplane tangent to A at a_0. Assume $\mathrm{supp}\, f$ lies on one side of T_{a_0}. Let $\alpha \neq 0$ and $r_0 > 0$ and let $x_0 \in S(a_0, r_0)$ and let ξ_0 be normal to $S(a_0, r_0)$ at x_0. Let $\xi_0 = (x_0 - a_0) \cdot \mathbf{dx}$

and let $\eta_0 = -(P_a(x_0 - a_0) \cdot \mathbf{dx} + r_0 \mathbf{dr})$ *where* P_a *is the orthogonal projection onto the hyperplane* $T_a - a$. *Then,*

$$(x_0; \alpha\xi_0) \in \mathrm{WF}(f) \text{ if and only if } (a_0, r_0; \alpha\eta_0) \in \mathrm{WF}(Rf) \tag{5}$$

Furthermore,

$$(x_0; \alpha\xi_0) \in \mathrm{WF}^s(f) \text{ if and only if } (a_0, r_0; \alpha\eta_0) \in \mathrm{WF}^{s+(n-1)/2}(Rf) \tag{6}$$

The covector $\xi_0 = (x_0 - a_0) \cdot \mathbf{dx}$ is conormal to the sphere $S(a_0, r_0)$ at x_0 (it corresponds to a vector normal to this sphere at x_0) so the theorem gives information about singularities of f conormal to $S(a_0, r_0)$. If Rf is smooth (or in $H^{s+(n-1)/2}$) in the direction η_0 in the theorem, then f is smooth (or in H^s) in direction ξ_0. So, smoothness of the spherical transform of f corresponds to smoothness of f in directions conormal to $S(a_0, r_0)$. More precisely, let $\mathcal{A} \subset A \times (0, \infty)$ be the open subset over which data are taken, then

$$\mathrm{WF}^s(f) \cap \left(\bigcup_{(a,r) \in \mathcal{A}} N^*(S(a, r)) \right) \tag{7}$$

is the set of H^s–stably reconstructed wavefront directions.

This is true because R satisfies (6) for data satisfying the condition of Theorem 3. Directions $(x_0, a\xi_0)$ satisfying (6) are the ones in the union in (7). These are directions conormal to the spheres $S(a, r)$ in the data set (for $(a, r) \in \mathcal{A}$).

This theorem says nothing about "invisible" singularities (ones not in (7)) but one can easily come up with functions f with singularities in directions not in (7) such that Rf is smooth; these singularities of f disappear in Rf.

This can be used to understand which boundaries of f (boundaries of objects in the ocean) are detectable from local sonar data. Let A be a smooth open set on the surface of the ocean. Let the reflector f lie below T_a for all $a \in A$. Let SONAR data be given on an open connected set $\mathcal{A} \subset A \times (0, \infty)$. Then, singularities of f conormal to $S(a, r)$ will be detectable from the given data for all $(a, r) \in \mathcal{A}$. But, singularities not conormal the sphere will not be stably detected by data near (a, r). For example, if $A = \mathcal{P}$ is a horizontal plane, then vertical boundaries will not be stably detected by any data with centers on $\mathcal{P}$ because no sphere centered on $\mathcal{P}$ has vertical conormals below the surface, $\mathcal{P}$.

Furthermore, according to (7), if $\mathcal{A} = A \times (0, R)$ for some $R > 0$, then more wavefront directions are stably visible near A than far away because the union in (7) includes more directions for points near A than far from A.

Note that Theorem 3 says nothing about points $x_0 \in S(a_0, r_0)$ that are on the equator $S(a_0, r_0) \cap T_{a_0}$. In fact, Theorem 3.3 of [3] makes no conclusion about such points.

This proof is related to Theorem 3.3 of [3], and it will be given in a future article. In particular, (5) and (6) follow from the fact that, for distributions f supported on one side of T_{a_0}, R is an elliptic Fourier integral operator that satisfies the Bolker Assumption [11].

Palamodov [18] has done a careful analysis of singularities of this operator in the plane when $A = P$ is a line. He has L^2 estimates even for the invisible directions (ones not conormal to spheres in the data set). This special structure lends itself to more precise information.

Our next theorem is a very general local uniqueness theorem.

Theorem 4 (Local Uniqueness for the Spherical Transform). *Let A be a real analytic hypersurface in $\mathbb{R}^n$ and let $\mathcal{A} \subset A \times (0, \infty)$ be open and connected. Let $f \in \mathcal{D}'(\mathbb{R}^n)$. Assume for all $(a, r) \in \mathcal{A}$, that f is supported on one side of T_a, the hyperplane tangent to A at a. Assume for some $(a_0, r_0) \in \mathcal{A}$ that $S(a_0, r_0)$ is disjoint from supp f. Then,*

$$f = 0 \quad on \quad \bigcup_{(a,r) \in \mathcal{A}} S(a, r).$$

In this theorem, we must assume A is real-analytic because there are counterexamples to uniqueness for C^∞ Radon transforms. Local uniqueness theorems are known if $A = P$ is a hyperplane. In [8] it is shown that if U is an open subset of a plane P, and f is zero on one side of P and $Rf(a, r) = 0$, for all $(a, r) \in \mathcal{A} = U \times (0, \infty)$ then $f \equiv 0$. In [4] uniqueness is shown if $\mathcal{A}$ is the set of all spheres centered on P and lying inside a given sphere $S(a_0, r_0)$. In this case, $f = 0$ inside $S(a_0, r_0)$. Theorem 4 is stronger than the ones in [8] and [4] since A is not restricted to be a plane and the sets of spheres is more general.

Here is how one could use this theorem as a guide in choosing which SONAR data to use in exploration. Let A be a small open connected set on the surface of the ocean. Assume A and the reflector in the ocean, f, satisfy the conditions of Theorem 4. Assume data are given on A for all spheres of radius less than some r_0. So, $\mathcal{A} = A \times (0, r_0)$. Then, f is determined on $\cup\{S(a, r) \,|\, (a, r) \in \mathcal{A}\}$ by SONAR data on spheres in $\mathcal{A}$. Furthermore, the set of wavefront directions in (7) are stably reconstructed.

The proof of Theorem 4 is similar in spirit to the proof in [6] and it will be given in a future article.

4 Discussion and Future Directions

There is some debate whether the Born approximation and spherical integrals are the right model for the SONAR problem when sources and detectors are at the same location. However, even if the model is inaccurate, as long as a reasonable assumption about singularities is valid, the analysis would still be valid. In particular, as long as singularities of the objects conormal to the spherical wavefronts result in singularities of the data (as described by (5) and (6)) then backprojection singularity detection algorithms would work. If real SONAR data of a scatterer f has the same singularities as Rf would have, then R^* of the data would reproduce the visible singularities of f. This is because the backprojection operator takes singularities of Rf (and so singularities of anything with the same singularities as Rf) to the visible singularities of f.

This analysis suggests that one consider local singularity detection methods. When $A = \mathcal{P}$ is a plane (or line in $\mathbb{R}^2$), Palamodov [18] and Denisjuk [9] have developed an inversion method for sonar data that reduces the problem to inversion of the classical Radon transform for functions supported in the unit disk, D. In order to get data over all lines in D, one needs sonar data over all spheres. They have proposed using limited angle inversion methods on this Radon data. One of the authors and, independently, Peter Kuchment have suggested using local Lambda tomography on this data. A student of the second author, Alexander Beltukov, is working on implementing this idea. One of the authors has proposed using a sort of local CT directly on the sonar data. Let R^* be a backprojection operator ($R^*g(x)$ is the average of $g(a,r)$ over all $(a,r) \in \mathcal{A}$ with $x \in S(a,r)$ in a smooth weight that is zero near the boundary of $\mathcal{A}$). Then, the singularities of $\Delta R^* R f$ will give the visible singularities of f, at least theoretically. Mr. Beltukov will investigate these methods, too.

One advantage of using local methods on the sonar data directly (as opposed to mapping to the classical Radon transform), is that one does not have to assume the surface of the ocean is planar; it can have waves. These methods will be presented in a future article.

Acknowledgement

Lively conversations with Richard Barakat, Margaret Cheney, and Adel Faridani about the practical aspects of sonar and references were very helpful to the authors as this research was being developed. The second author is indebted to the Universität des Saarlandes for hospitality as the authors did research for this article. The second author was supported by the German Humboldt Stiftung and the US National Science Foundation.

References

1. M. Agranovsky, C. Berenstein, and P. Kuchment, Approximation by spherical waves in L^p spaces, J. Geom. Analysis 6(1996), 365-383.
2. M.L. Agranovsky and E.T. Quinto, Injectivity sets for the Radon transform over circles and complete systems of radial functions, J. Functional Anal., **139**(1996), 383-414.
3. M.L. Agranovsky and E.T. Quinto, Geometry of Stationary Sets for the Wave Equation in $\mathbb{R}^n$. The Case of Finitely Supported Initial Data, preprint, 1999.
4. L-E. Andersson, On the determination of a function from spherical averages, SIAM J. Math. Anal. **19**(1988), 214-232.
5. R. Burridge and G. Beylkin, On double integrals over spheres, Inverse Problems **4**(1988), 1-10.
6. J. Boman and E.T. Quinto, Support theorems for real analytic Radon transforms, Duke Math. J. **55**(1987), 943-948.
7. M. Cheney and J. Rose, Three-dimensional inverse scattering for the wave equation: weak scattering approximations with error estimates, Inverse Problems **4**(1988) 435-447.

8. R. Courant and D. Hilbert, *Methods of Mathematical Physics, II*, Wiley-Interscience, New York 1962.

9. A. Denisjuk, Integral Geometry on the family of semi-spheres, Fractional Calculus and Applied Analysis, **2**(1999), 31-46.

10. J.A. Fawcett, inversion of N−dimensional Spherical Averages, SIAM J. Appl. Math **42**(1985), 336-341.

11. V. Guillemin and S. Sternberg, *Geometric Asymptotics,* Amer. Math. Soc., Providence, RI 1977.

12. S. Helgason, A duality in integral geometry, some generalizations of the Radon transform, Bull. Amer. Math. Soc. **70**(1964), 435-446.

13. L. Hörmander, *The analysis of linear partial differential operators I,* Springer-Verlag, 1983.

14. F.B. Jensen, W.A. Kuperman, M.B. Porter, H. Schmidt, *Computational Ocean Acoustics,* AIP Press, New York.

15. F. John, *Plane waves and spherical means,* Interscience, 1955.

16. M. Lavrent'ev, V. Romanov, and V. Vasiliev, *Multidimensional Inverse Problems for Differential Equations,* Lecture Notes in Mathematics 167, Springer Verlag, 1970.

17. S.J. Norton, Reconstruction of a two-dimensional reflecting medium over a circular domain: Exact Solution, J. Acoust. Soc. Am. **64**(1980), 1266-1273.

18. V. Palamodov, Reconstruction from limited data of arc means, preprint, 1998.

19. B. Petersen, *Introduction to the Fourier Transform and Pseudo-Differential Operators,* Pittman Boston, 1983.

20. E.T. Quinto, Pompeiu transforms on geodesic spheres in real analytic manifolds, Israel J. Math. **84**(1993), 353-363.

21. V.G. Romanov, *Integral Geometry and inverse Problems for Hyperbolic Equations,* Springer Tracts in Natural Philosophy, **26**, 1974.

22. R. Schneider, Functions on a sphere with vanishing integrals over certain subspheres, J. Math. Anal. Appl. **26**(1969), 381–384.

Efficient Methods in Hyperthermia Treatment Planning

T. Köhler[1], P. Maass[1], P. Wust[2]

[1] Universität Bremen, Zentrum für Technomathematik, 28359 Bremen
[2] Rudolf Virchow Klinikum Berlin, SFB 273

Abstract. The aim of this paper is to describe and analyse functionals which can be used for computing hyperthermia treatment plans. All
these functionals have in common that they can be optimised by efficient numerical methods. These methods have been implemented and
tested with real data from the Rudolf Virchow Klinikum, Berlin. The
results obtained by these methods are comparable to those obtained by
comparatively expansive global optimisation techniques.

1 Introduction

Regional hyperthermia is an emerging technology in cancer therapy. The basic
idea is to heat up the tumor region Ω as much as possible while keeping the
temperature in the surrounding healthy tissue G/Ω below a critical temperature.
This heating significantly improves the success of a subsequent chemo or radio
therapy.

The heating is achieved with a set of N microwave antennae surrounding the
patient. The latest generation of hyperthermia equipment uses up to $N = 24$
antennae, which are in fixed positions and operate at the same frequency. The
free parameters are the phase $\varphi_j, j = 1, .., N$ and amplitudes $a_j, j = 1, .., N$ of
the emitted microwaves.

A basic hyperthermia treatment plan therefore consists of N complex numbers $p_j = a_j e^{-ij\varphi_j}$ which determine the emitted microwaves and the resulting
temperature distribution in the body. More precisely the forward problem of
computing the steady state temperature distribution inside the body G for a
given set of parameters $p = (p_1, .., p_N)^t \in \mathbb{C}^N$ consists if two steps:

- compute the resulting electrical field $E(x)$ inside the inhomogeneous body
  by solving Maxwell equations,
- compute the resulting temperature distribution $T(x)$ by solving the bio-heat-
  transfer equation.

The quality of a hyperthermia treatment plan p is measured by the T_{90} temperature, which is the temperature achieved in at least 90% of the tumor region.

Now we can state the inverse or optimisation problem of hyperthermia treatment planning in more detail. Let us assume that a critical temperature $T_c(x)$

has been assigned by medical experience for every point $x \in G/\Omega$ in the healthy region:

$$\max_{p \in C} T_{90} \quad \text{subject to} \quad T(x) \le T_c(x) \quad \forall x \in G/\Omega \ . \tag{1}$$

This is a high-dimensional nonlinear optimisation problem in a Banach space formulation which requires substantial computations. Most of this paper deals with adequate simplifications of this functional which can be solved efficiently.

In Sect. 2 the basic equations describing the inhomogeneous electromagnetic fields as well as the equation describing the transition from an electrical field to a resulting temperature distribution are stated. Section 3 first describes a very much simplified but elegant and efficient approach introduced by [3] for computing hyperthermia treatment plans. This 2D-approach is based on just optimising a simple functional for the weighted electrical field.

On the other side of the spectrum of numerical schemes stands the approach described at the end of Sect. 3. It is based on optimising another functional by a global optimisation routine, see [2]. The results of this comparatively expensive global optimisation approach is used as a general comparison for the quality of other approaches throughout this paper.

More precisely we aim at generalizing the approach of [3] step by step to a realistic setting (3D-model and weight functions (Sect. 4.1), optimizing the temperature distribution instead of the electrical field (Sect. 4.2), time dependent hyperthermia treatment plans (Sect. 5).

All steps of this approach are illustrated by computations with real data obtained at the Rudolf Virchow Klinikum, Berlin. In this paper we can only briefly sketch the medical, physical and mathematical foundations of hyperthermia treatment planning. There exists a fast growing literature on these topics, for more detailed description of various aspects see e.g. [12, 3, 13, 14].

2 Basic mathematical models

We consider N pairs of antennae in fixed positions which operate with frequency ω. The free parameters consists of N complex phases $p_j = a_j e^{-i\varphi_j}$ which are combined in the phase vector p. In case of a homogeneous medium with dielectricity ϵ_0 each of these antennae generates an electric field $p_j E_{0_j}(x)$, where E_{0_j} is called the j-th antenna characteristics. The total electrical field in a homogeneous medium is then generated by a superposition of these basic electrical fields:

$$E_0(p)(x) = \sum_{j=1}^{N} p_j E_{0_j}(x) \ . \tag{2}$$

Due to the presence of the inhomogeneous body this electrical field and the related magnetic field change according to Maxwell's equations. A usual simplification assumes that the magnetic properties of a human body are very similar to those of water, i.e. we can still assume magnetic homogeneity μ_0. The inhomogeneous dielectricity $\epsilon(x)$ then leads to an electric field $E(x)$ which can e.g.

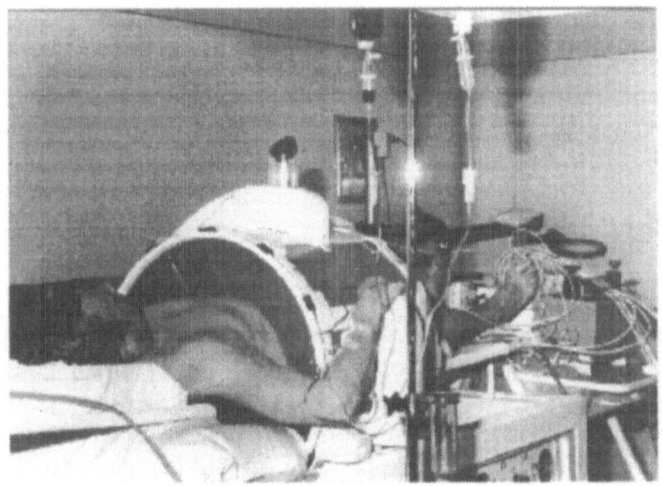

Fig. 1. Regional hyperthermia, hyperthermia-system BSD 2000, sigma 60

be determined as a solution of [14]

$$E_0(x) = E(x) + \int_G (\omega^2 \mu_0 (\epsilon_0 - \epsilon(y)) E(y) \Phi(x,y)$$

(3)

$$+ \frac{1}{\epsilon(y)} (E(y) \cdot \nabla \epsilon(y)) \nabla_y \Phi(x,y)) dy,$$

here Φ denotes the 3D-Greens function

$$\Phi(x,y) \;=\; \frac{e^{i\omega \sqrt{\mu_0 \epsilon_0} |x-y|}}{4\pi |x-y|} \quad.$$

Again the total electrical field $E(x)$ is the superposition of the primary inhomogeneous electrical fields generated by the antennae:

$$E(x) \;=\; \sum_{j=1}^{N} p_j \, E_j(x) \quad,$$

(4)

where each E_j can be computed via (3). The relevant quantity for hyperthermia treatment is not the electrical field itself but rather the absorbed energy (ARD, absorbed rate density), which also depends on the specific conductivity σ:

$$ARD(p)(x) \;=\; \sigma(x) \, |E(x)|^2 \quad.$$

First mathematical approaches in hyperthermia treatment planning concentrated on optimising the ARD distribution, see e.g. [3]. We will describe this approach in more detail in Sect. 3.

The ultimate aim of hyperthermia treatment plannig is to optimise the resulting temperature distribution. So far no complete and efficient model exists, which incorporates all aspects of the heat transfer (diffusion, perfusion, large vessels, temperature depending material constants etc.). However a standard simplified model, the stationary bio-heat-transfer-equation, governs all major aspects and produces results which agree sufficiently with reality. More precisely its solution determines the increase in temperature, $T_{hyp}(x)$, achieved by a given ARD-distribution:

$$\text{div}(\kappa \text{grad} T_{hyp}) - cW(T_{hyp} - T_a) + ARD = 0 , \tag{5}$$

here T_a denotes the temperature of the arterial blood, c combines several quantities describing the material properties of the blood flow, $W(x)$ models the local perfusion, $\kappa(x)$ denotes thermal conductivity.

The computation of the forward problem therefore consists of two steps: Given a set of parameters p compute E by solving (3), compute T by solving (5).

The inverse or optimisation problem consists of finding a vector p s.t. (1) is maximised.

3 Functionals for hyperthermia treatment planning

As already stated, optimising the functional in (1) requires to solve a high-dimensional, non-linear optimisation problem. This approach is not suitable for clinical practice. A first approach for an efficient optimisation procedure was introduced in [3]. The idea of this paper is to consider a functional which compares the absorbed energy in the tumor region Ω with the absorbed energy in the healthy part G/Ω:

$$\max_{p \in \mathbb{C}^N, ||p||=1} \frac{\int_\Omega \sigma(x)|E(x)|^2 \, dx}{\int_{G/\Omega} \sigma(x)|E(x)|^2 \, dx} = \max_{p \in \mathbb{C}^N, ||p||=1} \frac{||E||_{L^2(\Omega,\sigma)}}{||E||_{L^2(G/\Omega,\sigma)}} . \tag{6}$$

This functional differs substantially from (1). However, from a mathematical point of view, the advantage of this new functional lies in its Hilbert space structure, which allows an efficient computation: Using (4) this functional is equivalent to.

$$\max_{p \in \mathbb{C}^N, ||p||=1} \frac{<p, Ap>_{\mathbb{C}^N}}{<p, Bp>_{\mathbb{C}^N}} \tag{7}$$

with $N \times N$ matrices A, B,

$$A_{ij} = \int_\Omega \sigma(x)E_i(x)\overline{E_j(x)} \, dx , \quad B_{ij} = \int_{G/\Omega} \sigma(x)E_i(x)\overline{E_j(x)} \, dx . \tag{8}$$

This problem is solved by the normalised eigenvector corresponding to the largest generalised eigenvalue of

$$Av = \lambda Bv .$$

The normalised largest eigenvector only gives the relative differences between the different antenna parameters p_j, hence, the final treatment plan is achieved by multiplying p with an additional amplitude factor a, s.t. all restrictions in the healthy tissue are met.

In its original form this elegant approach was realised in a 2D setting, this can be easily generalised to higher dimensions, but it has several additional severe drawbacks:

1. averaging over G/Ω gives little weight to locally overheated areas in the healthy tissue, i.e. the resulting hyperthermia treatment plans produced pronounced hot spots,
2. the approach is limited to optimising the ARD distribution, the bio-heat-transfer equation and the temperature distribution are neglected.

The main purpose of this paper is to find a functional, which has the same structure as (6) but which better models clinical demands. I.e. we want to keep the efficient way of minimising the resulting functional by a generalised eigenvalue problem. Of course we would like to compare the computed hyperthermia treatment plans with the global optimum obtained by (1). However this functional is too costly to maximise. Hence we use for comparison the results obtained by the hyperthermia group of the Konrad-Zuse-Centre in Berlin. They solve

$$\min_{p \in \mathbb{C}^N} \left\{ \int_\Omega f_1(x) \, dx + w_2 \int_{G/\Omega} f_2(x) \, dx + w_3 \int_{G/\Omega} f_3(x) \, dx \right\}, \qquad (9)$$

$$f_1(x) = \begin{cases} (43 - T(x))^2 & \text{for } T(x) < 43^\circ C \\ 0 & \text{otherwise} \end{cases}$$

$$f_2(x) = \begin{cases} (T(x) - T_c(x))^2 & \text{for } T(x) > T_c(x) \\ 0 & \text{otherwise} \end{cases}$$

$$f_3(x) = \begin{cases} (T(x) - T_c(x) - \epsilon)^2 & \text{for } T(x) > T_c(x) - \epsilon \\ 0 & \text{otherwise} \end{cases}$$

by an iterative method (damped Gauss-Newton method). This approach is a compromise between numerical efficiency and medical/clinical demands. We would like to stress, that approximating the global maximum of this functional can be done with reasonable numerical effort, but each single step of the Gauss-Newton-iteration already requires substantially more computation time than maximising the functional (7).

4 Efficient functionals for optimising hyperthermia treatment plans

The starting point for this section is the basic functional (6). We will discuss various improvements of this functional leading to a medically relevant and compatible methodology for computing hyperthermia treatment plans. All of these functionals have the same structure as (6), hence they can be maximised by a simple eigenvalue computation.

More precisely we will follow a three step process:

- *ARD* optimisation: introducing adaptive weight functions w and Sobolev-spaces allows to control hot spots,
- Temperature optimisation: we introduce a functional, which allows to incorporate the temperature solution of the bio-heat-equation in a suitable Hilbert space functional,
- Combination of hyperthermia treatment plans: every single hyperthermia treatment plan exhibits regions of healthy tissue where the temperature reaches critical values; this restricts the total energy which can be emitted by the antennae and therefore also limits the energy transported into the tumor region; an optimal, time varying combination of different hyperthermia treatment plans, which are computed online with different adaptive weight functions, allows a further improvement of the overall hyperthermia treatment.

4.1 Weighted functionals for *ARD* optimisation

A first natural step to improve the clinical relevance of (6) is to introduce Sobolev spaces and weight functions. Sobolev norms of order $s > 0$ can be used to give higher weight to localised structures with large derivatives like hot spots. Hence we tested various combinations of Sobolev norms in the following functional

$$\max_{p \in \mathbb{C}^N, ||p||=1} \frac{||E||_{H^s(\Omega,\sigma)}}{||E||_{H^t(G/\Omega,\sigma)}} . \tag{10}$$

The best results were obtained for $s = 0, t = 1$. However, the success was limited, because the electrical field E has discontinuities at every boundary between organs, bones, muscles etc., hence not only hot spots are weighted by introducing Sobolev norms.

The introduction of suitable weight functions $w(x)$, which give high values to

1. sensitive healthy regions
2. regions where hot spots are expected by clinical experience

lead to significant improvements.

This leads to functionals of type

$$\max_{p \in \mathbb{C}^N, ||p||=1} \frac{||E||_{H^s(\Omega,w\sigma)}}{||E||_{H^t(G/\Omega,w\sigma)}} . \tag{11}$$

We list some results obtained with these different functionals. All functionals were used to optimise the ARD-distribution. Nevertheless we plot the resulting temperature distribution after solving the bio- heat-equation in order to compare the results with those obtained by the temeperature optimisation introduced in Sects. 4.2 and 5. The treatment plan "synchron" refers to a naive, constant p with $p_i = p_j$, "ARD-L^2" ist obtained with the unweighted L^2- functional, the other two treatment plans were obtained with adapted weight functions for an L^2-functional and a Sobolev-functional ($s = 0, t = 1$).

Functional	T_{90}	T_D	D_{42}
"synchron"	39.00°C	40.61°C	9.94%
"ARD-L^2"	39.38°C	41.43°C	37.15%
"ARD-L^2 $\downarrow$EG"	39.47°C	41.68°C	43.48%
"ARD-H^1 $\downarrow$EG"	39.50°C	41.74°C	44.94%

Table 1. Results obtained with hyperthermia treatment plans. These treatment plans were optimised according to different functionals as described in Sect. 4.1.

Three different cross-sections of a male patient are displayed. Table 1 displays three different quality measures for hyperthermia treatment plans: T_{90} has been explained already, T_D measures the average temperature in the tumor, D_{42} gives the percentage of the tumor, where the therapeutically relevant temperature of 42^0C has been reached.

4.2 Weighted functionals for temperature optimisation

So far we have described different functionals for optimising the ARD-distribution

$$ARD(x) = \sigma(x)|E(x)|^2 = \sum_{i,j=1}^{N} p_i \overline{p_j} \sigma(x) E_i(x) E_j(x) .$$

The advantage of e.g. (11) is the possibility to optimise this functional by an equivalent matrix description $< p, Ap > / < p, Bp >$, see (7).

Now we turn to optimising the temperature distribution itself. The main problem is to find adequate function spaces and functionals, which also allow an optimisation by a generalised eigenvalue problem. The temperature $T(x)$, which is finally reached by the hyperthermia treatment plan p, has two components: the basal temperature T_{bas}, which describes the temperature prior to the treatment, and T_{hyp}, the temperature increase due to p and the related ARD- distribution, T_{hyp} is computed via (5), i.e.

$$T(x) = T_{bas}(x) + T_{hyp}(x) .$$

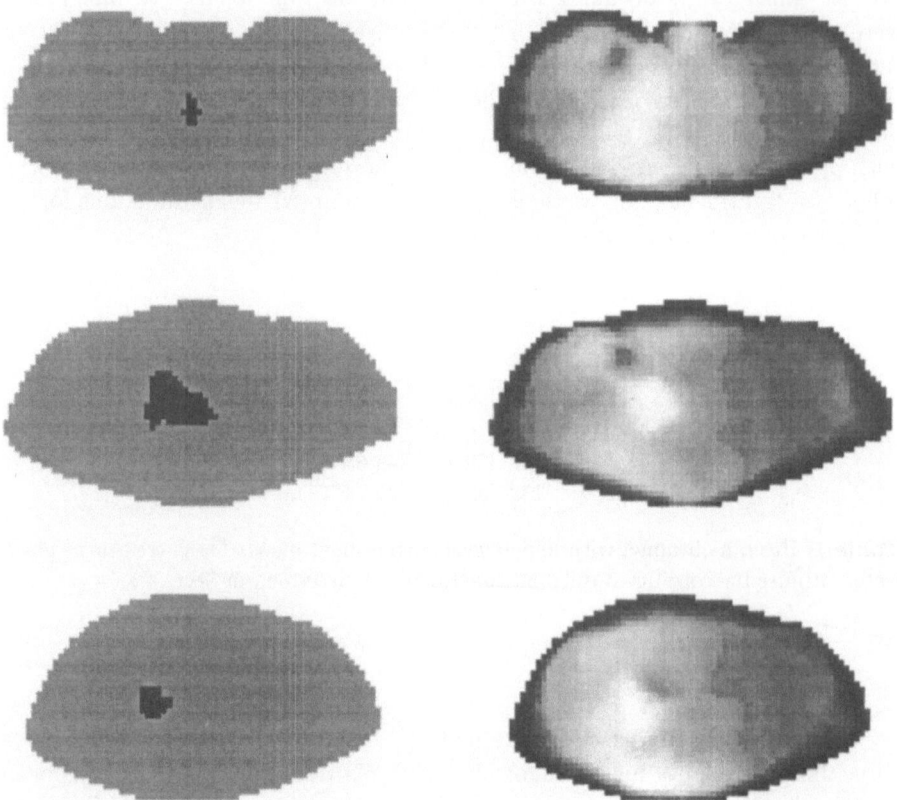

Fig. 2. The left column displays crossesections through the tumor in different slices, the right column displays the corresponding temperature distribution obtained with the functional "$ARD\text{-}H^1{\downarrow}EG$" (slices 53, 63, 78)

The bio-heat-transfer equation describes a linear dependence of T_{hyp} on ARD, i.e.

$$T_{hyp}(ARD)(x) \;=\; \sum_{i,j=1}^{N} p_i \overline{p_j} T_{hyp}(\sigma E_i \cdot E_j)(x) \;=\; <p, M(x)p> ,$$

where the coefficients of the matrix $M(x) = (M(x))_{ij})_{i,j=1}^{N}$ are the solutions of (5) with $ARD(x) = \sigma(x)E_i(x) \cdot E_j(x)$, "$\cdot$" denotes the scalar product in $\mathbb{C}^3$. Hence, we restrict ourselves to optimise the increase in temperature T_{hyp} in an L^1-setting. This is reasonable, because the basal temperature is rather homogeneous in the relevant interior of the body. Again we introduce weight functions w and define:

$$\max_{p \in \mathbb{C}^N, \|p\|=1} \frac{\int\limits_{\Omega} T_{hyp}(x)w(x)\ dx}{\int\limits_{G/\Omega} T_{hyp}(x)w(x)\ dx} \quad . \tag{12}$$

Theorem 1. *The functional (12) is maximised by the normalised eigenvector p_{opt} of the largest eigenvalue λ of the (positive definite) eigenvalue problem*

$$Ap = \lambda Bp \quad ,$$

where A, B are $N \times N$ matrices with

$$A_{ij} = \int\limits_{\Omega} M_{ij}(x)w(x)\ dx \ , \quad B_{ij} = \int\limits_{G/\Omega} M_{ij}(x)w(x)\ dx \quad .$$

For a proof see [7].

One of the main advantages of introducing a weight w in connection with the efficient eigenvalue optimisation is its adaptiveness: starting with a uniform weight function one obtains an *ARD*- or temperature distribution which exhibits hot spots, this information can be used to formulate an adapted weight function for a second optimisation step.

Again we list some results obtained with these different functionals. All functionals were used to optimise the temperature distribution. "T_{hyp}" is obtained with the unweighted functional, the other two treatment plans were obtained with different adapted weight functions, for details see [7]. Table 2 displays three different quality measures for these hyperthermia treatment plans. Figure 3 displays the temperature distribution in slice 63 for different optimisation strategies.

5 Optimal combinations of hyperthermia treatment plans

So far we have optimised single hyperthermia treatment plans p. The aim was to determine a phase and amplitude vector for the antennae steering, s.t. the resulting electrical field/temperature distribution reaches an optimal steady state.

Functional	T_{90}	T_D	D_{42}
"T_{hyp}"	39.43°C	41.39°C	35.51%
"$T_{hyp}\downarrow$EG"	39.59°C	41.82°C	46.20%
"$T_{hyp}\downarrow$EG&HS"	39.90°C	41.89°C	48.61%

Table 2. Results for optimising the temperature distribution with different functionals.

This approach has been generally used by hyperthermia research groups over the last years. However, considering only the steady state solution for the temperature distribution neglects the process of how the temperature increases from its original temperature $T_{bas}(x)$ to the steady state temperature $T(x) = T_{bas}(x) + T_{hyp}(x)$.

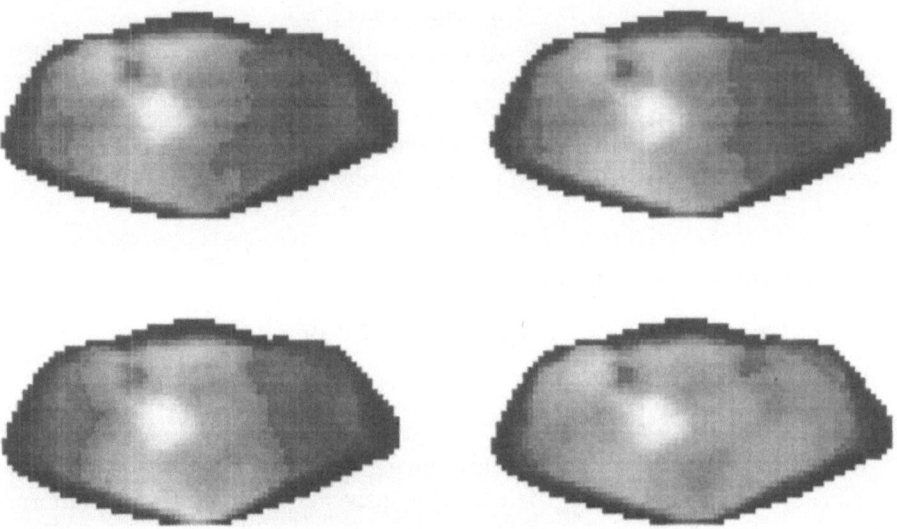

Fig. 3. Temperature distributions in slice 63 for different functionals: "T_{hyp}" upper left, "$T_{hyp}\!\downarrow\!EG$" upper right, "$T_{hyp}\!\downarrow\!EG\&HS$" lower left, "globOpt" obtained by global optimisation upper right.

Modelling this process opens new directions for optimising hyperthermia treatment plans. We will do this in the final chapter for two main reasons:

1. If we control and stop the heating process accordingly, then we can originally choose hyperthermia treatment plans which exhibit severe hot spots in their steady states.
2. Let us assume, that we have a set of hyperthermia treatment plans, which in their steady states exhibit hot spots in different areas. This allows to adjust higher amplitudes for these plans if we control the heating process and switch between these plans accordingly.

In order to put this in a mathematical framework we need to model the heating process and we need optimality criteria for the switching points.

Of course, an heuristic approach of this sort is used in clinical reality already. A typical hyperthermia treatment session lasts approximately 75 min. and the medical supervisor modifies the original p according to the patients reactions. Our aim is to describe and implement some basic ideas for a strict mathematical optimisation.

We use a simple rule for the heating (or cooling) process. Assume that a hyperthermia treatment plan p will achieve an increase in temperature $T_{hyp}(x)$. Then according to Newton's law and clinical feasibility the basic model for the temperature at time t in x is given by

$$T(x,t) \;=\; T_{bas} \;+\; T_{hyp}(x)(1 - e^{-\xi(x)t}) \quad.$$

The exponent $\xi(x)$ varies locally in reality, we have neglected that in our computations by choosing a sensible average value $\xi(x) = 0.5$.

The above described basic model has to be modified slightly if we switch between different hyperthermia treatment plans, for details see [7]. Figure 4 displays the temperature at a fixed point x for a combination of two hyperthermia treatment plans. Figure 4a treats a combination of treatment plans were the second achieves a higher T_{hyp}, Figure 4b describes a situation were the second plan actually leads to a slight cooling in position x during the second phase of the treatment.

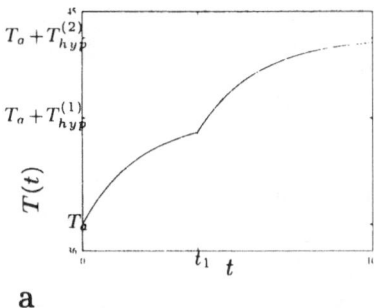
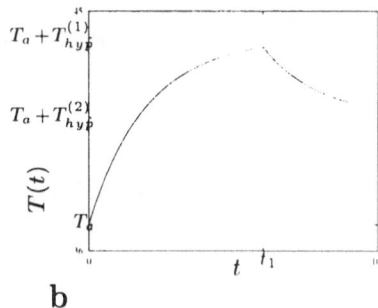

a   b

Fig. 4. Pointwise heating/cooling profile for a combination of two hyperthermia treatment plans.

In the following we describe the results obtained by combining two hyperthermia treatment plans p^1 and p^2. The weight function w used in the computation of p^2 has been defined adaptively depending on the temperature distribution of p^1. By controlling the heating process we can amplify these treatment plans by factors a^1 and a^2. Moreover we have to specify the duration Δt_1 and Δt_2 for each treatment plan. Due to our simple but realistic heating model, this low-dimensional optimisation problem can be solved by a direct method, for details and implementational issues see [7].

We want to stress, that the computation of a set of suitable hyperthermia treatment plans can be done by the methods of Sect. 3 (choosing weight functions according to the resulting temperature distribution of the previous hyperthermia treatment plans, computing the optimal plans by the generalised eigenvalue approach) very efficiently. All computations can be done online during the treatment itself in virtually no time. In combination with external control measurement, this opens new directions for the clinical application of hyperthermia treatment.

In Table 3 we listed the quality parameters T_{90}, T_D, D_{42} obtained by different methods. The values "globOpt" refer to the results obtained by the Konrad-Zuse-Center with a global optimisation routine. Table 3 demonstrates, that - within the accuracy of the used mathematical models - the method of combining just two treatment plans by the efficient procedure sketched above, yields results which are comparable with those obtained by the comparatively expansive global optimisation.

Functional	T_{90}	T_D	D_{42}
"synchron"	$39.00°$ C	$40.61°$ C	9.94%
"$ARD\text{-}L^2$"	$39.38°$ C	$41.43°$ C	37.15%
"$ARD\text{-}L^2 \downarrow EG$"	$39.47°$ C	$41.68°$ C	43.48%
"$ARD\text{-}H^1 \downarrow EG$"	$39.50°$ C	$41.74°$ C	44.94%
"T_{hyp}"	$39.43°$ C	$41.39°$ C	35.51%
"$T_{hyp} \downarrow EG$"	$39.59°$ C	$41.82°$ C	46.20%
"$T_{hyp} \downarrow EG\&HS$"	$39.90°$ C	$41.89°$ C	48.61%
"comb"	$40.00°$ C	$41.95°$ C	49.56%
"globOpt"	$40.15°$ C	$42.06°$ C	50.70%

Table 3. Summary of optimisation results.

6   Conclusions

We have described various functionals for optimising hyperthermia treatment plans. All functionals have in common, that they can be maximised by an efficient eigenvalue approach. The basic ingredient is a simple weight function w, which can easily and adaptively be chosen online during the treatment.

Moreover, the optimisation procedure has been extended to time varying treatment plans. Again, all these computations can be done online during the treatment itself in virtually no time. The results obtained by these methods are comparable to those obtained by global optimisation routines.

Acknowledgement

The work on this project of the first author has been supported by Graduiertenkolleg "Temperaturabhängige Effekte für Diagnostik und Therapie", Rudolf Virchow Klinikum, the second author has been supported by DFG grant MA 1657/1-3.

References

1. F. Bardati, A. Borrani, A. Gerardino, G.A. Lovisolo, SAR Optimization in a Phased Array Radiofrequency Hyperthermia System, *IEEE Trans. on Biomed. Eng.* **42** (1995), S.1201-1207.

2. R. Beck, P. Deuflhard, H.C. Hege, M. Seebass, D. Stalling, Numerical Algorithms and Visualization in Medical Treatment Planning, in H.C. Hege, K. Poltier (Hrsg.), Visualization and Mathematics, Berlin 1997, S.303-325.
3. M. Böhm, J. Kremer, A.K. Louis, Efficient algorithm for computing optimal control of antennas in hyperthermia, *Surv. Math. Ind* **3** (1993), S.233-251.
4. C. Großmann, J. Terno, Numerik der Optimierung, Stuttgart 1993.
5. J.D. Jackson, Klassische Elektrodynamik, Stuttgart 1983.
6. G. Joos, Lehrbuch der Theoretischen Physik, Wiesbaden 1989.
7. T. Köhler, Effiziente Algorithmen für die Optimierung der Therapie-Planung zur regionalen Hyperthermie, Dissertation, Universität Potsdam (1999).
8. J. Kremer, A.K. Louis, On the mathematical foundation of hyperthermia treatment, *Math. Meth. Appl. Sci.* **13** (1990), S.467-479.
9. J. Lang, B. Erdmann, M. Seebass, Impact of Nonlinear heat Transfer on Temperature Control in Regional Hyperthermia, Preprint SC 97-73, Konrad-Zuse-Zentrum fuer Informationstechnik Berlin (1997).
10. C. Müller, Grundprobleme der mathematischen Theorie elektromagnetischer Schwingungen, Berlin - Göttingen - Heidelberg 1957.
11. K.S. Nikita, N.G. Maratos, N.K. Uzunoglu, Optimal Steady-State Temperature Distribution for a Phased Array Hyperthermia System, *IEEE Trans. on Biomed. Eng.* **40** (1993), S.1299-1306.
12. R.B. Roemer, Inverse Techniques for Estimating Complete Temperature Distributions during Hyperthermia Cancer Therapy, presented at French Heat Transfer Society (SFT) National Conference, Poitiers (1995).
13. P. Wust, K. Dieckmann, H. Stahl, M. Seebass, J. Nadobny, R. Felix, Technologische Grundlagen und Entwicklungsmöglichkeiten der regionalen Hyperthermie, *Minimal Invasive Medizin - MED TECH* **4** (1993), S.16-21.
14. P. Wust, M. Seebass, J. Nadobny, P. Deuflhard, G. Mönich, R. Felix, Simulation studies promote technological development of radiofrequency phased array hyperthermia, *Int. J. Hyperthermia* **12** (1996), S.477-494.

Solving Inverse Problems with Spectral Data

Joyce R. McLaughlin
Department of Mathematical Sciences
Rensselaer Polytechnic Institute
Troy, NY 12180
mclauj@rpi.edu

Abstract

We consider a two dimensional membrane. The goal is to find properties of the membrane or properties of a force on the membrane. The data is natural frequencies or mode shape measurements. As a result, the functional relationship between the data and the solution of our inverse problem is both indirect and nonlinear. In this paper we describe three distinct approaches to this problem. In the first approach the data is mode shape level sets and frequencies. Here formulas for approximate solutions are given based on perturbation results. In the second approach the data is frequencies and boundary mode shape measurements; uniqueness results are obtained using the boundary control method. In the third approach the data is frequencies for four boundary value problems. Local existence, uniqueness results are established together with numerical results for approximate solutions.

Introduction

We consider two dimensional membranes. Spectral data is measured for these membranes; that is natural frequencies and/or some specific measurements of the corresponding mode shapes. Choices for the mode shape measurements are level sets or nodal sets in the interior of the membrane or flux or displacement measurements on the boundary. We ask: What can we learn about the membrane from these measurements?

To obtain this data we excite the membrane at a sequence of natural frequencies, often driving the membrane with a time harmonic force at a single point. What results is a wave that travels across the membrane reflecting from the boundary, traveling back, reflecting again and so on. At most frequencies the initial and reflected waves interfere with each other producing a small response. At a natural frequency the initial and reflected waves reinforce each other producing a large response. This response, a combination of traveling waves, makes the membrane appear to be oscillating up and down, we call what we see "a standing wave," and the shape at any instant of time is called a mode shape. The shape is the same at all instances except for a multiplicative or amplitude factor.

When, for example, the edge of the membrane is fixed we can measure level sets of the mode shape by illuminating the membrane with two lasers. The interference pattern is a set of dark and light lines called a holographic image. Each line is a level set. Of course one of these level sets is the nodal set, the set of points that don't vibrate when the membrane is excited at a natural frequency. If, however, we want only the nodal set, a Doppler shift experiment may be considered. There, a single laser is used. As it scans the membrane the Doppler shift in the backscatter is measured and minimum Doppler shift is achieved at nodal points. See [McL2] for more discussion about this.

If, for example, the edge of the membrane is free and our data is displacement of the mode shape at the boundary points together with the corresponding natural frequencies a third experiment can be considered. We can impart an impulse at a point on the membrane. Displacement as a response of the impulse is measured at points around the boundary. We Fourier analyze the response at each point and obtain the natural frequencies and the corresponding displacements at the boundary.

To put these two approaches in perspective we recall what is known in one dimension for Sturm-Liouville problems with separable boundary conditions. The classic approach was initially successfully attacked by Borg [B], Gelfand-Levitan [GL], Hochstadt [Ho] and many others for the differential equation $y'' + (\lambda - q)y = 0$ on $0 \le x \le 1$ with separable boundary conditions and $q \in L^2(0,1)$, with a complete solution to the inverse spectral problem for this equation given by [PT], [IT], [IMT]. These results for the inverse problem require two sets of spectral data. In one approach to the problem the data set consists of two sequences of frequencies each for a different set of separable boundary conditions; another data set is one set of frequencies together with mode shape measurements on the boundary (at $x = 0$ and 1), that are different from the boundary conditions. Each data set produces a unique solution; in the work of Trubowitz, et. al., necessary and sufficient conditions that the data determines q together with the boundary conditions are established. An important extension of these results is given in [CMcL1], [CMcL2] where necessary and sufficient conditions, together with formulas for exact solutions, for the inverse spectral problem for $(py')' + \lambda py = 0$, $p > 0$, $p \in H^1(0,1)$ and with Dirichlet boundary conditions are established. This equation cannot be transformed to the one given above with q. For that transformation, $p \in H^2(0,1)$ is needed. Note that the weakening of the smoothness properties of p are important for applications. In fact for applications requiring p to be of bounded variation, i.e. $p \in BV[0,1]$, is often the more realistic assumption, especially when the material properties are expected to have discontinuities. A complete characterization as was established by [PT], [IT], [IMT], for q, and extended by Coleman and McLaughlin [CMcL1], [CMcL2], for $p \in H^1(0,1)$, is still an open problem for $p \in BV[0,1]$.

Note, however, that inroads for the $p \in BV[0,1]$ problem have been made by [C], [Ha1], [W1], [W2]. In [Ha2], [Ha3], the author presents an implementation of numerical methods intended for geophysical applications. See also [HMcL4] for new results on the asymptotic behavior of eigenvalues in the BV case.

In a second approach for solving the one dimensional inverse spectral problem, the data is natural frequencies and level set measurements for the corresponding mode shapes. Sufficient conditions for a unique $q \in L^2(0,1)$ are given in [McL1]. Sufficient conditions for a unique $p, \rho \in H^2(0,1)$, for $(py')' + \lambda \rho y = 0$ with separable boundary conditions or an even smoother pair are given in [HMcL2], [HMcL3]. In [HMcL3] numerical implementation of algorithms, that calculate an approximation to one coefficient when the others are known, and that require a natural frequency together with level set measurements from a single mode shape, are given. Error bounds for the difference between the computed approximation and the true solution are established. For these computations, the zero level set, or nodal set, is the one that is used. This work is extended to $p, \rho \in BV[0,L]$ in [HMcL4] (with a significant advance using number theory on the asymptotic behavior of the frequencies). Note that in all cases considered in [HMcL2], [HMcL3], [HMcL4] the algorithms produce piecewise constant approximates; each constant is the difference or ratio of the squares of two frequencies. One of the frequencies is measured and the other is calculated by a very simple formula using only adjacent zero level set data, or adjacent nodes.

This second one-dimensional approach using the zero level set data is extended by [ST], [S], [LY], [LSY]. There new algorithms for q are given and the smoothness of q is established from the actual positions of the zero level set data.

The mathematics for solving two dimensional inverse spectral problems has developed along a number of fronts. Some of the resultant theorems are remarkably similar to the one dimensional results; the techniques however are new. In all cases new results for the direct problem are required. In one case, described in the first section of this paper, perturbation methods are used. Here for the direct problem, perturbation expansions for almost all natural frequencies and mode shapes are developed. In the work completed to date by McLaughlin, Hald, Lee, and Portnoy, [HMcL1], [LMcL], [McLP], [McL3], a rectangular membrane is considered. The goal is to find the (nonconstant) density or a (nonconstant) coefficient in a restoring force. The problem is difficult because even when there is no restoring force and the density is constant the spacing of the eigenvalues, the squares of the natural frequencies, is very irregular, even when all the frequencies are distinct. Some eigenvalues are well-spaced relative to their neighbors while others are clustered together. Perturbation expansions are established when, in the no restoring force - constant density case at least two conditions are satisfied:

1. There is a bound, which decreases as the frequency increases, on the distance to the adjacent frequencies;
2. The distance to selected frequencies is large and increasing as the frequency increases. The selected frequencies have corresponding mode shapes with similar oscillation properties.

Number theory and analysis are used to establish these results. Further, a^2, the square of the ratio of the sides has favorable number theoretic properties; it is irrational and satisfies a Diophantine condition. This aids us to establish the spacing properties above and allows us to write down specific a^2, e.g. $\sqrt{2}$ or $\sqrt{5}$, where our analysis holds.

The hard work to obtain the perturbation expansions pays off. It is applied to the case where the data is a natural frequency and level set measurements of the mode shape. With this data, simple formulas are obtained for the coefficient, q, in the restoring force. In one formula the value of q at selected points is approximated by the difference of two eigenvalues; one is measured, the other is calculated from the data near the selected point. This formula generalizes to two dimensions a corresponding one-dimensional formula. In another formula, q is approximated at selected points by four data points chosen near the selected point. There is no one-dimensional counterpart of this formula, see [HMcL3], [McL3], [McL4]. For each formula error estimates for the difference between q and its approximate value are given.

Even for this data set, which is the eigenfrequency plus the level set measurements, it's possible to eliminate the extensive mathematics needed for the perturbation results if the error bounds are given in terms of additional displacement measurements. The implementation of such an algorithm is given in [LMcL]. The algorithm produces a piecewise constant approximation to the density ρ. Similar to the one-dimensional problem, each constant is the ratio of two frequencies, one that is measured and one that is calculated from zero level set data measured in the neighborhood where that constant approximates ρ.

Note that in all of these solutions the formulas yield local information; i.e. if the solution is required in only part of the membrane, level set data need only be measured in the same region. The convergence of the piecewise constant approximates yields the uniqueness result for q when the domain is a rectangle.

Fewer results have been established when the data, natural frequencies and level set measurements, are given for domains other than rectangular membranes. In [G], however, it is shown that in the q identically equal to zero case, when the domain is a circular disk, if all the nodal lines for all natural frequencies are the same as for the ρ equal constant case, then ρ can only be a constant.

We turn now to other data sets for the two dimensional inverse spectral problem. Both kinds of data sets are suggested by analogous one dimensional inverse problems. For one set of results the data is the eigenfrequencies plus boundary measurements. The boundary measurements, together with the boundary condition, yield full Cauchy data on the boundary of the domain for each eigenfunction. Uniqueness results, see [NSU], [KK], are known and in one case error bounds have been established, see [AS], when only partial data or noisy data is given. Two very different approaches are applied to achieve results. In one approach the spectral data is shown to be sufficient to define the Dirichlet to Neumann, DtN, map. Then properties of the DtN maps are used to establish the uniqueness and error bound results, see [NSU] and [AS]. In the second approach, the model for the membrane is much more general. The proofs do not depend on perturbation results but use the boundary control method of Belishev [Be1], [Be2]. The allowable models are divided into classes defined by groups of gauge transformations. A single element in each class of models is determined by the spectral data. This nonuniqueness when the model class is more general is also seen for one-dimensional problems. These results are described in the second section of this paper. Note that, so far, no numerical algorithms and numerical computations, using this two dimensional data set, have been presented in the literature.

The final inverse spectral problem discussed in this paper uses only eigenvalues as data for the inverse problem. The conjecture is that if there is a single unknown coefficient in the membrane model then the eigenvalues for four eigenvalue problems, each with distinct but related boundary conditions, are enough to determine the unknown coefficient. The last section of this paper describes existence, uniqueness results when the unknown coefficient is close to a constant, is in a finite dimensional space, the domain is a rectangle, and the data is a finite number of eigenvalues. Results from numerical reconstruction in two separate cases (one for q and one for ρ) using this data, are given. This discussion is presented in the last section of this paper.

The inverse spectral problem–using frequency and level set data

We concentrate here on two dimensional results. Under the assumption that the motion of the vibrating membrane is time harmonic we concentrate on the resultant elliptic equation with Dirichlet boundary conditions

$$-T\Delta u + qu = \lambda\rho u, \qquad \underset{\sim}{x} \in R, \qquad (1)$$

$$u = 0, \qquad \underset{\sim}{x} \in \partial R. \qquad (2)$$

where $R = [0, \pi/a] \times [0, \pi]$ and the eigenvalue $\lambda = \omega^2$ were ω is the natural frequency. Here T is the (constant) tension, q is the (often nonconstant) amplitude of a restoring force and ρ is the (often nonconstant) density per unit area. The two inverse problems we will consider in this section are: (1) recover q/T when ρ/T is a known constant; or (2) recover ρ/T when q/T is identically zero. Note that in one dimension, with enough smoothness, using the Liouville change of variables (see, [BR]) these problems can be transformed one to the other. In two dimensions this is not the case.

We begin with the case where q/T is identically zero. Here we will give a numerical algorithm and demonstrate that it gives rather good results. The method for calculating a piecewise constant approximate for ρ/T when q/T is identically zero when the data set is the zero level set of a mode shape together with the corresponding natural frequency is as follows. Using a graphical display we let Figure 1, 2, and 3 be the nodal set when ρ/T is a constant, the nodal set when ρ/T is not constant, and the division of the nonconstant ρ/T nodal domains into the same number of subdomains, Ω_j', as in the ρ/T equal constant case.

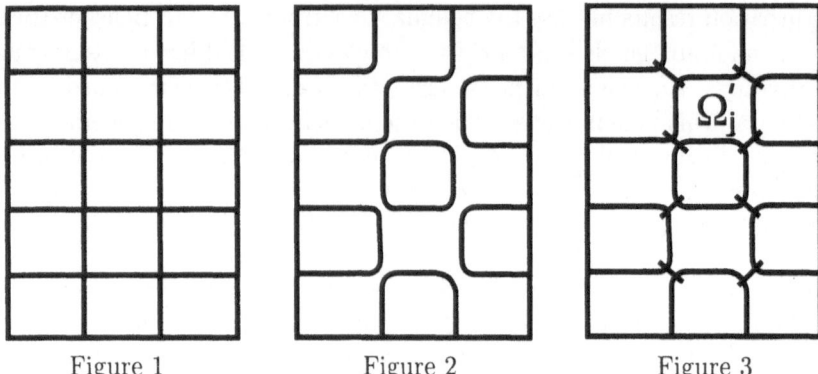

Figure 1   Figure 2   Figure 3

Note that Figure 2 shows a typical pattern when ρ/T is not constant. Some atypical zero level set patterns occasionally also occur. The method below is not used for these atypical patterns. Further for typical zero level set patterns the diagonal cuts of Figure 3 are known a priori. The piecewise constant approximate is, see [LMcL],

$$(\rho/T)_a = \left\{ \frac{\lambda_{10}(\Omega'_j)}{\lambda_n}, \qquad \underset{\sim}{x} \in \Omega'_j, \right.$$

where $\lambda_n = (\omega_n)^2$ is the square of the natural frequency for the displayed mode shape in Figures 2 and 3 and $\lambda_{10}(\Omega'_j)$ is the smallest eigenvalue for

$$-\Delta u = \lambda u, \qquad \underset{\sim}{x} \in \Omega'_j,$$
$$u = 0, \qquad \underset{\sim}{x} \in \partial\Omega'_j.$$

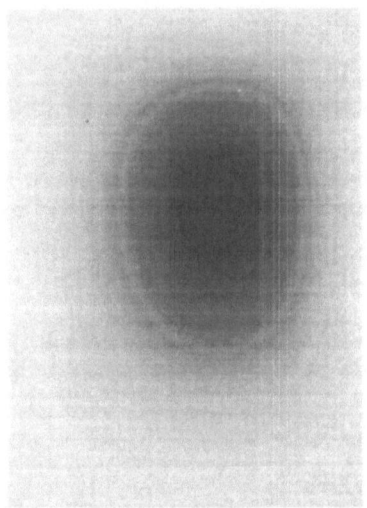

Figure 4   Figure 5

Figures 4 and 5 show a nonconstant ρ/T and one calculated approximate $(\rho/T)_a$ calculated from the data from one of the mode shapes.

Note that for this example $R = [0,1] \times [0, e/2]$ and the eigenvalues and nodal position synthetic data are calculated using a spectral method where

$$
\rho(x,y) = \begin{cases} 1 + \exp\left(\frac{(0.125)^2 \ln 0.125}{(0.125)^2 - (x - 0.75)^2 - (y-1)^2}\right) & \text{for } (x - 0.75)^2 + \\ & (y - 1)^2 \leq (0.125)^2, \\ \\ 1 & \text{otherwise.} \end{cases}
$$

The lowest eigenvalue $\lambda_{10}(\Omega_j')$ for the cut subdomain Ω_j' is calculated using a finite element method. See [Lee] for additional details. Note also that in [Lee] error bounds for the difference between the true $\rho(x,y)$ and its approximate are given. These bounds depend on measurements of the mode shape along the cuts.

Changing now to the problem where ρ/T is constant and q/T is nonconstant we start with the development of perturbation results for the eigenvalues, λ, and the corresponding mode shapes (eigenfunctions). These results are valid for almost all eigenfunctions, including arbitrarily large ones. This job, which is more or less straight forward in one dimension, is made more difficult by the fact that the eigenvalues in two dimensions, which are real, are on average equally spaced but in actual fact are quite irregularly spaced on the real line. Further under sufficient smoothness assumptions on q/T we can show that the biggest change in the mode shape, and hence change in the mode shape level sets, due to a change from constant q/T to nonconstant q/T is made from an interaction with other mode shapes with similar oscillation properties. In addition the eigenvalues corresponding to those mode shapes with similar oscillation properties are not the nearest neighbors on the number line.

To make this more clear we begin now to state some of the hypotheses. The first goal will be to establish results for the spacing of the eigenvalues when q/T is identically zero and ρ/T is a constant where we absorb the constant into the eigenvalues creating a normalization of ρ/T, that is $\rho/T \equiv 1$. Further from now on we'll simply relabel q/T as q. Second we make a choice for a^2; it will be irrational so that all eigenvalues are distinct and satisfy a Diophantine condition. The latter hypothesis allows the utilization of number theoretic arguments in the proofs, considerably shortening the arguments and further allows, because of a fundamental result of Roth [R], that specific a^2, in particular algebraic numbers, can be given for which the theory holds. Specifically the condition required is

The Diophantine Condition:
Let $J = (1, a_0)$ and $0 < \epsilon_0 < 2$ be given. Let Z be the set of integers and define $V = \{a \in J \mid \text{there exists } 0 < \delta < \epsilon_0/6 \text{ and } K > 0 \text{ such that for all } p, q \in Z$ with $q > 0 : |a^2 - p/q| > K/q^{2+\delta}\}$.

Then V is of full measure in J.
Examples of numbers that don't satisfy this condition are given in [HMcL1].

With this assumption the eigenvalue problem

$$-\Delta u = \lambda u, \qquad\qquad \underset{\sim}{x} \in R, \qquad\qquad (3)$$
$$u = 0, \qquad\qquad \underset{\sim}{x} \in \partial R, \qquad\qquad (4)$$

is considered; it has eigenvalue, (normalized) eigenfunction pairs

$$\lambda_{\alpha 0} = a^2 n^2 + m^2,$$
$$u_{\alpha 0} = \frac{2\sqrt{a}}{\pi} \sin anx \sin my,$$

for each $\alpha = (an, m)$, n, m positive integers. Notice that α is an element in a lattice plane $L = \{\alpha = (an, m) \mid n, m > 0, n, m \in Z\}$ while $\lambda_{\alpha 0} = |\alpha|^2$ is a point on the real line. Requiring spacing properties for $\lambda_{\alpha 0}$ on the number line, sometimes relative to the position of the corresponding α in the lattice plane, is an essential part in the perturbation expansion. Specifically it is established that

Lemma 1:
For $0 < \delta < \epsilon_0/6$, and a^2 satisfying the Diophantine condition, that is $a^2 \in V$, the set

$$M_{10}(a) = \{\alpha \in L \mid \text{there exists } \beta \in L, \ \beta \neq \alpha, \| \ |\alpha|^2 - |\beta|^2 | < 4 \, |\alpha|^{-\epsilon_0}\}$$

satisfies

$$\lim_{r \to 0} \frac{\# M_{10}(a) \cap \{\alpha \in L \mid \ |\alpha| < r\}}{\# \{\alpha \in L \mid \ |\alpha| < r\}} = 0.$$

That is $M_{10}(a)$ has density zero in L.

Note that this lemma says that for almost all $\alpha \in L$ there is a lower bound, that slowly decreases as $|\alpha|$ gets large, on the distance from $\lambda_{\alpha 0} = |\alpha|^2$ to the nearest neighbor, $\lambda_{\beta 0} = |\beta|^2$. This is illustrated in the graph in Figure 6.

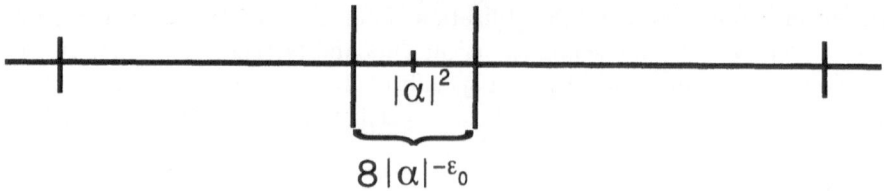

Figure 6

The second spacing lemma is:

Lemma 2:
For $0 < \epsilon_1 + 5\delta + \epsilon_1\delta < \epsilon_2 < \frac{1}{2}$, $C_1.C_2 \geq 1$ and $a^2 \in V$, the set

$$M_{11}(a) = \{\alpha \in L \mid \text{there exists } \beta \in L, \text{ with } \beta \neq \alpha, \text{ and}$$
$$|\alpha - \beta| < C_1 |\alpha|^{\epsilon_1},$$
$$|\lambda_{\alpha 0} - \lambda_{\beta 0}| < C_2 |\alpha|^{1-\epsilon_2}\}$$

has density zero in L, that is

$$\lim_{r \to \infty} \frac{\#\{\alpha \in M_{11}(a) \mid |\alpha| < r\}}{\#\{\alpha \in L \mid |\alpha| < r\}} = 0.$$

This lemma establishes the fact that for almost all α, the lattice points β, that are near α in the lattice plane, and so have the property that the corresponding mode shapes $u_{\alpha 0}, u_{\beta 0}$ have similar oscillation properties, also have the property that $\lambda_{\alpha 0} = |\alpha|^2$ and $\lambda_{\beta 0} = |\beta|^2$ are a large distance apart on the number line. A graphical illustration of this is in Figure 7.

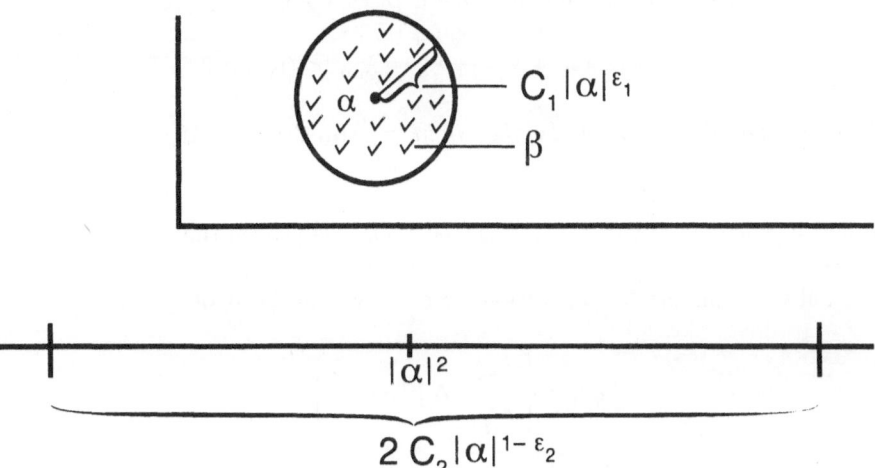

Figure 7

For this lemma we comment that it is slightly different from the corresponding Lemma 1.2 in [HMcL1]. The proof for the above result is shorter, using number theoretic arguments, while the hypothesis on ϵ_1 and ϵ_2 is slightly stronger.

One additional condition is needed to establish the perturbation results. In [HMcL1], since an inverse problem is solved, it is assumed that the location of the eigenvalues $\{\lambda_{jq}\}_{j=1}^{\infty}$ for

$$-\Delta u + qu = \lambda u \qquad\qquad \underset{\sim}{x} \in R, \qquad\qquad (5)$$
$$u = 0, \qquad\qquad \underset{\sim}{x} \in \partial R, \qquad\qquad (6)$$

are known. Then, without loss of generality, it is assumed that $\int_R q = 0$. The third condition becomes

Lemma 3:

Let

$$L\backslash M = \{\alpha \in L\backslash M_{10} \cup M_{11} \mid \text{ exactly one } \lambda_{kq} \text{ is in the interval}$$
$$(\mid \alpha \mid^2 - 2 \mid \alpha \mid^{-\epsilon_0}, \mid \alpha \mid^2 + 2 \mid \alpha \mid^{-\epsilon_0}\}.$$

Then M has density zero, that is

$$\lim_{r \to \infty} \frac{\{\alpha \in M \mid \mid \alpha \mid < r\}}{\{\alpha \in L \mid \mid \alpha \mid < r\}} = 0.$$

While Lemma 3 is satisfactory for the inverse problem, it would be stronger if the condition did not depend on q. Indeed in [McLP], this improvement is accomplished. The alternate lemma is:

Lemma 3A:

Let $0 < \epsilon_0 < (\epsilon_2 - \epsilon_1)/2$, $0 < \epsilon_3 < \epsilon_2 - \epsilon_1 - 2\epsilon_0$. Let $a \in V$. Define

$$M_{13}(a) = \{\alpha \in L\backslash M_{10} \mid \text{ there exists } \beta, \gamma \in L, \ \beta, \gamma \neq \alpha,$$
$$0 < \mid \beta - \gamma \mid < C_1 \mid \alpha \mid^{\epsilon_1},$$
$$\mid \lambda_{\beta 0} - \lambda_{\alpha 0} \mid \mid \lambda_{\gamma 0} - \lambda_{\alpha 0} \mid < C_3 \mid \alpha \mid^{\epsilon_3}\}.$$

Then M_{13} has density zero. Further if $\check{M} = M_{10} \cup M_{11} \cup M_{13}$ then $\check{M}$ is of density zero in L.

Note that when $\alpha \in L\backslash M_{10} \cup M_{11}, \cup M_{13}$, then it is proved that $\alpha \in L\backslash M$.

Finally the smoothness condition for q is given in terms of

$$\mid q \mid_\ell = \left\{ \sum_{\alpha \in L'} \mid \alpha \mid^{2\ell} \mid a_\alpha \mid^2 \right\}^{1/2}$$

where $L' = \{\alpha = (an, m) \mid n, m \geq 0, n, m \in Z\}$, where a_α are the Fourier coefficients for q for the basis $v_\alpha = c_\alpha \cos(anx) \cos my$, $\alpha \in L'$, and where

$$c_\alpha = \begin{cases} \frac{2\sqrt{a}}{\pi} & \text{if } n \neq 0, \ m \neq 0 \\[2ex] \frac{\sqrt{2a}}{\pi} & \text{if } n = 0, \ m \neq 0 \text{ or } n \neq 0, \ m = 0 \\[2ex] \frac{\sqrt{a}}{\pi} & \text{if } n = 0, \ m = 0 \end{cases}$$

The precise conditions on ℓ, C_1, C_2 and C_3 are given in [HMcL1], [McLP], as well as the perturbation expansions. Note that for each $\alpha \in L\backslash \check{M}$, or $\alpha \in L\backslash M$, there is a unique eigenvalue, eigenfunction pair $\{\lambda_{\alpha q}, u_{\alpha q}\}$ that is a perturbation of $\{\lambda_{\alpha 0}, u_{\alpha 0}\}$. Further, in [HMcL1] the exact error estimates needed for the inverse problem solution are derived while in [McLP] a full perturbation series is established.

Turning now to the solution of the inverse problem, formulas are given in the following three theorems. To frame the first result we show graphically how we subdivide the membrane. When $\alpha \in L \backslash M$ (or $\alpha \in L \backslash \check{M}$) satisfies our conditions, the zero level set for $u_{\alpha 0}$ and the zero level set for $u_{\alpha q}$, when q is not identically zero, are shown in Figures 8 and 9; in Figure 10 we show the subdivision of the domain into nm subregions Ω'_j using the zero level set in Figure 9 and diagonal straight line cuts. The straight lines for the diagonal cuts are known apriori and are described in [HMcL1]. Notice the similarity with the Figures 1, 2 and 3.

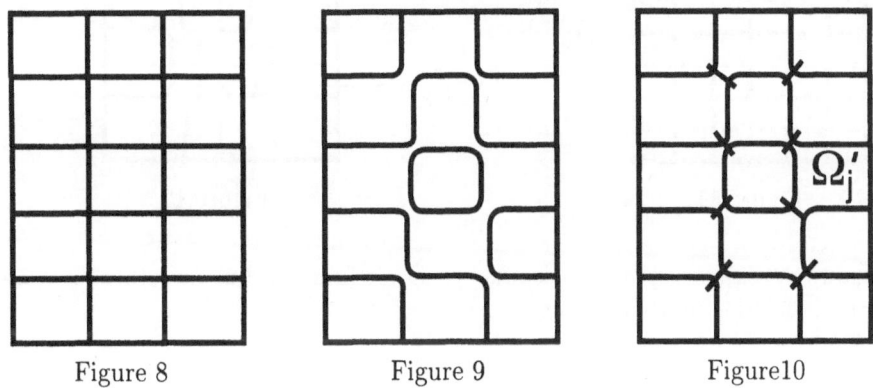

| Figure 8 | Figure 9 | Figure10 |

Then we form the piecewise constant approximate q_a

$$q_a = \left\{ \lambda_{\alpha q} - \lambda_{10}(\Omega'_j) \qquad \underset{\sim}{x} \in \Omega'_j, \ j = 1, ..., nm \right.$$

where $\lambda_{10}(\Omega'_j)$ is the smallest eigenvalue for

$$-\Delta u = \lambda u, \qquad \underset{\sim}{x} \in \Omega'_j,$$
$$u = 0, \qquad \underset{\sim}{x} \in \partial \Omega'_j,$$

$j = 1, 2, ..., m$. Then it is proved that,

Theorem 1:
Let $\alpha \in L \backslash M$ (or $\alpha \in L \backslash \check{M}$). Then in each Ω'_j, $j = 1, 2, ..., nm$ there exists an x'_j with

$$| q(x'_j) - q_\alpha(x'_j) | < \frac{1}{9 \, | \, \alpha \, |^{2-4a_2}}.$$

The error estimate in Theorem 1 is valid even when $\int_R q \neq 0$. Note that similar results have been obtained, see [VA], when the Dirichlet boundary conditions are replaced by mixed boundary conditions.

While this is an elegant formula, extensive numerical computations may be needed to compute each $\lambda_{10}(\Omega'_j)$ to obtain $q_\alpha(x'_j)$. Surprisingly, we can simplify this a great deal. Again we show the idea graphically. In Figure 11 we repeat Figure 8 adding the dashed midlines; in Figure 12 we repeat Figure 9 adding the midlines of Figure 11.

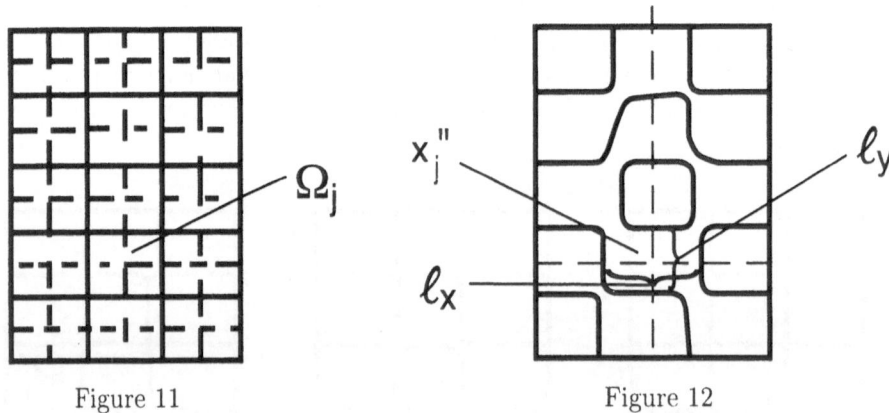

Figure 11   Figure 12

Again approximate q by a piecewise constant function but now there is a new formula for the approximate q_{aa}

$$q_{aa} = \left\{ \lambda_{aq} - 3\lambda_{a0} - \frac{2}{\pi}[(an)^3 \ell_x + m^3 \ell_y], \quad \underset{\sim}{x} \in \Omega_j \right.,$$

$j = 1, ..., nm$, where Ω_j is any subdomain in Figure 8 (or 11) and ℓ_x and ℓ_y are chosen for the same subdomain. Note that the calculation here is very straight forward requiring only simple differences of the measured data. To choose the lengths ℓ_x and ℓ_y let x''_j be the point of intersection of the dashed midlines in Ω_j. Then ℓ_y is the length of the vertical dashed midline passing through x''_j and measured from the closest nodal point below x''_j to the closest nodal point above x''_j. The length ℓ_x is the corresponding horizontal distance. We can establish

Theorem 2:

Let $\alpha \in L\backslash M$ (or $\alpha \in L\backslash \check{M}$). Then there exists a constant C with

$$| q(x''_j) - q_{aa}(x''_j) | \le C \, |\alpha|^{-2+(3/2)\epsilon_2} .$$

Finally there are several similar formulas we can establish using nonzero level sets. We give only one here again using a graphical representation. Figure 13 shows a non zero level set.

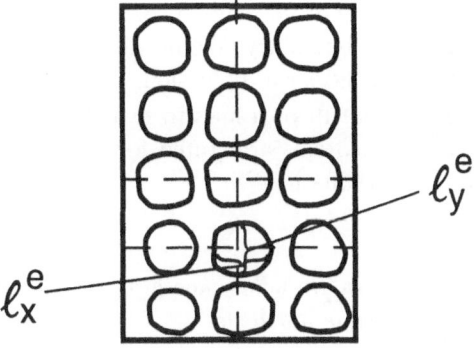

Figure 13

The piecewise constant approximate

$$q_{aaa} = \lambda_{\alpha q} - 3\lambda_{\alpha 0} + 2\left((an)^2 2x^e + m^2 2y^e\right)$$
$$+ \frac{2}{\pi}\left((an)^3 \ell_x^e + m^3 \ell_y^e\right) \qquad \underset{\sim}{x} \in \Omega_j,$$

where x^e and y^e are given apriori and do not depend on the data; error bounds for $q - q_{aaa}$ at x_j'', similar to those in Theorem 2, can be established, see [McL4].

The inverse spectral problem–using boundary, spectral data

In this section we briefly review results for inverse spectral problems where the data is eigenvalues and boundary data for the eigenmodes. The boundary data is chosen so that full Cauchy data is known for the eigenmodes on the entire boundary of the region. Two approaches for establishing results have been used. In one approach, see [NSU], the spectral data is shown to establish the Dirichlet to Neumann (DtN) map and then results for DtN maps are used to achieve results. This is a clever use of existing results. Further an important estimate of the error, see [AS], when only partial spectral data is known and when the data may contain an order ϵ error is established. A second approach yielding an extensive set of results, see e.g. [BK] and [KK], and relies on the boundary control method first put forth by Belishev, [Bel], for isotropic inverse problems. This work has required the development of significant new mathematics and has been generalized to mathematical models that can include anisotropy.

Note that the full set of data considered here, eigenvalues plus boundary data is more than is needed to achieve a solution to the inverse problem. The richness of this data is evident in the fact that a full matrix of coefficients which could represent anisotropy in a physical medium can be determined by the data, even when a finite number of eigenvalues and the boundary eigenmode data for the corresponding eigenmodes is omitted. Note also, in some cases it is possible to use only partial data on the boundary. We do not give these results here but refer the reader to [Be2, p. R39] for a discussion.

Here we restrict the statement of results to two dimensions even though the original statements of the theorems are stated for dimension $n \geq 2$. We begin by quoting one of the first results in [NSU].

Theorem 3:

Let Ω be a bounded domain in R^2 with smooth boundary. Let $q_i \in L^\infty(\Omega), i = 1, 2$, and consider the eigenvalue problems

$$-\Delta u + q_i u = \mu u, \qquad x \in \Omega,$$

$$u = 0, \qquad x \in \partial\Omega$$

Denote $\{\mu_j(q_i), \phi_j(x; q_i)\}_{j=1}^\infty$ as the eigenvalue, eigenfunction pairs for the above eigenvalue problems, $i = 1, 2$, and suppose that

$$\mu_j(q_1) = \mu_j(q_2), \qquad j = 1, 2, \dots,$$

$$\frac{\partial\phi_j(x; q_1)}{\partial\nu} = \frac{\partial\phi_j(x; q_2)}{\partial\nu}, \qquad j = 1, 2, \dots, \qquad x \in \partial\Omega$$

where ν is the unit outward normal to $\partial\Omega$. Then $q_1(x) = q_2(x)$ for all $x \in \Omega$.

Note that although the statement of the above theorem has a Dirichlet boundary condition in the eigenvalue problem, this condition is not necessary for the result; the boundary condition can be changed to $\partial u/\partial\nu + \alpha u = 0$ for $x \in \partial\Omega$ and α a smooth real function in $\partial\Omega$. In this latter case the condition $\partial\phi_j(x; q_1)/\partial\nu = \partial\phi_j(x; q_2)/\partial\nu$ for $x \in \partial\Omega$ is changed to $\phi_j(x; q_1) = \phi_j(x; q_2)$ for $x \in \partial\Omega, j = 1, 2, \dots$ With these changes, the conclusion of the theorem is unchanged.

Uniqueness results suggest that stability results can follow. Such results are established in [AS] where the main theorem is (again stated only for two dimensions)

Theorem 4:

Let Ω be a bounded domain in R^2 with smooth boundary. Let $q_i, i = 1, 2$ be bounded Hölder continuous functions in $\bar{\Omega}$ with

$$\|q_i\|_{L^\infty(\Omega)} \leq M,$$

$$|q_i(x) - q_i(y)| \leq E |x - y|^\alpha, \quad x, y \in \bar{\Omega}$$

for some $M, E > 0$ and $0 < \alpha < 1$. Let $\{\lambda_j^i, \phi_j(x; q_i)\}_{j=1}^\infty$, be the eigenvalue, eigenfunction pairs for

$$-\Delta u + q_i u = \mu u, \qquad \underset{\sim}{x} \in \Omega,$$

$$u = 0, \qquad \underset{\sim}{x} \in \partial\Omega,$$

$i = 1, 2$. *Then there exist positive constants A, B, C and $0 < \sigma < 1$ such that for every $N > 0$,*

$$\|q_1 - q_2\|_{L^\infty(\Omega)} \le C(N^A \epsilon^\sigma + N^{-B}),$$

where

$$\epsilon = \sup_{j \le N} |\lambda_j^1 - \lambda_j^2| + \sup_{j \le N} \left\| \frac{\partial\phi_j(x; q_1)}{\partial\nu} - \frac{\partial\phi_j(x; q_2)}{\partial\nu} \right\|_{L^\infty(\partial\Omega)}.$$

The authors point out that if the eigenfunctions are not chosen carefully in the multiple eigenvalue case then it is possible to have $\epsilon > 0$ even when $q_1 \equiv q_2$. Note also that the constants A, B and σ in the theorem depend only on the Hölder constant α and the space dimension. The proof of this result exploits the connection with the DtN map.

We turn now to another set of results where the mathematics that provides the proofs of these results is based in differential geometry. Instead, then, of speaking about anisotropic media, the authors of these results speak about a compact, connected, oriented differentiable (C^∞) manifold, M, with dim $M = 2$ (again we restrict our statements to two dimensions) and smooth non-zero (one) dimensional manifold boundary $S = \partial M$. The Riemannian metric is denoted by g on M with associated measure $dV = dV_g$. What is considered then is the operator

$$\mathcal{A}u = a(x, D)u = -g^{1/2}\left(\frac{\partial}{\partial x_k} + ib_k\right)\left(g^{1/2}g^{k\ell}\mu\left(\frac{\partial}{\partial x_\ell} + ib_\ell\right)u\right) + qu, \quad (7)$$

where a sum over k and ℓ is understood implicitly. Note that the metric tensor $(g^{k\ell})_{k,\ell,=1,2}$ is symmetric,

$$g(x) = \{\det[g^{k\ell}]\}^{-1}, \quad dV_g = g^{1/2}dx_1 dx_2,$$

and $\mu > 0$, with $q, b_j, j = 1, 2$, being real valued and C^∞ smooth, and $b_j, j = 1, 2$ forming a differential 1 - form on M. The above operator is considered together with the boundary condition

$$\mathcal{B}u = \left(\frac{\partial}{\partial\nu} + ib \cdot \nu + \sigma\right)u = 0, \qquad \underset{\sim}{x} \in S, \tag{8}$$

where σ is a real valued smooth function defined on S.

The goal of the inverse problem is to recover the coefficients in $\mathcal{A}$ and $\mathcal{B}$ from incomplete boundary spectral data (IBSD) defined as follows:

Definition:
Let N be the set of positive integers and $K' \subset N$ be a finite subset. Let $\{\lambda_k, \phi_k\}_{k \in N}$ be the eigenvalue, eigenfunction pairs for

$$\begin{aligned}
\mathcal{A}u &= \lambda u, &\quad \text{on } M,\\
\mathcal{B}u &= 0 &\quad \text{on } S.
\end{aligned} \tag{9}$$

Then the collection $(S, \{\lambda_k\}_{k \in N - K'}, \{\phi_k \mid_S\}_{k \in N - K'})$ is called the incomplete boundary spectral data (IBSD) for the operator $\mathcal{A}$ (together with $\mathcal{B}$).

Optimistically one might expect that the IBSD would determine all the co-efficients in $\mathcal{A}$ and $\mathcal{B}$ but such is not the case. It happens that there is a set of transformations that leave the eigenvalues fixed, that multiply each eigenfunction on the boundary by the same function, that leave the manifold unchanged but change $\mathcal{A}$ and $\mathcal{B}$; this is the group $\mathcal{G}$ of generalized gauge transformations. In any orbit of $\mathcal{G}$, see[KK], there is a unique canonical representation called the Schrödinger operator (with magnetic potential) (see [KK]). The uniqueness result that can then be obtained is (without specifically defining the canonical Schrödinger operator explicitly) is

Theorem 5:
Let $\mathcal{A}$ together with $\mathcal{B}$ be a canonical Schrödinger operator. Then its IBSD $(S, \{\lambda_k\}_{k \in N - K'}, \{\phi_k \mid_S\}_{k \in N - K'})$ determines $\mathcal{A}$ and $\mathcal{B}$, i.e. $(M, g), q, \sigma$, and b uniquely.

The proof, which is quite extensive, does not rely on perturbation methods. Rather detailed results about Gaussian beams, which are rapidly oscillating solutions of $\mathcal{A}u(x,t) + \frac{\partial^2}{\partial t^2}u(x,t) = 0$, concentrated near a space-time ray, are needed to recover coefficients, such as q, which have a low order effect on the spectral data.

Note that the need to restrict to a canonical problem in order to achieve uniqueness is mirrored in one dimension. There the situation is this. We consider the eigenvalue problem

$$\begin{aligned}
(pu_x) - q + \lambda \rho u &= 0, &\quad 0 < x < 1,\\
(u_x + au) \mid_{x=0} = 0, &\quad (u_x + bu) \mid_{x=1} = 0,
\end{aligned} \tag{10}$$

where $p_{xx}, \rho_{xx}, q \in L^2(0, 1), p, \rho > 0$. Then multiply λ by a constant and divide ρ by the same constant so that the resultant p, ρ satisfy $\int_0^1 \sqrt{\rho/p}\,dx = 1$. Following that make the change of dependent and independent variables (the Liouville transformation)

$$s = \int_0^x \sqrt{\rho/p}\,dx,$$

$$v = (p\rho)^{1/4}u,$$

to obtain that v satisfies the equations

$$v_{ss} - ((q/\rho) + [((p\rho)^{1/4})_{ss}/(p\rho)^{1/4}])v + \lambda v = 0, \qquad 0 < s < 1,$$

$$(v_s\sqrt{\rho/p} + av)\,|_{s=0}, \quad (v_s\sqrt{\rho/p} + bv)\,|_{s=1} = 0. \tag{11}$$

The boundary spectral data becomes

$$\{\lambda_j, v_j(0), v_j(1)\}_{j=1}^\infty$$

where $\{\lambda_j, v_j\}_{j=1}^\infty$ are the eigenvalue, eigenfunction pairs for (11). Since each v_j can be multiplied by a constant, the information content in $v_j(0)$ and $v_j(1)$ is contained in $v_j(1)/v_j(0)$. It is known see e.g. [IT], [IMT] that the data

$$\{\lambda_j, v_j(1)/v_j(0)\}_{j=1}^\infty$$

is exactly the right amount of data to uniquely determine the triple

$$\left\{ \begin{array}{c} \hat{a} \\ \hat{b} \\ \hat{q}(s) \end{array} \right\} = \left\{ \begin{array}{l} a/\sqrt{\rho/p(0)} \\ b/\sqrt{\rho/p(1)} \\ (q/\rho) + [((p\rho)^{1/4})_{ss}/(p\rho)^{1/4}] \end{array} \right\} \tag{12}$$

It is not possible, however, to recover the three functions q, p, ρ from (12) uniquely. In fact there is a whole class of Liouville transformations that could be applied to obtain problems of the form (10) from the given data. Three possibilities include:

(I) $p, \rho \equiv 1, \quad s = x, \quad a = \hat{a}, \quad b = \hat{b}, \quad q = \hat{q}$;

(II) $q = c\rho$ with $c < \min_j \lambda_j$, $p \equiv \rho$, $s = x$, $a = \hat{a}$, $b = \hat{b}$, and p is a positive solution of $(p^{1/2})_{ss} - (\hat{q} - c)p^{1/2} = 0$;

(III) $q \equiv c < \min_j \lambda_j$, $\rho \equiv 1$, with p a positive solution of the equation $(p^{1/4})_{ss} - (\hat{q} - c)p^{1/4} = 0$, satisfying $\int_0^1 \sqrt{p(s)}ds = 1$ and the original independent variable $x = \int_0^s \sqrt{p(s)}ds$, also $a = \hat{a}\sqrt{p(0)}, b = \hat{b}\sqrt{p(1)}$.

The inverse spectral problem–using only eigenvalues

It is important to include one more set of results and these results address the problem where one recovers material properties using only eigenvalues. To address this challenging problem we first recall that, in one dimension, if we know the eigenvalues for the following two problems, with the same $q \in L^2(0, 1/2)$,

$$y'' + (\lambda - q)y = 0 \qquad 0 < x < \frac{1}{2} \tag{13}$$

$$y(0) = y(\frac{1}{2}) = 0$$

with eigenvalues $\lambda_1, \lambda_2, \ldots$ and

$$z'' + (\mu - q)z = 0 \qquad 0 < x < \frac{1}{2} \tag{14}$$

$$z(0) = z'(\frac{1}{2}) = 0$$

with eigenvalues $\mu_1, \mu_2, \ldots$

then there is at most one $q \in L^2(0, 1/2)$ with these sets of eigenvalues. This can be established in a straight forward manner from the results in [PT] and is addressed in [L]. It is also equivalent to the following. First extend q to q_e, defined on $0 < x < 1$, by making an even reflection of q about $x = 1/2$. Then the combined set $\{\lambda_i\}_{i=1}^{\infty} \cup \{\mu_i\}_{i=1}^{\infty}$ is the set of eigenvalues for

$$w'' + (\eta - q_e)w = 0, \qquad 0 < x < 1, \tag{15}$$

$$w(0) = w(1) = 0.$$

An independent proof for (15), see [PT], shows that there is at most one symmetric $q_e \in L^2(0, 1)$ for which the combined set $\{\lambda_i\}_{i=1}^{\infty}, \cup \{\mu_i\}_{i=1}^{\infty}$ are the eigenvalues of (15).

These one dimensional results suggest two related two dimensional inverse spectral problems. The basic idea is this. If in one dimension two sets of eigenvalues provide a uniqueness result for a nonsymmetric q then are four sets of eigenvalues enough to provide a uniqueness result in two dimensions? To illustrate, consider the following example. Let $\hat{R}$ be the rectangle $\hat{R} = [0, \pi/2a] \times [0, \pi/2]$ with $\partial_1 \hat{R} = \{(\pi/2a, y) \mid y \in [0, \pi/2]\}, \partial_2 \hat{R} = \{(x, \pi/2) \mid x \in [0, \pi/2a]\}, \partial_3 \hat{R} = \{(0, y) \mid y \in [0, \pi/2]\}$ and $\partial_4 \hat{R} = \{(x, 0) \mid x \in [0, \pi/2a]\}$ subsets of ∂R. Consider the four eigenvalue problems with the same $q \in L^{\infty}(\hat{R})$ and $\underset{\sim}{n}$ the unit outward normal to points on the $\partial \hat{R}$:

$$\Delta u + (\lambda - q)u = 0, \qquad \underset{\sim}{x} \in \hat{R}, \tag{16}$$

$$u = 0, \qquad \underset{\sim}{x} \in \partial \hat{R},$$

with eigenvalues $\lambda_1 < \lambda_2 \leq \lambda_3 \leq \ldots,$

$$\Delta u + (\lambda - q)u = 0, \qquad \underset{\sim}{x} \in \hat{R}, \tag{17}$$

$$u = 0, \qquad \underset{\sim}{x} \in \partial_1 \hat{R} \cup \partial_3 \hat{R} \cup \partial_4 \hat{R},$$

$$\nabla u \cdot \underset{\sim}{n} = u_y = 0 \qquad \underset{\sim}{x} \in \partial_2 \hat{R},$$

with eigenvalues $\mu_1 < \mu_2 \leq \mu_3 \leq \ldots,$

$$\Delta u + (\nu - q)u = 0 \qquad \underset{\sim}{x} \in \hat{R}, \tag{18}$$

$$u = 0 \qquad \underset{\sim}{x} \in \partial_2 \hat{R} \cup \partial_3 \hat{R} \cup \partial_4 \hat{R},$$

$$\nabla u \cdot \underset{\sim}{n} = u_x = 0, \qquad \underset{\sim}{x} \in \partial_1 \hat{R},$$

with eigenvalues $\nu_1 < \nu_2 \leq \nu_3 \leq ...$, and

$$\Delta u + (\eta - q)u = 0, \qquad \underset{\sim}{x} \in \hat{R}, \tag{19}$$
$$u = 0, \qquad \underset{\sim}{x} \in \partial_3 \hat{R} \cup \partial_4 \hat{R},$$
$$\nabla u \cdot \underset{\sim}{n} = 0 \qquad \underset{\sim}{x} \in \partial_1 \hat{R} \cup \partial_2 \hat{R},$$

with eigenvalues $\eta_1 < \eta_2 \leq \eta_3 \leq ...$

By evenly reflecting q about $\partial_1 \hat{R}$ and then evenly reflecting the resultant q about the resultant extension of $\partial_2 \hat{R}$ then we obtain the eigenvalue problem

$$\Delta u + (\lambda - q_e)u = 0, \qquad \underset{\sim}{x} \in R = [0, \pi/a] \times [0, \pi], \tag{20}$$
$$u = 0, \qquad \underset{\sim}{x} \in \partial R,$$

with eigenvalues $\{\lambda_i\}_{i=1}^{\infty} \cup \{\mu_i\}_{i=1}^{\infty} \cup \{\nu_i\}_{i=1}^{\infty} \cup \{\eta_i\}_{i=1}^{\infty}$ and where q_e is the even extension of q to all of R. Hence for this particular problem, solving the inverse problem of finding a symmetric $q_e \in L^{\infty}(R)$ from the eigenvalues of (20) is equivalent to finding a nonsymmetric $q \in L^{\infty}(\hat{R})$ from the eigenvalues of (16), (17), (18), (19).

To illustrate the known results then we consider only (20) and a related problem

$$\frac{1}{\lambda} \Delta u + \rho u = 0, \qquad \underset{\sim}{x} \in R, \tag{21}$$
$$u = 0, \qquad \underset{\sim}{x} \in \partial R,$$

where in this related problem the goal is to solve the inverse problem: find a symmetric $\rho > 0$ from the eigenvalues of (21).

Two local results are known. Both produce functions q or ρ for which (20), or respectively (21), have a given finite number, say m, eigenvalues. Both extended a method first developed by Hald, [Ha1], for the one dimensional inverse spectral problem. The basic idea is to establish an approximate matrix eigenvalue problem for (20) (and similarly (21)) where the approximate problem is derived using spectral approximations. For each matrix problem q (or ρ) is assumed to be in the span of m given basis functions and each eigenfunction is in the span of N (possibly different) basis functions, $N \geq m$. With m fixed for each N a function q_N (or ρ_N is determined as a solution of the corresponding $N \times N$ matrix inverse eigenvalue problem. As $N \to \infty, q_N \to q$ (or $\rho_N \to \rho$) a function in the span of the m basis functions. For that q or ρ the eigenvalue problem (20) (and similarly (21)) has the given finite set of eigenvalues.

Specifically in [KMcL] and [McC], the following are proved. For (20),

Theorem 6:

Let $\{\lambda_n^0\}_{n=1}^m = \Lambda^0$ be the first m eigenvalues for

$$\Delta u + \lambda u = 0, \qquad \underset{\sim}{x} \in R,$$
$$u = 0, \qquad \underset{\sim}{x} \in \partial R,$$

with $R = [0, \pi/a] \times [0, \pi], a > 0$ and with $\min_{1 \le n \ne n' \le m} | \lambda_n^0 - \lambda_{n'}^0 | = \delta > 0$. Let $\Lambda = \{\lambda_n\}_{n=1}^m$, all distinct, be given along with a set of symmetric, orthonormal basis functions $\{\psi_n\}_{n=1}^m$, each symmetric on R. Then there exists $\delta_1, \delta_2 > 0$, $\{\beta_n\}_{n=1}^m$, and symmetric $q = \sum_{n=1}^m \beta_n \psi_n$ with $\| \Lambda - \Lambda^0 \|_{\ell^2} < \delta_1, \| q \|_{L_\infty} < \delta_2$ and with the property that $\Lambda = \{\lambda_n\}_{n=1}^m$ are the first m eigenvalues of

$$\Delta u + (\lambda - q)u = 0, \quad \underset{\sim}{x} \in R,$$
$$u = 0, \quad \underset{\sim}{x} \in \partial R.$$

And for (21),

Theorem 7:

Let $\{1/\lambda_n^0\}_{n=1}^m = \Gamma^0$ be the first m eigenvalues for

$$\frac{1}{\lambda} \Delta u + u = 0, \qquad \underset{\sim}{x} \in R,$$
$$u = 0, \qquad \underset{\sim}{x} \in \partial R.$$

with $\min_{1 \le n \ne n' \le m} | 1/\lambda_n^0 - 1/\lambda_{n'}^0 | = \delta > 0$. Let $\Gamma = \{1/\lambda_n\}_{n=1}^m$, all distinct be given along with a set of symmetric orthonormal basis functions $\{\psi_n\}_{n=1}^m$. Then there exist $\delta_1, \delta_2 > 0, \{\beta_n\}_{n=1}^m$, and symmetric $\rho = 1 + \sum_{n=1}^m \beta_n \psi_n$ with $\| \Gamma - \Gamma^0 \|_{\ell^2} < \delta_1, \| \rho - 1 \|_{L_\infty} < \delta_2$ such that $\Gamma = \{1/\lambda_n\}_{n=1}^m$ are the first m eigenvalues for

$$\frac{1}{\lambda} \Delta u + \rho u = 0, \qquad \underset{\sim}{x} \in R,$$
$$u = 0 \qquad \underset{\sim}{x} \in \partial R.$$

Note that in [KMcL] and [McC], the constants δ_1 and δ_2 are given explicitly in terms of δ. Note also that both theorems are proved using contraction mappings and the same idea is used for the numerical algorithms. Examples to show achieved results from the numerical computations are contained in Figures 14, 15, 16 and Figure 17. Figure 14 exhibits

$$q_1(x, y) = \begin{cases} \exp\left(\frac{-1}{d(x,y)}\right) & \text{if } d(x, y) = 1 - (x - \frac{\pi}{2a})^2 - 3(y - \frac{\pi}{2})^2 > 0, \\ \\ 0 & \text{otherwise,} \end{cases}$$

with $a = \sqrt{0.9}$, used to compute the eigenvalues $\{\lambda_n\}_{n=1}^8$. The eigenvalues are calculated using a Matlab finite element package. Figure 15 shows the projection of q_1 onto the span of $\{\psi_n\}_{n=1}^8$ where each

$$\psi_n = (2\sqrt{a}/\pi)\sin((2s_n - 1)ax)\sin((2t_n - 1)y)$$

with (s_n, t_n) distinct pairs of integers for $n = 1, ..., 8$; Figure 16 exhibits the reconstruction of the approximate $q_1 = \sum_{n=1}^{8} \beta_n \psi_n$ using the matrix approximation of (20), the data $\{\lambda_n\}_{n=1}^{8}$ and the resultant fixed point iteration to solve the matrix inverse problem for $\{\beta_n\}_{n=1}^{8}$ when $N = 64$.

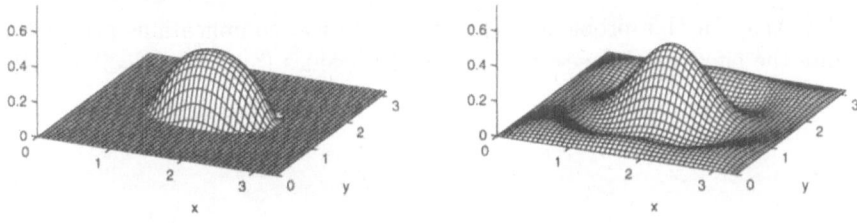

Figure 14   Figure 15

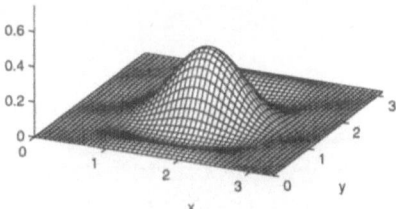

Figure 16

190

Figure 17 shows the results of the computation for approximate $\rho = 1 + \sum_{n=1}^{10} \beta_n \psi_n$ in (21) from given $\Gamma = \{1/\lambda_n\}_{n=1}^{10}$. Here the eigenvalue data is again calculated with a Matlab finite element tool box and with

$$\rho = \begin{cases} 1 + \exp\left(\frac{-1}{d(x,t)}\right) & \text{if } d(x,y) = \left(\frac{\pi}{3}\right)^2 - 4\left(x - \frac{\pi}{2a}\right)^2 - \left(y - \frac{\pi}{2}\right)^2 > 0, \\ 1 & \text{otherwise,} \end{cases}$$

and with $a = \sqrt{0.85}$.

Note that for this problem (21), new theoretical complications arise partly because the effect of changes in ρ on the eigenvalues (and vice versa) is rather strong.

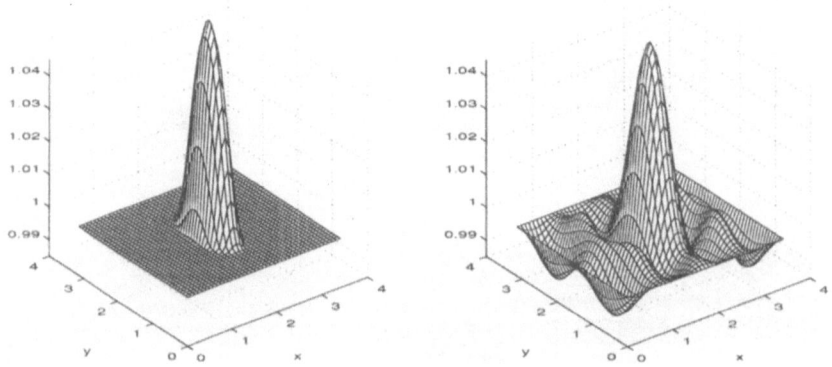

Figure 17

Determining whether four sets of eigenvalues is enough to determine q (or ρ) for general domains, for q (or ρ) in an infinite dimensional space, and for q (or ρ) not sufficiently close to a constant is an open problem.

Acknowledgement

The author is grateful to C. J. Lee, Roger Knobel and Maeve McCarthy for allowing for the inclusion of their numerical calculations in this paper.

References

[A] G. Alessandrini, "Stable Determination of Conductivity by Boundary Measurements", *Applicable Analysis*, Vol. 27, (1988), pp. 153-172.

[AS] G. Alessandrini and J. Sylvester, "Stability for a Multidimensional Inverse Spectral Theorem", *Comm. Math. Phys.*, Vol. 5, No. 5, (1990), pp. 711-736.

[Be1] M.I. Belishev, "On an approach to Multidimensional Inverse Problems for the wave equation", *Dokl. Akad. Nauk SSSR* (in Russian), Vol. 297, No. 3, (1987), pp. 524-527.

[Be2] M.I. Belishev, "Boundary control in reconstruction of manifolds and metrics (the BC method)," *Inverse Problems*, Vol. 13 (1997), No. 5 pp. R1-R45.

[Bo] G. Borg, *Eine Umkehrung der Sturm-Liouvilleschen Eigenwertaufgabe, Acta Math.* Vol. 78, (1946), pp. 1-96.

[BB] V.M. Babich and V.S. Buldyrev, *Short-wavelength Diffraction Theory*, Springer Verlag, New York, 1991.

[BK] M.J. Belishev and Ya. Kurylev, "To the Reconstruction of a Riemannian Manifold via its Spectral Data (BC-method)", *Comm. P.D.E.*, Vol. 17, No. 5-6, (1992), pp. 767-804.

[BR] G. Birkhoff and G-C. Rota, *Ordinary Differential Equations*, Wiley, New York, 1989.

[C] R. Carlson, "An Inverse Spectral Problem for Sturm-Liouville Operators with Discontinuous Coefficients," *Proc. Amer. Math. Soc.*, Vol. 120, (1994) pp. 475-484.

[CMcL1] C.F. Coleman and J.R. McLaughlin, "Solution of the Inverse Spectral Problem for an Impedance with Integrable Derivative, Part I." *Comm. Pure and Appl. Math.*, Vol. 46 (1993), pp. 145-184.

[CMcL2] C.F. Coleman and J.R. McLaughlin, "Solution of the Inverse Spectral Problem for an Impedance with Integrable Derivative, Part II." *Comm. Pure and Appl. Math.*, Vol. 46 (1993), pp. 185-212.

[GL] I.M. Gel'fand and B.M. Levitan, "On the Determination of a Differential Equation from its Spectrum," *Ivz. Akad. Nauk SSSR Ser. Math.*, Vol. 15 (1951), pp. 309-360; *Amer. Math. Trans.*, Vol. 1 (1955), pp. 233-304.

[Ha1] O.H. Hald, "The Inverse Sturm-Liouville Problem and the Rayleigh-Ritz Method," *Math. Comp.*, Vol. 32, No. 143, (1978), pp. 687-705.

[Ha2] O. H. Hald, "Inverse Eigenvalue Problems for the mantle," *Geophysical Journal of the Royal Astronomical Society*, Vol. 62 (1980), pp. 41-48.

[Ha3] O. H. Hald, "Inverse eigenvalue problems for the mantel-II," *Geophysical Journal of the Royal Astronomical Society*, Vol. 72 (1983), pp. 139-164.

[Ho] H. Hochstadt, "The Inverse Sturm Liouville Problem", *CPAM*, Vol. XXVI, (1973), pp. 715-729.

[HMcL1] O.H. Hald and J.R. McLaughlin, *Inverse Nodal Problems: Finding the Potential from Nodal Lines*, AMS Memoir, January 1996.

[HMcL2] O.H. Hald and J.R. McLaughlin, "Inverse Problems Using Nodal Position Data - Uniqueness Results, Algorithms and Bounds," *Proceedings of the Centre for Mathematical Analysis*, edited by R.S. Anderson and G.N. Newsam, Australian National University, Vol. 17 (1988), pp. 32-59.

[HMcL3] O.H. Hald and J.R. McLaughlin, "Solutions of Inverse Nodal Problems," *Inverse Problems*, Vol. 5 (1989), pp. 307-347.

[HMcL4] O.H. Hald and J.R. McLaughlin, "Inverse Problems: Recovery of BV Coefficients from Nodes," *Inverse Problems*, Vol. 14 (1998), pp. 245-273.

[IT] E.L. Isaacson and E. Trubowitz, "The Inverse Sturm-Liouville Problem I," *Comm. Pure and Appl. Math*, Vol. 36 (1983), pp. 767-783.

[IMT] E.L. Isaacson, H.P. McKean and E. Trubowitz, "The Inverse Sturm-Liouville Problem II," *Comm. Pure and Appl. Math.* Vol. 36 (1983), pp. 767-783.

[KK] A. Katchalov and Ya. Kurylev, "Multidimensional Inverse Problem with Incomplete Boundary Spectral Data", *Comm. P.D.E.*, Vol. 23(1&2), (1998), pp. 55-95.

[KMcL] R. Knobel and J.R. McLaughlin, "A Reconstruction Method for a Two-Dimensional Inverse Eigenvalue Problem." *ZAMP*, Vol. 45 (1994), pp. 794-826.

[L] B. M. Levitan, "On the determination of a Sturm-Liouville equation by two spectra," *Izv. Akad. Nauk, SSSR Ser. Mat.* Vol. 28 (1964) pp. 63-78; *Amer. Math. Soc. transl.* Vol. 68 (1968) pp. 1-20.

[Lee] C. A. Lee, "An inverse nodal problem of a membrane," Ph. D. thesis, Rensselaer Polytechnic Institute, 1995.

[LMcL] C.J. Lee and J.R. McLaughlin, "Finding the Density for a Membrane from Nodal Lines," *Inverse Problems in Wave Propagation*, eds. G. Chavent, G. Papanicolaou, P. Sacks, W.W. Symes, (1997), Springer Verlag, pp. 325-345.

[LSY] C.K. Law, C-L Shen and C-F Yang, "The Inverse Nodal Problem on the Smoothness of the Potential Function," *Inverse Problems*, Vol. 15 (1999), pp. 253-263.

[LY] C.K. Law and C.F. Yang, "Reconstructing the Potential Function and its Derivative Using Nodal Data," *Inverse Problems*, Vol. 14, (1998), pp. 299-312.

[McC] C.M. McCarthy, "The Inverse Eigenvalue Problem for a Weighted Helmholtz Equations," (to appear *Applicable Analysis*).

[McL1] J.R. McLaughlin, "Inverse Spectral Theory Using Nodal Points as Data - A Uniqueness Result." *J. Diff. Eq.*, Vol. 73. (1988), pp. 354-362.

[McL2] J.R. McLaughlin, "Good Vibrations," *American Scientist*, Vol. 86, No. 4, July-August (1998), pp. 342-349.

[McL3] J.R. McLaughlin, "Formulas for Finding Coefficients from Nodes/Nodal Lines," *Proceedings of the International Congress of Mathematicians, Zürich, Switzerland 1994*, Birkhäuser Verlag, (1995), pp. 1494-1501.

[McL4] J.R. McLaughlin, "Using Level Sets of Mode Shapes to Solve Inverse Problems." (to appear).

[McLP] J.R. McLaughlin and A. Portnoy, "Perturbation Expansions for Eigenvalues and Eigenvectors for a Rectangular Membrane Subject to a Restorative Force," *Comm. P.D.E.*, Vol. 23 (1&2), (1998), pp. 243-285.

[NSU] A. Nachman, J. Sylvester and G. Uhlmann, "An n-Dimensional Borg-Levinson Theorem," *Comm. Math. Phys.* Vol. 115, No. 4 (1988), pp. 595-605.

[PT] J. Pöschel and E. Trubowitz, *Inverse Spectral Theory*, Academic Press, Orlando, (1987).

[S] C.L. Shen, "On the Nodal Sets of the Eigenfunctions of the String Equations," *SIAM J. Math. Anal.*, Vol. 19, (1998), pp. 1419-1424.

[ST] C.L. Shen and T.M. Tsai, "On Uniform Approximation of the Density Function of a String Equation Using Eigenvalues and Nodal Points and Some Related Inverse Nodal Problems," *Inverse Problems*, Vol. 11, (1995), pp. 1113-1123.

[VA] C. Vetter Alvarez, "Inverse Nodal Problems with Mixed Boundary Conditions," Ph. D. thesis, University of California, Berkeley, 1998.

Low Frequency Electromagnetic Fields in High Contrast Media

Liliana Borcea[1] and George C. Papanicolaou[2]

[1] Computational and Applied Mathematics, MS 134, Rice University, 6100 Main Street, Houston, TX 77005-1892 `borcea@caam.rice.edu`
[2] Department of Mathematics, Stanford University, Stanford, CA 94305 `papanicolaou@stanford,edu`

Abstract. Using variational principles we construct discrete network approximations for the Dirichlet to Neumann or Neumann to Dirichlet maps of high contrast, low frequency electromagnetic media.

1 Introduction

Imaging of the electrical conductivity and permittivity of a heterogeneous body by means of low-frequency electrical or electromagnetic field measurements is an inverse problem, often called "impedance tomography", "electromagnetic induction tomography", "magnetotellurics" and so on. Applications arise in many areas, for example in medicine with diagnostic imaging, in nondestructive testing, in oil recovery, in subsurface flow monitoring, in underground contaminant detection, etc. In this paper we will focus attention on imaging heterogeneous media with large variations in the magnitude of their electrical properties. This is relevant in many geophysical applications where the conductivity can vary over several orders of magnitude. For example, a dry rock matrix is insulating compared to liquid filled pores. Some pore liquids, such as hydrocarbons, are poor conductors in comparison with other pore liquids, such as brines [4, 33]. Thus, the subsurface conductivity can have very large variations, even at macroscopic length scales where some averaging is built into the model.

Mathematically, we have to consider direct and inverse problems for elliptic systems of linear partial differential equations with high contrast coefficients. Inverse problems for such partial differential equations are highly nonlinear and pose difficult analytical and computational questions. In particular, standard imaging methods that use some form of linearization about a reference medium (Born approximation) [1, 15, 21, 43] are not appropriate for imaging high contrast media. Nonlinear output least squares [47, 32] and nonlinear variational methods [37] can fail as well.

For imaging high contrast media we take a different approach. We combine asymptotic methods and recently introduced variational principles to give an accurate and efficient description of the behavior of the fields in the region of interest. We show in particular that transport properties of high contrast media

can be well approximated by electrical networks. Discrete circuit approximations of continuum conductivity problems have been considered in the past [20,
18, 19], where the resistor networks are numerical discretizations of the partial
differential equations. The networks considered in this study are radically different. They arise from the strong channeling of currents in materials that have
large variations of electrical conductivity and/or permittivity. In this paper, we
review and build upon our results in [9, 11, 10, 8] to justify rigorously network
approximations of static and quasi-static electromagnetic transport in high contrast media. In the static case, we have resistor networks whereas at nonzero
frequency, we have resistor-capacitor or resistor-inductor networks, depending
on the properties of the high contrast media. Data available in inversion applications in high contrast media contain information mostly about the asymptotic
networks. Therefore imaging high contrast media reduces mainly to imaging
these networks.

We begin with the time-harmonic Maxwell's equations [31]

$$\nabla \times \mathbf{H}(\mathbf{x}, \omega) = [\sigma(\mathbf{x}) - i\omega\varepsilon(\mathbf{x})] \mathbf{E}(\mathbf{x}, \omega),$$
$$\nabla \times \mathbf{E}(\mathbf{x}, \omega) = i\omega\mu(\mathbf{x})\mathbf{H}(\mathbf{x}, \omega), \tag{1}$$
$$\nabla \cdot [\varepsilon(\mathbf{x})\mathbf{E}(\mathbf{x}, \omega)] = 0,$$
$$\nabla \cdot [\mu(\mathbf{x})\mathbf{H}(\mathbf{x}, \omega)] = 0 \ ,$$

where $\mathbf{H}$ and $\mathbf{E}$ are the complex magnetic and electric fields, $i = \sqrt{-1}$, σ is
the electrical conductivity, ε is the electric permittivity, μ is the magnetic permeability and ω is the frequency. The excitation is given by some form of non-
homogeneous boundary conditions. For example, in impedance tomography the
excitation consists of the current flux $\mathbf{j} = \nabla \times \mathbf{H}$ normal to the boundary. In electromagnetic induction tomography the excitation consists of magnetic dipoles at
the boundary. We consider only low frequencies and distinguish three different
elliptic problems to study:

1. Static ($\omega = 0$) conducting media. We study the *d.c. impedance tomography
   problem.*
2. Quasistatic dielectric media, where the magnetic term $i\omega\mu\mathbf{H}$ in (1) is negligible. The resulting problem has a: *complex conductivity.*
3. Quasistatic conductive media, where the displacement current $i\omega\varepsilon\mathbf{E}$ is negligible. The resulting problem is: *inductive.*

Our theory applies to a simply connected domain $\Omega \subset \mathbb{R}^2$. Extensions to three
dimensions can be done in some cases (see for example [34] for a local, asymptotic
analysis).

2 High Contrast D.C. Impedance Tomography

2.1 Formulation of the Problem

At zero frequency, equations (1) reduce to

$$\nabla \cdot [\sigma(\mathbf{x})\nabla\phi(\mathbf{x})] = 0 \quad \text{in } \Omega \ , \tag{2}$$

which we consider with Neumann boundary conditions

$$\sigma \frac{\partial \phi(\mathbf{x})}{\partial n} = I(\mathbf{x}) \quad \text{on } \partial\Omega, \quad \int_{\partial\Omega} I(s)ds = 0 , \tag{3}$$

Here, $\mathbf{E}(\mathbf{x}) = -\nabla\phi(\mathbf{x})$, n is the outward normal to the boundary $\partial\Omega$ and I is the normal current density given at the boundary. In impedance tomography $\sigma(\mathbf{x})$ is unknown and it is to be found from simultaneous measurements of currents and voltages at the boundary. Thus, for a given current excitation I we overspecify problem (2) by requiring that

$$\phi(\mathbf{x}) = \psi(\mathbf{x}) \quad \text{on } \partial\Omega , \tag{4}$$

where ψ is the measured voltage at the boundary. When all possible excitations and measurements at the boundary are available then we know the Neumann to Dirichlet (NtD) map which maps current I into voltages ψ at the boundary. The mathematical problem of impedance tomography is to find σ in the interior of Ω, from the NtD map. In practice, we rarely have the full NtD map available, so the imaging has to be done with partial, noisy information about it. The inverse problem can also be formulated in terms of the Dirichlet to Neumann (DtN) map which maps voltages into currents. In this case, we design our data gathering experiments so that we specify the voltage at the boundary and measure currents.

We assume that σ has high contrast, which means that the ratio of its maximum to its minimum value is large. There are many ways in which high contrast may arise in conducting media. We can have, for example, media with insulating or highly conducting inclusions in a smooth background. Since in most applications we do not have detailed information about the medium, such as the shape of inclusions, we model high contrast conductivity as a continuous function given by

$$\sigma(\mathbf{x}) = \sigma_0 e^{-S(\mathbf{x})/\epsilon} , \tag{5}$$

where σ_0 is constant, $S(\mathbf{x})$ is a smooth function with isolated, nondegenerate critical points (a Morse function) and ϵ is a small and positive parameter. Thus, as ϵ decreases, the contrast of σ becomes exponentially large.

The NtD and DtN maps have been studied extensively. See for example [2, 45, 29] for discussion of important questions such as injectivity, continuity of the maps and their inverse, etc. In this section, we address a new question: How do the NtD and DtN maps behave in the asymptotic limit of infinitely high contrast or, equivalently, $\epsilon \to 0$? The DtN map $\Lambda^\epsilon : H^{\frac{1}{2}}(\partial\Omega) \to H^{-\frac{1}{2}}(\partial\Omega)$, is defined by

$$\Lambda^\epsilon \psi = \sigma_0 e^{-S(\mathbf{x})/\epsilon} \left.\frac{\partial\phi}{\partial n}\right|_{\partial\Omega} = I(s) , \tag{6}$$

where $\mathbf{n}$ is the unit outer normal to $\partial\Omega$ and ϕ is the solution of (2) with Dirichlet boundary conditions (4), for a given $\psi \in H^{\frac{1}{2}}(\partial\Omega)$. Λ^ϵ is selfadjoint, positive semidefinite [45] and its quadratic form has the Dirichlet variational formulation [17]

$$(\psi, \Lambda^\epsilon\psi) = \int_{\partial\Omega} I(s)\psi(s)ds = \min_{\phi|_{\partial\Omega}=\psi} \int_\Omega \sigma_0 e^{-S(\mathbf{x})/\epsilon} \nabla\phi(\mathbf{x}) \cdot \nabla\phi(\mathbf{x}) \, d\mathbf{x} . \tag{7}$$

The generalized inverse of the map Λ^ϵ, the NtD map, $(\Lambda^\epsilon)^{-1}I = \psi$, is defined on the restricted space of currents $I \in H^{-\frac{1}{2}}(\partial\Omega)$ that satisfy $\int_{\partial\Omega} I(s)ds = 0$. The NtD map is selfadjoint and positive definite and its quadratic form has the Thomson variational formulation [17]

$$(I, (\Lambda^\epsilon)^{-1} I) = \int_{\partial\Omega} I(s)\psi(s)ds = \min_{\substack{\nabla \cdot \mathbf{j} = 0 \\ -\mathbf{j} \cdot \mathbf{n}\,|_{\partial\Omega} = I}} \int_\Omega \frac{1}{\sigma_0} e^{S(\mathbf{x})/\epsilon}\,\mathbf{j}(\mathbf{x}) \cdot \mathbf{j}(\mathbf{x})\,d\mathbf{x} \ ,$$

$$(8)$$

where $\mathbf{j}$ is the current density. We conclude this section with the observation that at their minima, both variational principles (7) and (8) give the same result, $\int_{\partial\Omega} I(s)\psi(s)ds$, which is physically the power dissipated into heat in the conducting medium.

2.2   Asymptotic Resistor Network Approximation

In [11, 10] we carry out an asymptotic analysis of the elliptic problem (2) in a high contrast medium with conductivity (5). We show that static transport in such a high contrast continuum can be approximated by current flow through a resistor network. The network topology is uniquely defined by ridges of maximal conductivity (5), in the domain of the solution. Suppose that the high contrast conductivity (5) has the set $\mathcal{N}_I$ of N_I maxima in the interior of Ω. These are the interior nodes of the resistor network. Suppose that the ridges of maximal conductivity intersect the boundary at a set $\mathcal{N}_B$ of N_B points. These are the peripheral nodes of the network. A branch of the network connects two adjacent nodes through a saddle point, say $\mathbf{x}_S$ with resistance

$$R(\mathbf{x}_s) = \frac{1}{\sigma(\mathbf{x}_s)} \sqrt{\frac{k_1}{k_2}} \ . \tag{9}$$

To get this asymptotic resistance at the saddle we use the local Taylor series expansion of the scaled logarithm of σ

$$S(\mathbf{x}) = S(\mathbf{x}_s) + \frac{k_1}{2}y^2 - \frac{k_2}{2}x^2 + \dots \ , \tag{10}$$

where we consider a system of coordinates centered at $\mathbf{x}_s$, with axis x pointing along the ridge of maximal σ passing through $\mathbf{x}_s$. We complete the definition of the asymptotic network by defining the potential and current excitations, given an arbitrary potential $\psi(s)$ and normal current $I(s)$ measured at arclength s along the boundary $\partial\Omega$ of the high contrast continuum. Consider a peripheral node $j \in \mathcal{N}_B$ that is located at s_j on $\partial\Omega$. In [10], we show that the network potential at node j is

$$\Psi_j = \psi(s_j) \ . \tag{11}$$

To calculate the current $\mathcal{I}_j$, flowing into node j, we focus attention on the maximum of σ, say $\mathbf{x}_M$ that is adjacent to s_j. We define the basin of attraction of

$\mathbf{x}_M$ as the region in Ω that contains $\mathbf{x}_M$ and lies between $\partial\Omega$ and the nearby ridges of minimal σ. Suppose that these ridges intersect $\partial\Omega$ at points s_{j_1} and s_{j_2}. Clearly, s_j lies between s_{j_1} and s_{j_2} and, as shown in [10],

$$\mathcal{I}_j = \int_{s_{j_1}}^{s_{j_2}} I(s)ds \; . \tag{12}$$

This means that the input current at node $j \in \mathcal{N}_B$ in the asymptotic network is the net current flowing into the basin of attraction of the maximum $\mathbf{x}_M$, adjacent to peripheral node j. Results (12) and (11) are proved rigorously in [10] and they allow us to calculate all peripheral network currents $\mathcal{I} = (\mathcal{I}_1, \mathcal{I}_2, \dots \mathcal{I}_{N_B})^T$ and potentials $\boldsymbol{\Psi} = (\Psi_1, \Psi_2, \dots \Psi_{N_B})^T$.

In figure 1, we show an example of how the asymptotic resistor network is constructed. We consider a continuum with a high contrast conductivity that has four maxima and six minima shown in the figure by $x_a, \dots x_d$ and $x_{a'}, \dots x_{f'}$, respectively. There are also five saddle points of σ denoted in the figure by $1, \dots 5$. The current flows along paths of maximal conductivity, shown in figure 1 with a full line. Each maximum of σ has a basin of attraction delimited in Ω by the ridge of minimal conductivity passing through the neighboring saddle points (the dotted curves in figure 1). Because of the external driving, the current must flow from one maximum of σ to another with the least resistive path goes through the saddle points. When the contrast of σ is high, the current is strongly concentrated along the paths of maximal conductivity and so it flows like current in a resistor network. The branches of the network connect adjacent maxima of σ through the saddle points. Thus, in figure 1 we have five branches, each with resistance R_i, $i = 1, \dots 5$. We also have four peripheral nodes a, b, c and d, where $\partial\Omega$ intersects ridges of maximal σ. The asymptotic resistor network is shown in figure 1. The network excitation at the peripheral node a is defined as in (11) and (12): $\Psi_a = \psi(s_a)$ and $\mathcal{I}_a = \int_{s_{a'}}^{s_{b'}} I(s)ds$, respectively.

The discrete DtN map $\Lambda^{D,\epsilon}$ is a symmetric $N_B \times N_B$ positive semidefinite matrix with a null space spanned by the vector $(1, 1, \dots 1)^T \in \mathbb{R}^{N_B \times 1}$. This matrix gives the input/output current in terms of the boundary potential,

$$\mathcal{I}_j = \sum_{k \in \mathcal{N}_B} \Lambda_{jk}^{D,\epsilon} \Psi_k, \quad \text{for all } j \in \mathcal{N}_B \; . \tag{13}$$

Its quadratic form has the variational formulation

$$< \boldsymbol{\Psi}, \Lambda^{D,\epsilon} \boldsymbol{\Psi} > = \sum_{j \in \mathcal{N}_B} \mathcal{I}_j \Psi_j = \min_{\Phi_k = \Psi_k \text{ if } k \in \mathcal{N}_B} \frac{1}{2} \sum_{j \in \mathcal{N}} \sum_{k \in \mathcal{V}_j} \frac{1}{R_{jk}^\epsilon} (\Phi_j - \Phi_k)^2 \; . \tag{14}$$

Here, $\mathcal{V}_j$ is the set of neighbors of node j in the set $\mathcal{N} = \mathcal{N}_I \bigcup \mathcal{N}_B$. The resistance R_{jk}^ϵ is given by (9), where $\mathbf{x}_s$ is the saddle point that connects the maxima $\mathbf{x}_j$ and $\mathbf{x}_k$ of σ or, equivalently, the nodes j and k in the network. The superscript ϵ reminds us that the resistance depends on the small parameter ϵ because of

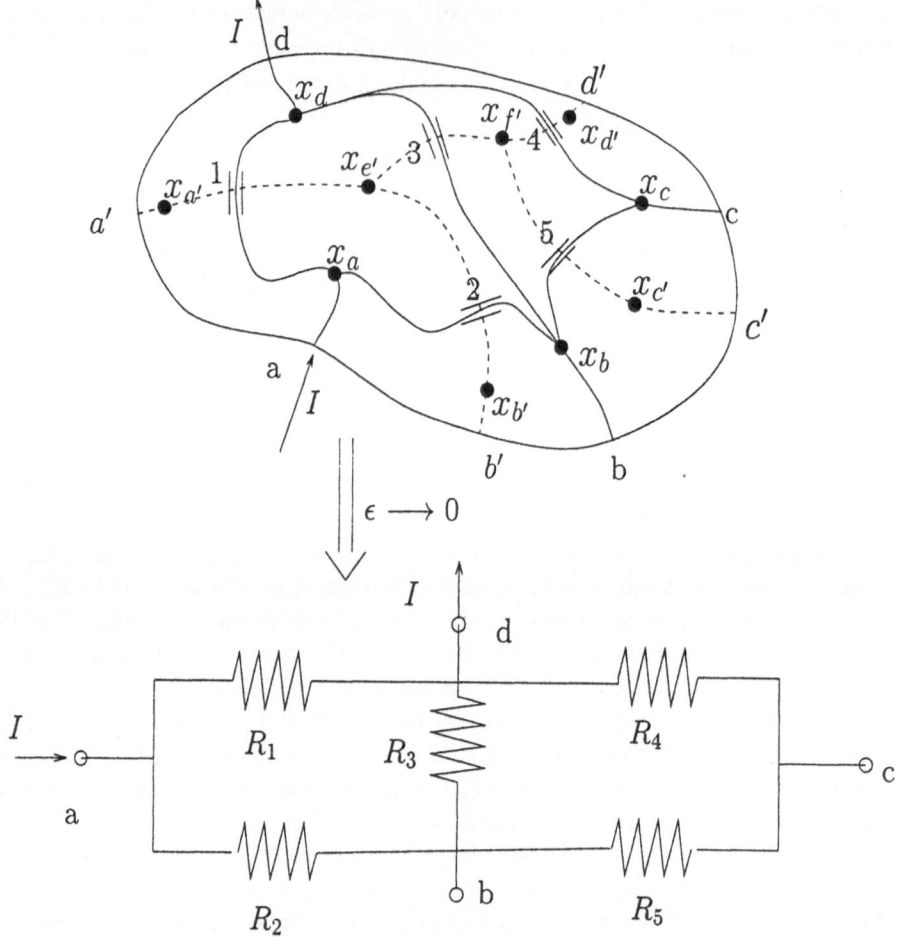

Fig. 1. Example of an asymptotically equivalent resistor network

the factor $\sigma(\mathbf{x}_s) = \sigma_0 e^{-S(\xi,\eta)/\epsilon}$. The pseudoinverse of $\Lambda^{D,\epsilon}$ is the NtD map that gives the boundary potentials in terms of the currents,

$$\Psi_j = \sum_{k \in \mathcal{N}_B} (\Lambda^{D,\epsilon})^{-1}_{jk} \mathcal{I}_k, \quad \text{for all } j \in \mathcal{N}_B . \tag{15}$$

The domain of the map $(\Lambda^{D,\epsilon})^{-1}$ is the $N_B - 1$ dimensional space spanned by current vectors $\mathcal{I} = (\mathcal{I}_1, \mathcal{I}_2, \dots \mathcal{I}_{N_B})^T \in R^{N_B \times 1}$ that satisfy the constraint (13). $(\Lambda^{D,\epsilon})^{-1}$ is symmetric, positive definite and its quadratic form has the following variational formulation

$$< \mathcal{I}, (\Lambda^{D,\epsilon})^{-1} \mathcal{I} > = \sum_{j \in \mathcal{N}_B} \mathcal{I}_j \Psi_j = \min_{J_{jk}} \frac{1}{2} \sum_{j \in \mathcal{N}} \sum_{k \in \mathcal{V}_j} R^{\epsilon}_{jk} J^2_{jk} . \tag{16}$$

Here the currents J_{jk} satisfy Kirchhoff's nodal law

$$\sum_{k \in \mathcal{V}_j} J_{jk} = \begin{cases} 0 & \text{if } j \in \mathcal{N}_I, \\ \mathcal{I}_j & \text{if } j \in \mathcal{N}_B . \end{cases} \tag{17}$$

2.3 Asymptotics and the Variational Principles

We will analyze briefly the quadratic forms $(\psi, \Lambda^{\epsilon}\psi)$ and $(I, (\Lambda^{\epsilon})^{-1}I)$ of the DtN and NtD maps in media with conductivity (5), in the asymptotic limit $\epsilon \to 0$. We show that, as $\epsilon \to 0$, we have

$$(\psi, \Lambda^{\epsilon}\psi) = < \Psi, \Lambda^{D,\epsilon}\Psi > [1 + o(1)] , \tag{18}$$
$$(I, (\Lambda^{\epsilon})^{-1}I) = < \mathcal{I}, (\Lambda^{D,\epsilon})^{-1}\mathcal{I} > [1 + o(1)] .$$

The proof of (18) relies on the variational principles (7) and (8). We obtain upper and lower bounds on the quadratic forms $(\psi, \Lambda^{\epsilon}\psi)$ and $(I, (\Lambda^{\epsilon})^{-1}I)$ and show that these bounds match asymptotically and are equal to the right hand sides in (18). To simplify the exposition here, let us assume that we have a boundary current I that is concentrated precisely at the peripheral nodes of the asymptotic network. Of course, in a real experiment the current excitation is unlikely to satisfy this assumption. In [10], we show how to handle the case of arbitrary $I \in H^{-\frac{1}{2}}(\partial\Omega)$. We introduce a thin layer near the boundary, where the current adjusts from the given I along $\partial\Omega$ to the current concentrated along ridges of maximal σ in the interior of Ω. We show that the contribution of this layer to $(I, (\Lambda^{\epsilon})^{-1}I)$ is negligible as $\epsilon \to 0$. The analysis is done in [10] and shall not be repeated here.

To get an upper bound on $(I, (\Lambda^{\epsilon})^{-1}I)$ we take in (8) a trial field

$$\mathbf{j}(x) = \left(-\frac{\partial}{\partial y}, \frac{\partial}{\partial x} \right) H(x, y) , \tag{19}$$

such that, at the boundary, $-\mathbf{j}(s) \cdot \mathbf{n}(s) = I(s) = \frac{\partial H}{\partial s}$, where we assume that I is concentrated at $s_j \in \partial\Omega$, for $j \in \mathcal{N}_B$. Thus, we have the Dirichlet boundary condition

$$H(s) = \int^s I(t)dt, \quad \text{on } \partial\Omega , \tag{20}$$

where H is constant along pieces of $\partial\Omega$ that lie between two adjacent peripheral nodes of the network. In the interior of Ω, away from the ridges of maximal conductivity, we take H to be a constant.

Along the ridges of maximal σ we expect strong, concentrated currents. Consider such a ridge of maximal conductivity and introduce curvilinear coordinates (ξ, η), with ξ arclength along the ridge. For $|\eta| \le \delta \ll 1$, such that $\frac{\delta^2}{\epsilon} \to \infty$ as $\epsilon \to 0$, we have $\sigma(\xi, \eta) = \sigma_0 e^{-S(\xi,\eta)/\epsilon}$, where

$$S(\xi, \eta) = S(\xi, 0) + \frac{k(\xi)}{2}\eta^2 + \frac{1}{6}\frac{\partial^3 S}{\partial \eta^3}(\xi, 0)\eta^3 + \frac{1}{24}\frac{\partial^4 S}{\partial \eta^4}(\xi, 0)\eta^4 + \cdots . \qquad (21)$$

Here $k(\xi)$ is the curvature in the direction η and, by our assumption of a conductivity function with non degenerate critical points, $k(\xi) > 0$. We take

$$H(\xi, \eta) = -\frac{f}{2}\,\mathrm{erf}\left(\frac{\eta}{\sqrt{\frac{2\epsilon}{k(\xi)}}}\right) + \text{constant} , \qquad (22)$$

where, f is the net current flowing along the ridge. Then, as shown in [10],

$$\begin{aligned}
\mathbf{j}(\xi, \eta) &= \frac{f}{\sqrt{2\pi\epsilon}}e^{-\frac{k(\xi)\eta^2}{2\epsilon}}\left[\sqrt{k(\xi)}\hat{\xi} - \frac{\eta}{2\sqrt{k(\xi)}}\frac{dk(\xi)}{d\xi}(1 + O(\delta))\,\hat{\eta}\right]\\
&\approx \frac{f\sqrt{k(\xi)}}{\sqrt{2\pi\epsilon}}e^{-\frac{k(\xi)\eta^2}{2\epsilon}}\hat{\xi} ,
\end{aligned} \qquad (23)$$

where $\hat{\xi}$ and $\hat{\eta}$ are the unit tangential and normal vectors along the ridge, respectively. We have

$$\int_\Omega \frac{1}{\sigma(\mathbf{x})}|\mathbf{j}(\mathbf{x})|^2 dx = \sum_{\text{ridges in }\Omega}\int_{\text{ridge}}\frac{1}{\sigma_0}e^{S(\xi,\eta)/\epsilon}\,|\,\mathbf{j}(\xi,\eta)\,|^2\,d\xi d\eta , \qquad (24)$$

where we sum over all ridges of maximal conductivity in Ω. The technical details of the calculation of the integrals in (24) are given in [10]. After integration in η, we have

$$\int_{\text{ridge}}\frac{1}{\sigma_0}e^{S(\xi,\eta)/\epsilon}\,|\,\mathbf{j}(\xi,\eta)\,|^2\,d\xi d\eta = \frac{1 + O(\delta)}{\sigma_0}\int_{\xi_{in}}^{\xi_{out}}\sqrt{\frac{k(\xi)}{2\pi\epsilon}}[f(\xi)]^2 e^{\frac{S(\xi,0)}{\epsilon}}\,d\xi . \qquad (25)$$

The integral (25) is of Laplace type [5] and the main contribution comes from the maxima of $S(\xi, 0)$ or, equivalently, the saddle points of σ. Consider a saddle point $\mathbf{x}_S = (\xi_S, 0)$ along the ridge. In the vicinity of $\mathbf{x}_S$, $S(\xi, 0)$ is given by

$$S(\xi, 0) = S(\mathbf{x}_S) - \frac{p(\mathbf{x}_S)(\xi - \xi_S)^2}{2} + \frac{(\xi - \xi_S)^3}{6}\frac{\partial^3 S}{\partial \xi^3}(\xi_S, 0) + \cdots , \qquad (26)$$

where $p(\mathbf{x}_S) > 0$ is the curvature of the function S at the saddle point in the direction $\hat{\xi}$. The contribution of this saddle point to the integral in (25) is

$$\frac{(1 + O(\delta))}{\sigma_0} \int_{\xi - \xi_S}^{\xi + \xi_S} e^{\frac{S(\mathbf{x}_S)}{\epsilon} - \frac{p(\mathbf{x}_S)(\xi - \xi_S)^2}{2\epsilon} + \frac{(\xi - \xi_S)^3}{6\epsilon} \frac{\partial^3 g}{\partial \xi^3}(\xi_S) + \cdots} \sqrt{\frac{k(\xi)}{2\pi\epsilon}} [f(\xi)]^2 d\xi$$

$$= \frac{[f(\mathbf{x}_S)]^2}{\sigma(\mathbf{x}_S)} \sqrt{\frac{k(\mathbf{x}_S)}{p(\mathbf{x}_S)}} (1 + O(\delta)) ,$$

where $f(\mathbf{x}_S)$ is the net current through the saddle. We now add the contribution of all saddle points along the ridge and, since $\delta \ll 1$, we obtain

$$\int_{\text{ridge}} \frac{1}{\sigma_0} e^{S(\xi,\eta)/\epsilon} \, |\, \mathbf{j}(\xi,\eta)\,|^2 \, d\xi d\eta = \sum_{\mathbf{x}_S \in \text{ridge}} [f(\mathbf{x}_S)]^2 R(\mathbf{x}_S)(1 + o(1)) , \quad (27)$$

where $R(\mathbf{x}_S) = \frac{1}{\sigma(\mathbf{x}_S)} \sqrt{\frac{k(\mathbf{x}_S)}{p(\mathbf{x}_S)}}$ is the effective resistance of saddle $\mathbf{x}_S$.

The upper bound is thus

$$(I, (\Lambda^\epsilon)^{-1} I) \leq \min_f \sum_{\mathbf{x}_S} [f(\mathbf{x}_S)]^2 R(\mathbf{x}_S)(1 + o(1)) = < \mathcal{I}, (\Lambda^{D,\epsilon})^{-1} \mathcal{I} > [1 + o(1)] .$$

$$(28)$$

Note that because of conservation of currents in Ω the minimization in (28) is done over all fluxes f that satisfy Kirchhoff's nodal law for currents. A similar calculation leads to the upper bound

$$(\psi, \Lambda^\epsilon \psi) \leq < \Psi, \Lambda^{D,\epsilon} \Psi > [1 + o(1)] , \quad (29)$$

where we use variational principle (7). Using the convex duality relation

$$(\psi, \Lambda^\epsilon \psi) = \sup_{I \in H^{-\frac{1}{2}}(\partial\Omega)} [2(I, \psi) - (I, (\Lambda^\epsilon)^{-1} I)] , \quad (30)$$

we obtain lower bounds on $(\psi, \Lambda^\epsilon \psi)$ and $(I, (\Lambda^\epsilon)^{-1} I)$. These lower bounds match asymptotically the upper bounds (28) and (29), respectively and the proof is complete. The details are given in [10].

2.4 How to Chose the Test Functions

The results presented in section 2.3 show that static conductive transport in high contrast media is asymptotically equivalent to current flow in a resistor network, uniquely determined by the conductivity (5). This asymptotic network approximation comes from the strong channeling of currents along ridges of maximal σ, given by (5). The proof of the asymptotic resistor network approximation, given in [10] and in section 2.3, relies entirely on the variational principles (7) and (8) and a careful choice of trial fields. We now give a brief explanation of how we found these trial fields by looking directly at the governing equations in Ω.

The current density

$$\mathbf{j}(\mathbf{x}) = -\sigma(\mathbf{x})\nabla\phi(\mathbf{x}) = \nabla^\perp H(x, y) \ , \tag{31}$$

where $\nabla^\perp = \left(-\frac{\partial}{\partial y}, \frac{\partial}{\partial x}\right)$, is the minimizer of (8) if

$$\nabla^\perp \cdot \left[\frac{1}{\sigma_0} e^{S(\mathbf{x})/\epsilon} \nabla^\perp H(\mathbf{x})\right] = 0 \quad \text{in } \Omega,$$

$$H(s) = \int^s I(s)ds, \quad \text{on } \partial\Omega \ . \tag{32}$$

Thus, we have the singularly perturbed problem [36]

$$\Delta H(\mathbf{x}) + \frac{1}{\epsilon}\nabla S(\mathbf{x}) \cdot \nabla H(\mathbf{x}) = 0 \quad \text{in } \Omega \ . \tag{33}$$

Consider $\mathbf{x} \in \Omega$ that is neither a minimum nor a maximum or a saddle point of σ that is, $\nabla S(\mathbf{x}) \neq \mathbf{0}$. Then, $H(\mathbf{x})$ satisfies approximately,

$$\nabla S(\mathbf{x}) \cdot \nabla H(\mathbf{x}) = 0 \ , \tag{34}$$

or, equivalently, H varies in directions orthogonal to $\nabla S(\mathbf{x})$. The current $\mathbf{j}(\mathbf{x}) = \nabla^\perp H(\mathbf{x})$ is orthogonal to $\nabla H(\mathbf{x})$ so, it is parallel to $\nabla S(\mathbf{x})$. In fact, $\mathbf{j}(\mathbf{x})$ points in direction $-\nabla S(\mathbf{x})$, towards higher conductivity, in order to achieve the minimum in (8).

Let us begin with points near the boundary and justify the result (12). To fix ideas we take the situation of figure 1 and focus attention on the basin of attraction of x_a, a maximum of σ. From (34) the current density $\nabla^\perp H$ flows from the boundary downhill towards x_a, the closest point of minimum resistance $\rho(\mathbf{x}) = \frac{1}{\sigma_0} e^{S(\mathbf{x})/\epsilon}$ (see figure 2). The net current flowing into x_a is $\mathcal{I}_a = \int_{s_{a'}}^{s_{b'}} I(s)ds$, as given by formula (12). At x_a, $\nabla S = 0$ and equation (34) is not valid. Here, we have an inner layer [36] of width $\delta \gg \sqrt{\epsilon}$, where H changes rapidly, as shown in [11]. Because of the external driving we have current flow from one point of minimal resistance to another. To achieve the minimum in (8) the flow goes along the least resistive paths in Ω. These paths are precisely the ridges of maximal σ that define the topology of the asymptotic network. Take for example the ridge in (21). Here,

$$\nabla S(\xi, \eta) = \left[\frac{\partial S}{\partial \xi}(\xi, 0) + \frac{k'(\xi)\eta^2}{2} + \dots\right]\hat{\xi} + k(\xi)\eta\hat{\eta} \approx \frac{\partial S}{\partial \xi}(\xi, 0)\hat{\xi}$$

and, by (34), $\nabla^\perp H$ is in direction $\hat{\xi}$ tangent to the ridge. Clearly, as seen from (25), most power is dissipated as current passes through the points of highest resistance along ridges of maximal conductivity, the saddle points of σ.

In the asymptotic limit $\epsilon \to 0$ the solution of (32) is now as follows: In Ω, H changes rapidly, on a length scale of order $\sqrt{\epsilon}$, across ridges of maximal

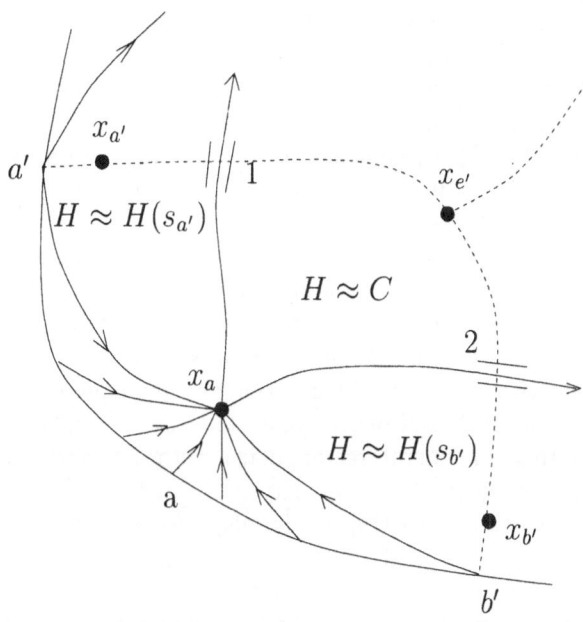

Fig. 2. Current flows from the boundary into the maximum x_a of σ, along paths of steepest descent. In the interior of Ω, current concentrates along paths of maximal conductivity

conductivity. Note that in [11] we do the local asymptotic analysis of (32) near saddle points of σ and find that H is of the form (22). In the rest of Ω, except possibly for thin boundary layers, H is approximately a constant, as shown in figure 2. Here, to the left of the network branch that passes through saddle point 1 we have $H \approx H(s_{a'})$, where $s_{a'}$ is arclength along $\partial\Omega$ at a', the intersection of the boundary with the ridge of minimal conductivity through $x_{a'}$. Similarly, below the branch through the saddle point 2 we have $H \approx H(s_{b'})$. In the interior of Ω between the two branches of figure 2 $H \approx C$, a constant, and so on. The actual constant values of H in the interior of Ω cannot be calculated from a local asymptotic analysis. For this, we need to study the global problem and the tools for that are variational principles (7) and (8).

Singular perturbation methods [36] applied to (32) give us a physically clear picture of static transport in high contrast media. Our proof in section 2.3 relies entirely on the variational principles (7) and (8) and it needs no further justification. However, the key to the successful use of the variational principles is, of course, having good trial fields. The local asymptotic analysis presented here is precisely the guide to finding these trial fields.

2.5 Imaging High Contrast Media

The results summarized in section 2.2 show that imaging high contrast σ of the form (5) is asymptotically equivalent to the identification of a resistor network from measurements of currents and voltages at the peripheral nodes. Therefore, in high contrast inversion the most important features of the conductivity are near the saddle points. Each saddle point of σ corresponds to a resistor in the asymptotic network and makes a significant contribution to the quadratic forms $(\psi, \Lambda^\epsilon \psi)$ and $(I, (\Lambda^\epsilon)^{-1} I)$ or, equivalently, to the eigenvalues of the DtN and NtD maps. The location of maxima and minima of σ in Ω determine the current flow topology and so they influence the spectra of the DtN and NtD maps. The actual value of σ at the maxima and minima is not so important in the asymptotic limit. Therefore, when imaging a high contrast conductivity (5) we cannot expect a good estimate of the value of σ at the minima or maxima. We should get, however, a good image at the saddle points as well as a fair estimate of the location of all critical points. The question of the unique recovery of resistor networks from the discrete DtN or NtD maps has been considered in [18, 24, 19]. It is shown there that, in general, rectangular resistor networks can be uniquely recovered. Even more general resistor networks can be uniquely recovered up to $Y - \Delta$ transformations (see [19] for details). However, the question of how to image these networks in practice does not have a satisfactory answer so far, especially when the network topology is not known a priori. In [9, 10], we propose imaging asymptotic resistor networks with the method of matching pursuit [41]. We show with extensive numerical computations that matching pursuit is effective in imaging high contrast conductive media if the library of functions is carefully constructed to capture the features of σ that come from the asymptotic theory.

3   Quasi-Static Approximation: Complex Conductivity

In dielectric media at low frequency the magnetic term $i\omega\mu\mathbf{H}$ in (2) can be neglected [11]. In two dimensions, the magnetic field $\mathbf{H}(\mathbf{x}) = H(\mathbf{x})\mathbf{e}_3$, $H = H_R + i\, H_I$, satisfies

$$\nabla^\perp \cdot \{[\rho(\mathbf{x}) + i\, C(\mathbf{x})]\, \nabla^\perp H(\mathbf{x})\} = 0, \quad \mathbf{x} \in \Omega \subset \mathbb{R}^2,$$

$$-\nabla^\perp H(\mathbf{x}) \cdot \mathbf{n}(\mathbf{x}) = I(\mathbf{x}) = I_R(\mathbf{x}) + i\, I_I(\mathbf{x}) \quad \text{on } \partial\Omega, \quad (35)$$

$$\int_{\partial\Omega} I(s)ds = 0 \;,$$

where $\rho + i\, C$ is the complex impedance consisting of the resistance(real part) $\rho = \sigma/(\sigma^2 + \omega^2\varepsilon^2)$ and the capacitive reactance (imaginary part) $C = \omega\varepsilon/(\sigma^2 + \omega^2\varepsilon^2)$. We define the current density

$$\mathbf{j}(\mathbf{x}) = \mathbf{j}_R(\mathbf{x}) + i\, \mathbf{j}_I(\mathbf{x}) = \nabla^\perp H(\mathbf{x}) \;, \tag{36}$$

such that $\nabla \cdot \mathbf{j}(\mathbf{x}) = 0$ for all $\mathbf{x} \in \Omega$ and, at the boundary, $-\mathbf{j}(\mathbf{x}) \cdot \mathbf{n}(\mathbf{x}) = I(\mathbf{x})$. Equation (35) implies that

$$\nabla\phi(\mathbf{x}) = -[\rho(\mathbf{x}) + i\, C(\mathbf{x})]\mathbf{j}(\mathbf{x}), \quad \mathbf{x} \in \Omega \;, \tag{37}$$

where ϕ is the complex electric potential. In the inverse problem of complex conductivity, $\rho(\mathbf{x})$ and $\mathcal{C}(\mathbf{x})$ are unknown and are to be found from simultaneous measurements of currents and potentials at the boundary. Thus, for a given current excitation I we overspecify problem (35) by asking that

$$\phi(\mathbf{x}) = \psi(\mathbf{x}) \quad \text{on } \partial\Omega \ , \tag{38}$$

where ψ is the complex potential measured at the boundary. When all possible excitations and measurements at the boundary are available, then we know the complex Neumann to Dirichlet (NtD) map which takes I and maps it into ψ. In most applications we do not have full knowledge of the NtD map and the inversion must be done with partial, noisy information about it.

We are interested in dielectric media with high contrast and we model the complex impedance by

$$\rho(\mathbf{x}) = \rho_0 e^{-S(\mathbf{x})/\epsilon}, \quad \mathcal{C}(\mathbf{x}) = \mathcal{C}_0 e^{-P(\mathbf{x})/\epsilon} \ , \tag{39}$$

where ρ_0 and $\mathcal{C}_0$ are constants, $S(\mathbf{x})$ and $P(\mathbf{x})$ are smooth functions with isolated, nondegenerate critical points and ϵ is a positive, small parameter. The complex NtD map $\Gamma^\epsilon = \Gamma_R^\epsilon + i\Gamma_I^\epsilon$ is defined by

$$\Gamma^\epsilon I(\mathbf{x}) = \psi(\mathbf{x}), \quad \mathbf{x} \in \partial\Omega \ . \tag{40}$$

The domain of the NtD map is the restricted space of currents $I = I_R + iI_I$ that satisfy I_R, $I_I \in \mathrm{H}^{-\frac{1}{2}}(\partial\Omega)$ and $\int_{\partial\Omega} I(s)ds = 0$. From Green's theorem, we have

$$(I^*, \Gamma^\epsilon I) = \int_{\partial\Omega} \psi(s)I^*(s)ds = \int_\Omega \left[\rho_0 e^{-S(\mathbf{x})/\epsilon} + i\, \mathcal{C}_0 e^{-P(\mathbf{x})/\epsilon} \right] |\,\mathbf{j}(\mathbf{x})\,|^2 \, d\mathbf{x} \ , \tag{41}$$

where I^* is the complex conjugate of I. Thus, the eigenvalues $\lambda^\epsilon = \lambda_R^\epsilon + i\lambda_I^\epsilon$ of Γ^ϵ have strictly positive real and imaginary parts.

In order to study the NtD map in the asymptotic limit $\epsilon \to 0$, it is important to have variational formulations for quadratic forms of Γ^ϵ. For this purpose we rewrite (35) as an elliptic system of equations

$$\nabla^\perp \cdot \left[\rho_0 e^{-S(\mathbf{x})/\epsilon} \mathbf{j}_R(\mathbf{x}) - \mathcal{C}_0 e^{-P(\mathbf{x})/\epsilon} \mathbf{j}_I(\mathbf{x}) \right] = 0,$$

$$\nabla^\perp \cdot \left[\rho_0 e^{-S(\mathbf{x})/\epsilon} \mathbf{j}_I(\mathbf{x}) + \mathcal{C}_0 e^{-P(\mathbf{x})/\epsilon} \mathbf{j}_R(\mathbf{x}) \right] = 0, \tag{42}$$

$$\mathbf{j}_R(\mathbf{x}) = \nabla^\perp H_R(\mathbf{x}), \ \mathbf{j}_I(\mathbf{x}) = \nabla^\perp H_I(\mathbf{x}), \quad \text{for } \mathbf{x} \in \Omega \ .$$

At the boundary, we give the normal current $-[\mathbf{j}_R(s) + i\mathbf{j}_I(s)] \cdot \mathbf{n}(s) = I(s) + i\, I_I(s)$ or, equivalently, the Dirichlet conditions

$$H_R(s) = \int^s I_R(s)ds,$$

$$H_I(s) = \int^s I_I(s)ds \ , \tag{43}$$

where s is arclength along $\partial\Omega$. Equations (42) with boundary conditions (43) have a unique (at least weak) solution H_R and $H_I \in H^1(\Omega)$, provided that $\rho(\mathbf{x}) > 0$ and $\rho(\mathbf{x})$ and $C(\mathbf{x})$ are uniformly bounded in Ω. Equations (42) are also the Euler equations for the following saddle point variational principles [16, 22, 11]

$$
\text{real}(I, \Gamma^\epsilon I)
$$

$$
= \min_{\substack{\nabla \cdot \mathbf{j}_R = 0 \\ -\mathbf{j}_R \cdot \mathbf{n} \,|_{\partial\Omega} = I_R}} \max_{\substack{\nabla \cdot \mathbf{j}_I = 0 \\ -\mathbf{j}_I \cdot \mathbf{n} \,|_{\partial\Omega} = I_I}} \int_\Omega \Big\{ \rho_0 e^{-S(\mathbf{x})/\epsilon} \big[|\, \mathbf{j}_R(\mathbf{x}) \,|^2 - |\, \mathbf{j}_I(\mathbf{x}) \,|^2 \big] -
$$

$$
2C_0 e^{-P(\mathbf{x})/\epsilon} \mathbf{j}_R(\mathbf{x}) \cdot \mathbf{j}_I(\mathbf{x}) \Big\} \, d\mathbf{x} \,, \qquad (44)
$$

$$
\text{imag}(I, \Gamma^\epsilon I)
$$

$$
= \min_{\substack{\nabla \cdot \mathbf{j}_R = 0 \\ -\mathbf{j}_R \cdot \mathbf{n} \,|_{\partial\Omega} = I_R}} \max_{\substack{\nabla \cdot \mathbf{j}_I = 0 \\ -\mathbf{j}_I \cdot \mathbf{n} \,|_{\partial\Omega} = I_I}} \int_\Omega \Big\{ C_0 e^{-P(\mathbf{x})/\epsilon} \big[|\, \mathbf{j}_R(\mathbf{x}) \,|^2 - |\, \mathbf{j}_I(\mathbf{x}) \,|^2 \big] +
$$

$$
2\rho_0 e^{-S(\mathbf{x})/\epsilon} \mathbf{j}_R(\mathbf{x}) \cdot \mathbf{j}_I(\mathbf{x}) \Big\} \, d\mathbf{x} \,. \qquad (45)
$$

3.1   Asymptotic Resistor-Capacitor Network Approximation

In [11], we carry out an asymptotic analysis of (35) for periodic dielectric media. In this section we review the results of [11] and we extend the asymptotic analysis to more general high contrast dielectrics, not just periodic ones. We show that in the asymptotic limit $\epsilon \to 0$ the solution $\mathbf{j} = \mathbf{j}_R + i\mathbf{j}_I$ of (42) and (43) can be approximated by current flow through a resistor-capacitor network. We show further that the NtD map Γ^ϵ is asymptotically equivalent to the discrete NtD map of the resistor-capacitor network.

Let us begin with the observation that since $\epsilon \ll 1$ the resistance $\rho(\mathbf{x})$ dominates $C(\mathbf{x})$ when its scaled logarithm $S(\mathbf{x})$ is less than $P(\mathbf{x})$. In fact, $\rho(\mathbf{x})$ and $C(\mathbf{x})$ are comparable in magnitude only when $S(\mathbf{x}) = P(\mathbf{x}) + O(\epsilon)$. In general, one does not expect S and P to be equal over large regions of the domain and we model the medium by alternating regions of dominant resistance and capacitance. Both ρ and C are smooth functions so these regions are separated by interfaces along which ρ is equal to C. In regions of dominant ρ or C, (42) is similar to the static problem (2). Thus, transport in these regions behaves like current flow in purely resistive or capacitive networks, respectively. The main question is how to connect these networks in order to get the global resistor-capacitor network that approximates the transport properties of the high contrast medium.

Suppose, for example, that the high contrast function $\rho(\mathbf{x})$ has the six minima denoted by $\circ$ in figure 3. Its maxima are denoted by $\bullet$ and the saddle points are numbered from 1 to 7. If we consider the static problem in a medium with resistance $\rho(\mathbf{x})$ the results of section 2.2 apply and the current flows along the ridges of minimal resistance, shown with a full line in figure 3. Similarly, in figure 4, we show an example of a high contrast function $C(\mathbf{x})$ and its associated asymptotic network.

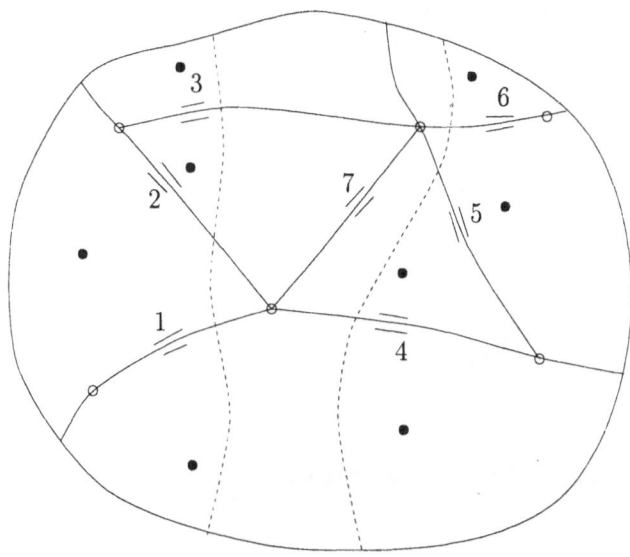

Fig. 3. The static network for an example of $\rho(\mathbf{x}) = \rho_0 e^{-S(\mathbf{x})/\epsilon}$

Consider now the complex impedance $\rho + iC$. As explained above, let us suppose that $C(\mathbf{x}) \gg \rho(\mathbf{x})$ for points between the dotted lines in figures 3 and 4. Along these lines (interfaces) ρ is equal to C and elsewhere in Ω, $\rho \gg C$. Away from the interfaces we already established that (42) is similar to the static problem of section 2.1, where σ in (2) is replaced by $\frac{1}{\rho}$ if $\rho \gg C$ and $\frac{1}{C}$ if $\rho \ll C$, respectively. Thus, in these regions, currents flow along ridges of minimal resistance and capacitive reactance, respectively. The ridges of minimal ρ and C cut the interfaces, as shown in figures 3 and 4, respectively. Along the interfaces, $\rho = C$ so ridges of minimal ρ and C must intersect the dotted lines at the same points. The network connection is now straightforward and the global network is shown in figure 5, where R_i and C_i correspond to the resistive/capacitive saddle points and they are given by formulas similar to (9). The construction illustrated in this example is general and it allows us to define the topology of the $R - C$ asymptotic network for any medium with impedance (39).

we now consider the quadratic forms $(I, \Gamma^\epsilon I)$ of the NtD map, for a high contrast impedance $\rho_0 e^{-S(\mathbf{x})/\epsilon} + i\, C_0 e^{-P(\mathbf{x})/\epsilon}$, in the asymptotic limit $\epsilon \to 0$. We show that Γ^ϵ is asymptotically equivalent to $\Gamma^{D,\epsilon}$, the NtD map of the asymptotic resistor-capacitor network. First we identify the topology of the network as explained in the example of figure 5. We divide Ω into resistive regions Ω_{ρ_j}, $j = 1, \ldots N_\rho$, where $\rho \gg C$ and capacitive regions Ω_{C_j}, $j = 1, \ldots N_C$, where $C \gg \rho$, respectively. These regions are separated by interfaces $\mathcal{L}_j$, $j = 1, \ldots N_\mathcal{L}$,

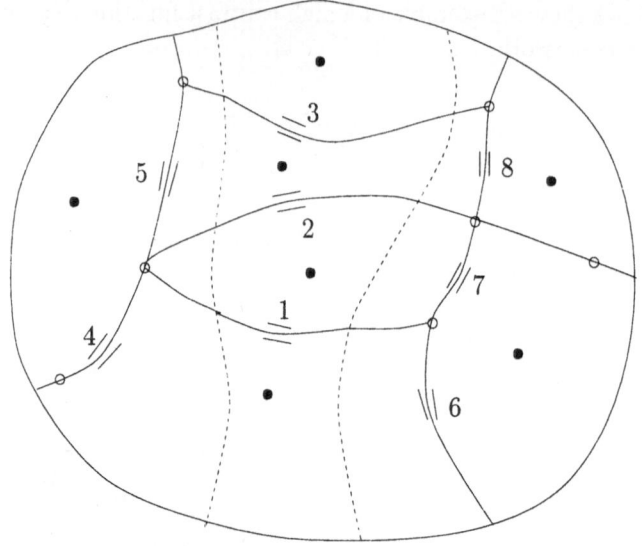

Fig. 4. The static network for an example of $\mathcal{C}(\mathbf{x}) = \mathcal{C}_0 e^{-P(\mathbf{x})/\epsilon}$

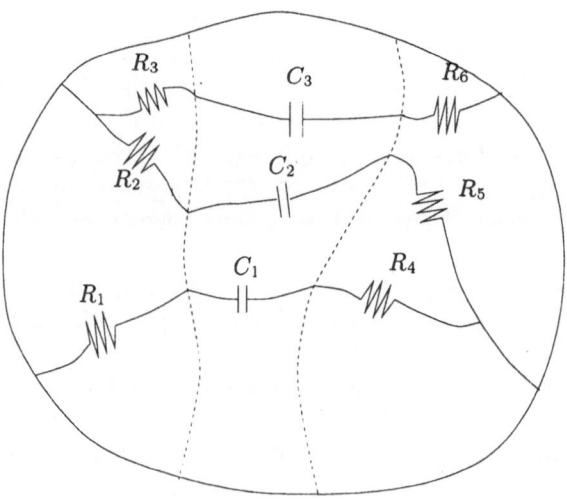

Fig. 5. The global R-C network

where $\rho = C$. Suppose that the global $R - C$ network has N_B peripheral nodes, where each node corresponds to a point of intersection of $\partial\Omega$ with a ridge of minimal ρ or C. The complex currents at the peripheral nodes are $\mathcal{I} = (\mathcal{I}_1, \mathcal{I}_2, \ldots \mathcal{I}_{N_B})^T$. The complex electric potentials at the peripheral nodes are $\Psi = (\Psi_1, \Psi_2, \ldots \Psi_{N_B})^T$. Take for example the point s_j of intersection of a ridge of minimal ρ with the boundary. The current flowing into this node is given by

$$\mathcal{I}_j = \int_{\partial\Omega \cap \overline{B}(\mathbf{x}_j)} I(s)ds \ , \tag{46}$$

where $\overline{B}(\mathbf{x}_j)$ stands for the closure of the basin of attraction of minimum $\mathbf{x}_j$ that is adjacent to s_j. The electric potential is

$$\Psi_j = \psi(s_j) \ . \tag{47}$$

The discrete NtD map, $\Gamma^{D,\epsilon} = \Gamma_R^{D,\epsilon} + i \ \Gamma_I^{D,\epsilon}$, is a complex $N_B \times N_B$ matrix that takes boundary currents and maps them into boundary potentials

$$\Psi_j = \sum_{k \in \mathcal{N}_B} \Gamma_{jk}^{D,\epsilon} \mathcal{I}_k, \quad \text{for all } j \in \mathcal{N}_B \ . \tag{48}$$

We show that, given any boundary current $I = I_R + iI_I$ such that $I_R, I_I \in \mathrm{H}^{-\frac{1}{2}}(\partial\Omega)$ and $\int_{\partial\Omega} I(\mathbf{s})ds = 0$,

$$(I, \Gamma^\epsilon I) = < \mathcal{I}, \Gamma^{D,\epsilon} \mathcal{I} > [1 + o(1)] \ , \tag{49}$$

where

$$< \mathcal{I}, \Gamma^{D,\epsilon} \mathcal{I} > = \sum_{j=1}^{N_B} \mathcal{I}_j \Psi_j \ . \tag{50}$$

3.2 Asymptotics Using the Variational Principles

The proof of (49) relies on the variational principles (44) and (45). To get an upper bound on real$(I, \Gamma^\epsilon I)$, we take a trial field

$$\mathbf{j}_R(\mathbf{x}) = \left(-\frac{\partial}{\partial y}, \frac{\partial}{\partial x} \right) H_R(x, y) \ , \tag{51}$$

so that at the boundary

$$H_R(s) = \int^s I_R(s)ds \ , \tag{52}$$

where s is arclength along $\partial\Omega$. To simplify the exposition let us assume that the excitation current $I(s) = I_R(s) + I_I(s)$ is concentrated at the peripheral nodes of the $R - C$ network, i.e. at the minima of $\max(\rho, C)$ along $\partial\Omega$. Then, H_R and H_I are constant along pieces of the boundary lying between two adjacent peripheral nodes. Of course, in a real experiment the current excitation is unlikely to satisfy this assumption. Then one must introduce a thin layer near the boundary where the current adjusts from the given I along $\partial\Omega$ to a current concentrated along ridges of minimal ρ or C. The analysis is identical to that of section 2.3 and [10] and shall not be repeated here.

We write the integral in (44) as a sum of integrals over the resistive regions Ω_{ρ_j}, $j = 1, \ldots N_\rho$, the capacitive regions Ω_{C_j}, $j = 1, \ldots N_C$ and the vicinities of interfaces $\mathcal{L}_j$, $j = 1, \ldots N_{\mathcal{L}}$. We analyze these three cases separately.

1. *The resistive region* $\mathbf{x} \in \Omega_{\rho_j}$ for some $j = 1, \ldots N_\rho$, such that $\rho(\mathbf{x}) \gg C(\mathbf{x})$. In Ω_{ρ_j}, we choose H_R to be a constant away from ridges of minimal ρ. In the vicinity of a ridge of minimal ρ consider coordinates (ξ, η) where ξ is arc length along the ridge. Here the scaled logarithm of ρ is given by

$$S(\xi,\eta) = S(\xi,0) - \frac{k(\xi)}{2}\eta^2 + \frac{1}{6}\frac{\partial^3 S}{\partial\eta^3}(\xi,0)\eta^3 + \cdots , \tag{53}$$

where $k(\xi) > 0$ and H_R is chosen as in section 2.2,

$$H_R(\xi,\eta) = -\frac{f_R}{2}\,\mathrm{erf}\left(\frac{\eta}{\sqrt{\frac{2\epsilon}{k(\xi)}}}\right) + \text{constant} , \tag{54}$$

where f_R is the real flux along the ridge. We have

$$\int_{\Omega_{\rho_j}} \left[\rho(\mathbf{x})\left(|\,\mathbf{j}_R(\mathbf{x})\,|^2 - |\,\mathbf{j}_I(\mathbf{x})\,|^2\right) - 2C(\mathbf{x})\mathbf{j}_R(\mathbf{x})\cdot\mathbf{j}_I(\mathbf{x})\right] dx \leq$$

$$\int_{\Omega_{\rho_j}} \left[(\rho(\mathbf{x}) + C(\mathbf{x}))\,|\,\mathbf{j}_R(\mathbf{x})\,|^2 - (\rho(\mathbf{x}) - C(\mathbf{x}))\,|\,\mathbf{j}_I(\mathbf{x})\,|^2\right] dx \approx \tag{55}$$

$$\sum_{\text{ridges in } \Omega_{\rho_j}} \int_{\text{ridge}} \rho(\mathbf{x})\,|\,\mathbf{j}_R(\mathbf{x})\,|^2\,dx - \int_{\Omega_{\rho_j}} \rho(\mathbf{x})\,|\,\mathbf{j}_I(\mathbf{x})\,|^2\,dx .$$

The maximum of (55) is achieved by imaginary currents that satisfy the Euler equation

$$\nabla^\perp \cdot \left[\rho_0 e^{-S(\mathbf{x})/\epsilon}\mathbf{j}_I(\mathbf{x})\right] = 0,$$
$$\nabla \cdot \mathbf{j}_I(\mathbf{x}) = 0, \quad \mathbf{x} \in \Omega_{\rho_j} . \tag{56}$$

This equation has been studied in section 2.2 and the result is that $\mathbf{j}_I$ is channeled along ridges of minimal resistance in Ω_{ρ_j}. Moreover,

$$\min_{\mathbf{j}_I} \int_{\Omega_{\rho_j}} \rho(\mathbf{x})\,|\,\mathbf{j}_I(\mathbf{x})\,|^2\,dx = \min_{f_I} \sum_{\mathbf{x}_S \in \Omega_{\rho_j}} R(\mathbf{x}_S)\,[f_I(\mathbf{x}_S)]^2\,[1 + o(1)] , \tag{57}$$

where $f_I(\mathbf{x}_S)$ is the net imaginary current through the saddle point $\mathbf{x}_S$ of $\rho(\mathbf{x})$ and $R(\mathbf{x}_S)$ is the effective resistance of the saddle, given by (9), where $\frac{1}{\sigma}$ is replaced by ρ.

Thus, the contribution of Ω_{ρ_j} to the upper bound is

$$\int_{\Omega_{\rho_j}} \left[\rho(\mathbf{x}) \left(|\mathbf{j}_R(\mathbf{x})|^2 - |\mathbf{j}_I(\mathbf{x})|^2\right) - 2C(\mathbf{x})\mathbf{j}_R(\mathbf{x}) \cdot \mathbf{j}_I(\mathbf{x})\right] d\mathbf{x} \le$$

$$\max_{f_I} \sum_{\mathbf{x}_S \in \Omega_{\rho_j}} R(\mathbf{x}_S) \left\{[f_R(\mathbf{x}_S)]^2 - [f_I(\mathbf{x}_S)]^2\right\} [1 + o(1)] . \tag{58}$$

2. *The capacitive region* $\mathbf{x} \in \Omega_{C_j}$ for some $j = 1, \ldots N_C$, such that $C(\mathbf{x}) \gg$ $\rho(\mathbf{x})$. In Ω_{C_j}, we choose $H_R(\mathbf{x})$ that minimizes $\int_{\Omega_{C_j}} C_0 e^{-P(\mathbf{x})/\epsilon} |\nabla^{\perp} H_R(\mathbf{x})|^2 d\mathbf{x}$. Clearly, H_R is the solution of

$$\nabla^{\perp} \cdot \left[C_0 e^{-P(\mathbf{x})/\epsilon} \nabla^{\perp} H_R(\mathbf{x})\right] = 0, \quad \text{for } \mathbf{x} \in \Omega_{C_j} . \tag{59}$$

From the results of sections 2.2 and 2.3, we know that, away from ridges of minimal C, $H_R(\mathbf{x})$ is approximately a constant. Take a ridge of minimal C and consider coordinates (ξ, η) where ξ is arc length along the ridge. In a neighborhood $|\eta| \le \delta \ll 1$ such that $\frac{\delta}{\sqrt{\epsilon}} \to \infty$ as $\epsilon \to 0$, we have

$$P(\xi, \eta) = P(\xi, 0) - \frac{k(\xi)}{2}\eta^2 + \frac{1}{6}\frac{\partial^3 P}{\partial \eta^3}(\xi, 0)\eta^3 + \cdots , \tag{60}$$

where $k(\xi) > 0$ and H_R is changing abruptly across the ridge, in such a way that $\mathbf{j}_R = \nabla^{\perp} H_R = O(\frac{1}{\sqrt{\epsilon}})\hat{\xi}$. The net current along the ridge is $f_R(\xi) = \int_{-\delta}^{\delta} \nabla^{\perp} H_R(\xi, t)dt$. Note that because of conservation of current $f_R(\xi)$ is either a constant or it changes along the ridge near minima of C. Thus,

$$\int_{\Omega_{C_j}} C_0 e^{-P(\mathbf{x})/\epsilon} |\nabla^{\perp} H_R(\mathbf{x})|^2 d\mathbf{x} = \min_{f_R} \sum_{\mathbf{x}_S \in \Omega_{C_j}} C(\mathbf{x}_S) [f_R(\mathbf{x}_S)]^2 [1 + o(1)] , \tag{61}$$

where $C(\mathbf{x}_S)$ is the effective capacity of the saddle, given by (9) with $\frac{1}{\sigma}$ is replaced by C.

We also study the imaginary current $\mathbf{j}_I = \nabla^{\perp} H_I(\mathbf{x})$ that maximizes

$$\int_{\Omega_{C_j}} \left[-2C(\mathbf{x})\nabla^{\perp} H_R(\mathbf{x}) \cdot \nabla^{\perp} H_I(\mathbf{x}) - \rho(\mathbf{x})|\nabla^{\perp} H_I(\mathbf{x})|^2\right] d\mathbf{x} . \tag{62}$$

The maximization is done over all $H_I(\mathbf{x})$ that satisfy boundary conditions

$$H_I(s) = \int^s \mathbf{j}_I(s) \cdot \mathbf{n}(s)ds , \tag{63}$$

where s is arclength along $\partial\Omega_{C_j}$. The boundary of Ω_{C_j} consists of some interfaces $\mathcal{L}_p$, $p = 1, \ldots N_{\mathcal{L}}$ and possibly a piece of $\partial\Omega$. In the resistive regions that are adjacent to Ω_{C_j} the imaginary current is concentrated along ridges of minimal

ρ, as shown in part 1 where we discussed the resistive regions. At the interfaces $\mathcal{L}_p$ these ridges meet the ridges of minimal C in Ω_{C_j}. Therefore, at the boundary of Ω_{C_j}, the imaginary current is concentrated at the peripheral nodes of the C (static) network and, as given by (63), H_I is a constant along a piece of $\partial\Omega_{C_j}$ that lies between two adjacent peripheral nodes.

We show next that the imaginary current that maximizes (62) is concentrated along the ridges of minimal $\mathcal{C}(\mathbf{x})$, in the interior of Ω_{C_j}. The Euler equation maximizing $\mathbf{j}_I = \nabla^\perp H_I$ satisfies is

$$\nabla^\perp \cdot [\rho(\mathbf{x})\nabla^\perp H_I(\mathbf{x}) + C(\mathbf{x})\nabla^\perp H_R(\mathbf{x})] = 0 , \tag{64}$$

where the resistance function is of the form

$$\rho(\mathbf{x}) = C(\mathbf{x})e^{-Q(\mathbf{x})/\epsilon} \ll C(\mathbf{x}), \quad \text{where } Q(\mathbf{x}) = S(\mathbf{x}) - P(\mathbf{x}) - \epsilon \log\frac{\rho_0}{C_0} . \tag{65}$$

We construct a function $\tilde{H}_I(\mathbf{x})$ such that

$$\tilde{H}_I(\mathbf{x}) = H_I(\mathbf{x}) \quad \text{on } \partial\Omega_{C_j} \tag{66}$$

and at $\mathbf{x}$ away from the ridges of minimal $\mathcal{C}(\mathbf{x})$, $\tilde{H}_I(\mathbf{x})$ is constant. In the vicinity of a ridge of minimal C we take

$$\nabla^\perp \tilde{H}_I(\xi, \eta) = \frac{f_I}{f_R}\nabla^\perp H_R(\xi, \eta) . \tag{67}$$

From (65)-(67), we have

$$\nabla^\perp \left[\rho(\mathbf{x})\nabla^\perp \tilde{H}_I(\mathbf{x}) + C(\mathbf{x})\nabla^\perp H_R(\mathbf{x})\right]$$
$$= \nabla^\perp \left\{ C(\xi, \eta)\left[1 + \frac{f_I}{f_R}e^{-Q(\xi,\eta)/\epsilon}\right]\nabla^\perp H_R(\xi, \eta)\right\}$$
$$\approx \nabla^\perp \left[C(\xi, \eta)\nabla^\perp H_R(\xi, \eta)\right] = 0 . \tag{68}$$

where $e^{-Q(\xi,\eta)/\epsilon} \ll 1$ so that $\rho \ll C$ holds. Thus, $\tilde{H}_I$ is the approximate solution of the Euler equation (64) and satisfies the same boundary conditions as H_I. By uniqueness of solutions H_I for the elliptic equation (64) we conclude that

$$\nabla^\perp H_I(\xi, \eta) = \frac{f_I}{f_R}\nabla^\perp H_R(\xi, \eta) . \tag{69}$$

From (65) and section 2.3 we now have

$$\max_{\nabla^\perp H_I} \int_{\Omega_{C_j}} [\rho(\mathbf{x}) \mid \nabla^\perp H_R(\mathbf{x}) \mid^2 - \rho(\mathbf{x}) \mid \nabla^\perp H_I(\mathbf{x}) \mid^2$$

$$-2C(\mathbf{x})\nabla^\perp H_R(\mathbf{x}) \cdot \nabla^\perp H_I(\mathbf{x})] \, d\mathbf{x}$$

$$\approx \sum_{\text{ridges in } \Omega_{C_j}} \int_{\text{ridge}} \left\{ C_0 e^{-P(\xi,\eta)/\epsilon} \left[e^{-Q(\xi,\eta)/\epsilon} (\mid \nabla^\perp H_R \mid^2 - \mid \nabla^\perp H_I \mid^2)\right.\right.$$

$$-2\nabla^\perp H_R \cdot \nabla^\perp H_I]\} \, d\xi d\eta$$

$$\approx -2 \sum_{\text{ridges in } \Omega_{C_j}} \int_{\text{ridge}} C_0 e^{-P(\xi,\eta)/\epsilon} \nabla^\perp H_R \cdot \nabla^\perp H_I d\xi d\eta$$

$$\approx \sum_{\text{ridges in } \Omega_{C_j}} \int_{\text{ridge}} C_0 e^{-P(\xi,0)/\epsilon} \times \frac{f_R(\xi) f_I(\xi) k(\xi)}{2\pi\epsilon} e^{-\frac{k(\xi)\eta^2}{2\epsilon}} d\xi d\eta$$

$$= \sum_{x_S \in \Omega_{C_j}} C(x_S) f_R(x_S) f_I(x_S) [1 + o(1)] \ . \tag{70}$$

3. *Resistive-capacitive region.* This is the case where x is in the neighborhood of an interface $\mathcal{L}_j$, $j = 1, \ldots N_{\mathcal{L}}$, where $\rho = C$. Consider a point $(\xi_{\mathcal{L}}, 0)$ along such an interface, where a ridge of minimal ρ and C intersects $\mathcal{L}_j$. We define a neighborhood of this point by

$$|\xi - \xi_{\mathcal{L}}| \le \alpha, \quad |\eta| \le \delta \ , \tag{71}$$

where (ξ, η) are coordinates along the ridge. Here, α and δ are small and $\frac{\alpha}{\epsilon} \to \infty$ and $\frac{\delta}{\sqrt{\epsilon}} \to \infty$ as $\epsilon \to 0$. Let us assume that there is no saddle point along $\mathcal{L}_j$ and that ρ decreases with ξ. Then C increases with ξ in such a way that $\rho(\xi_{\mathcal{L}}, 0) = C(\xi_{\mathcal{L}}, 0)$. We also have the expansions

$$S(\xi, \eta) \approx S(\xi_{\mathcal{L}} - \alpha, 0) + p(\xi - \xi_{\mathcal{L}} + \alpha) - \frac{k(\xi_{\mathcal{L}})}{2}\eta^2 \tag{72}$$

$$P(\xi, \eta) \approx P(\xi_{\mathcal{L}} + \alpha, 0) - q(\xi - \xi_{\mathcal{L}} - \alpha) - \frac{k(\xi_{\mathcal{L}})}{2}\eta^2 \ ,$$

where $p = \frac{\partial S}{\partial \xi}(\xi_{\mathcal{L}} - \alpha, 0) > 0$ and $q = \frac{\partial P}{\partial \xi}(\xi_{\mathcal{L}} + \alpha, 0) > 0$.

In the vicinity of $(\xi_{\mathcal{L}}, 0)$, we choose the trial field

$$H_R(\xi, \eta) \approx -\frac{f_R(\xi_{\mathcal{L}})}{2} \operatorname{erf}\left(\frac{\eta}{\sqrt{\frac{2\epsilon}{k(\xi_{\mathcal{L}})}}}\right) + \text{constant} \ , \tag{73}$$

whereas, away from points such as $(\xi_{\mathcal{L}}, 0)$, H_R is taken to be a constant. Note that our choice of $\dot{H}_R$ in a thin strip along the interface $\mathcal{L}_j$ agrees with the trial field H_R in the resistive and capacitive regions, separated by $\mathcal{L}_j$. Thus,

$$\nabla^\perp \cdot \left[C_0 e^{-P(\xi,\eta)/\epsilon} \nabla^\perp H_R(\xi, \eta)\right] \approx \frac{\partial}{\partial \eta}\left[C_0 e^{-P(\xi,\eta)/\epsilon} \frac{\partial H_R(\xi, \eta)}{\partial \eta}\right] \approx 0 \ . \tag{74}$$

We must calculate

$$\max_{j_I} \int_{\text{vicinity of } \mathcal{L}_j} \left[-\rho(x) |j_I(x)|^2 - 2C(x)\nabla^\perp H_R(x) \cdot j_I(x)\right] dx \ , \tag{75}$$

where we integrate in a thin strip along the interface $\mathcal{L}_j$. The Euler equation that the maximizing imaginary current $j_I = \nabla^\perp H_I$ satisfies is

$$\nabla^\perp \cdot \left[\rho(x)\nabla^\perp H_I(x)\right] = -\nabla^\perp \cdot \left[C(x)\nabla^\perp H_R(x)\right] \approx 0 \ . \tag{76}$$

Thus, H_I is the solution of the static elliptic equation considered in section 2.2, where σ is replaced by $1/\rho$ and $\nabla^\perp H_I$ is concentrated near minima, such as $(\xi_\mathcal{L}, 0)$, of ρ and C along $\mathcal{L}_j$. From (18) and the assumption that there are no saddle points lying on $\mathcal{L}_j$, we have

$$\int_{\text{vicinity of } \mathcal{L}_j} \rho\,|\nabla^\perp H_I|^2\,dx \ll R(\mathbf{x}_S)\,[f_I(\mathbf{x}_S)]^2 \,, \tag{77}$$

where $\mathbf{x}_S$ is the saddle point of ρ that is closest to $(\xi_\mathcal{L}, 0)$, along the ridge of minimal resistance.

From (74) we obtain

$$\int_{\text{vicinity of } \mathcal{L}_j} C\nabla^\perp H_R \cdot \nabla^\perp H_I\,dx$$

$$\approx \sum_{(\xi_\mathcal{L},0)\in\mathcal{L}_j} \int_{\xi_\mathcal{L}-\alpha}^{\xi_\mathcal{L}+\alpha} d\xi \int_{-\delta}^{\delta} d\eta \; C_0 e^{-P(\xi,\eta)/\epsilon}\frac{dH_R(\eta)}{d\eta}\frac{\partial H_I(\xi,\eta)}{\partial\eta}$$

$$\approx \int_{\xi_\mathcal{L}-\alpha}^{\xi_\mathcal{L}+\alpha} d\xi \int_{-\delta}^{\delta} d\eta \frac{\partial}{\partial\eta}\left[H_I(\xi,\eta)C_0 e^{-P(\xi,\eta)/\epsilon}\frac{dH_R(\eta)}{d\eta}\right]$$

$$\approx \int_{\xi_\mathcal{L}-\alpha}^{\xi_\mathcal{L}+\alpha} \sqrt{\frac{k(\xi_\mathcal{L})}{2\pi\epsilon}}\, f_R(\xi_\mathcal{L})f_I(\xi)\,C(\xi_\mathcal{L}+\alpha,0)e^{q(\xi-\xi_\mathcal{L}-\alpha)/\epsilon}\,d\xi$$

$$\approx \sqrt{\frac{\epsilon k(\xi_\mathcal{L})}{2\pi}}\,\frac{C(\xi_\mathcal{L}+\alpha,0)}{q}\,f_R(\xi_\mathcal{L})f_I(\xi_\mathcal{L}) \,, \tag{78}$$

where $f_I(\xi) = H_I(\xi, -\delta) - H_I(\xi, \delta)$ and we sum over all minima $(\xi_\mathcal{L}, 0) \in \mathcal{L}_j$. By assumption, C decreases with ξ so $C(\xi_\mathcal{L}+\alpha, 0) \le C(\mathbf{x}_S)$, where $\mathbf{x}_S$ is the location of the capacitive saddle adjacent to $(\xi_L, 0)$. Thus, (75) is negligible in comparison with the contribution of the saddle points of ρ in regions Ω_{ρ_j}, $j = 1, \ldots N_\rho$, and the saddle points of C in regions Ω_{C_j}, $j = 1, \ldots N_C$.

We conclude this calculation with the observation that the same result holds if ρ were increasing with ξ instead of decreasing as in (72). However, if $(\xi_\mathcal{L}, 0)$ is a saddle point its contribution to the upper bound must be taken into account. The calculation of (75) for $(\xi_\mathcal{L}, 0)$ a saddle point is identical to that of (27) in section 2.2 and shall not be repeated here.

We gather all the results in this section to obtain the upper bound

$$\text{real}(I, \Gamma^\epsilon I) \le \min_{f_R} \max_{f_R} \left\{ \sum_{j=1}^{N_\rho} \sum_{\mathbf{x}_S\in\Omega_{\rho_j}} R(\mathbf{x}_S)\left[(f_R(\mathbf{x}_S))^2 - (f_I(\mathbf{x}_S))^2\right] - \right.$$

$$\left. \sum_{j=1}^{N_C} \sum_{\mathbf{x}_S\in\Omega_{C_j}} 2C(\mathbf{x}_S)f_R(\mathbf{x}_S)f_I(\mathbf{x}_S) \right\} [1 + o(1)] \,, \tag{79}$$

where the min-max is taken over the fluxes f_R and f_I that satisfy Kirchhoff's nodal law for currents. We get a lower bound on $\text{real}(I, \Gamma^\epsilon I)$, by taking a trial imaginary current $\mathbf{j}_I$. The calculations are similar to the above and the result is

$$
\begin{aligned}
\text{real}(I, \Gamma^\epsilon I) \geq \min_{f_R} \ \max_{f_R} \Bigg\{ &\sum_{j=1}^{N_\rho} \sum_{\mathbf{x}_S \in \Omega_{\rho_j}} R(\mathbf{x}_S) \left[(f_R(\mathbf{x}_S))^2 - (f_I(\mathbf{x}_S))^2 \right] - \\
&\sum_{j=1}^{N_C} \sum_{\mathbf{x}_S \in \Omega_{C_j}} 2C(\mathbf{x}_S) f_R(\mathbf{x}_S) f_I(\mathbf{x}_S) \Bigg\} [1 + o(1)] \ .
\end{aligned}
\tag{80}
$$

The upper and lower bounds on $\text{real}(I, \Gamma^\epsilon I)$ match asymptotically and we conclude that

$$
\begin{aligned}
\text{real}(I, \Gamma^\epsilon I) = \min_{f_R} \ \max_{f_R} \Bigg\{ &\sum_{j=1}^{N_\rho} \sum_{\mathbf{x}_S \in \Omega_{\rho_j}} R(\mathbf{x}_S) \left[(f_R(\mathbf{x}_S))^2 - (f_I(\mathbf{x}_S))^2 \right] - \\
\sum_{j=1}^{N_C} \sum_{\mathbf{x}_S \in \Omega_{C_j}} 2C(\mathbf{x}_S) f_R(\mathbf{x}_S) f_I(\mathbf{x}_S) \Bigg\} [1 + o(1)] &= \text{real} < \mathcal{I}, \Gamma^{D,\epsilon} \mathcal{I} > [1 + o(1)] \ .
\end{aligned}
\tag{81}
$$

A similar calculation leads to

$$
\begin{aligned}
\text{imag}(I, \Gamma^\epsilon I) = \min_{f_R} \ \max_{f_R} \Bigg\{ &\sum_{j=1}^{N_C} \sum_{\mathbf{x}_S \in \Omega_{C_j}} C(\mathbf{x}_S) \left[(f_R(\mathbf{x}_S))^2 - (f_I(\mathbf{x}_S))^2 \right] + \\
\sum_{j=1}^{N_\rho} \sum_{\mathbf{x}_S \in \Omega_{\rho_j}} 2R(\mathbf{x}_S) f_R(\mathbf{x}_S) f_I(\mathbf{x}_S) \Bigg\} [1 + o(1)] &= \text{imag} < \mathcal{I}, \Gamma^{D,\epsilon} \mathcal{I} > [1 + o(1)] \ .
\end{aligned}
\tag{82}
$$

3.3 Remarks on the Variational Principles

It is clear that the min-max variational principles (44) and (45) are essential in the analysis even though they do not seem to have a direct physical meaning, as in the static problem of section 2. In fact, we have

$$
(I, \Gamma^\epsilon I) = (I^*, \Gamma^\epsilon I) + 2i \ (I_I, \Gamma^\epsilon I) \ ,
\tag{83}
$$

where $\text{real}(I^*, \Gamma^\epsilon I)$ and $\text{imag}(I^*, \Gamma^\epsilon I)$ are the power dissipated into heat and the electric energy stored in the system, respectively. Nevertheless, the results (81) and (82) give an approximation of the saddle functionals in (44) and (45) in the asymptotic limit of infinitely high contrast. The current $\mathbf{j}_R + i \ \mathbf{j}_I$ that achieves the min-max in (44) and (45) is the unique solution of equations (42), (43) and, implicitly, of (35). Thus, (81) and (82) show that the current density in a high contrast medium with impedance (39) is approximately given by current flow in an $R - C$ network. We also have

$$
\begin{aligned}
(I^*, \Gamma^\epsilon I) &= \int_\Omega \left[\rho_0 e^{-S(\mathbf{x})/\epsilon} + i \ C_0 e^{-P(\mathbf{x})/\epsilon} \right] |\, \mathbf{j}(\mathbf{x}) \,|^2 \ dx \\
&= < \mathcal{I}^*, \Gamma^{D,\epsilon} \mathcal{I} > [1 + o(1)] \ .
\end{aligned}
\tag{84}
$$

We have therefore shown that the NtD map Γ^ϵ of the high contrast continuum is asymptotically equivalent to the discrete NtD map of the $R - C$ circuit. This is important for inversions, where the data provide information about the NtD map Γ^ϵ and the asymptotic theory tells us that the first step should be the identification of the asymptotic $R - C$ network.

4   Quasi-Static Approximation: Inductive Approximation

At low frequency the displacement current $i\omega\varepsilon\mathbf{E}$ in (1) can be neglected in conductive media [3, 8]. We consider the transverse electric problem in a simply connected domain $\Omega \subset \mathbb{R}^2$, where $\mathbf{H}(\mathbf{x}) = (0, 0, H(\mathbf{x}))$ and $\mathbf{E}(\mathbf{x}) = (E_1(\mathbf{x}), E_2(\mathbf{x}), 0)$. Equations (1) reduce to

$$\mathbf{E}(\mathbf{x}) = -\frac{1}{\sigma(\mathbf{x})}\nabla^\perp H(\mathbf{x}),$$

$$\nabla^\perp \left[\frac{1}{\sigma(\mathbf{x})}\nabla^\perp H(\mathbf{x}) \right] = -i\omega\mu(\mathbf{x})H(\mathbf{x}), \quad \mathbf{x} \in \Omega . \tag{85}$$

At the boundary, we specify

$$H(\mathbf{x}) = h(\mathbf{x}) = h_R(\mathbf{x}) + i\, h_I(\mathbf{x}), \quad \mathbf{x} \in \partial\Omega , \tag{86}$$

for given h_R and h_I in $H^{\frac{1}{2}}(\partial\Omega)$. We are interested in high contrast conductive media, where we model the electrical conductivity as in (5). Typically, the variations in magnitude of the magnetic permeability $\mu(\mathbf{x})$ are much smaller than the variations of σ [3]. In fact, $\mu(\mathbf{x})$ is usually assumed to be a constant μ_0, the permeability of free space [3]. We take $\mu(\mathbf{x})$ to be a bounded, continuous function with variations of order one in Ω.

We define the complex map

$$\Gamma^\epsilon h(\mathbf{x}) = \mathbf{n}(\mathbf{x}) \times \mathbf{E}(\mathbf{x}), \quad \mathbf{x} \in \Omega , \tag{87}$$

where $\mathbf{n}(\mathbf{x})$ is the outer normal to the boundary. Given that $\sigma(\mathbf{x})$ is strictly positive and uniformly bounded in Ω, equation (85) with boundary conditions (86) has a unique solution $H(\mathbf{x}) = H_R(\mathbf{x}) + i\, H_I(\mathbf{x})$ (at least in the weak sense), where H_R and H_I are in $H^1(\Omega)$. Then, the real and imaginary parts of $\Gamma^\epsilon h$ are in $H^{-\frac{1}{2}}(\partial\Omega)$ and we define the inner product

$$(h^\star, \Gamma^\epsilon h) = \int_{\partial\Omega} h^\star(s)\mathbf{e}_3 \cdot [\mathbf{n}(s) \times \mathbf{E}(s)]\, ds , \tag{88}$$

where h^* is the complex conjugate of h and $\mathbf{e}_3$ is the unit vector orthogonal to the plane of Ω. We have

$$(h^*, \Upsilon^\epsilon h) = \int_{\partial\Omega} [\mathbf{E}(s) \times h^*(s)\mathbf{e}_3] \cdot \mathbf{n}(s) ds = \int_\Omega \nabla \cdot [\mathbf{E}(\mathbf{x}) \times H^*(\mathbf{x})\mathbf{e}_3] \, d\mathbf{x} =$$

$$\int_\Omega [H^*(\mathbf{x})\mathbf{e}_3 \cdot \nabla \times \mathbf{E}(\mathbf{x}) - \mathbf{E}(\mathbf{x}) \cdot \nabla \times H^*(\mathbf{x})\mathbf{e}_3] \, d\mathbf{x} =$$

$$-\int_\Omega \left[\tfrac{1}{\sigma(\mathbf{x})} \mid \nabla^\perp H(\mathbf{x}) \mid^2 -i \, \omega\mu(\mathbf{x}) \mid H(\mathbf{x}) \mid^2 \right] d\mathbf{x} \; .$$

$$(89)$$

Thus, the real and imaginary parts of $-(h^*, \Upsilon^\epsilon h)$ give the power dissipated into heat and the magnetic energy stored in the system, respectively.

We will analyze the complex map (87) in the asymptotic limit $\epsilon \to 0$ and for that we use the min-max variational principles introduced in [8] for quadratic forms of Υ^ϵ. Let us rewrite (85) as an elliptic system of equations

$$\nabla^\perp \left[\frac{1}{\sigma_0} e^{S(\mathbf{x})/\epsilon} \nabla^\perp H_R(\mathbf{x}) \right] - \omega\mu(\mathbf{x}) H_I(\mathbf{x}) = 0,$$

$$(90)$$

$$\nabla^\perp \left[\frac{1}{\sigma_0} e^{S(\mathbf{x})/\epsilon} \nabla^\perp H_I(\mathbf{x}) \right] + \omega\mu(\mathbf{x}) H_R(\mathbf{x}) = 0 \; ,$$

with Dirichlet boundary conditions (86). Equations (90) can be viewed as Euler equations for some real valued functionals. These functionals do not have a direct physical interpretation but do lead to variational principles that characterize the inductive problem. We have

$$-\mathrm{real}(h, \Upsilon^\epsilon h)$$

$$= \min_{H_R|_{\partial\Omega}=h_R} \max_{H_I|_{\partial\Omega}=h_I} \int_\Omega \left\{ \frac{1}{\sigma_0} e^{S(\mathbf{x})/\epsilon} \left[\mid \nabla^\perp H_R(\mathbf{x}) \mid^2 - \mid \nabla^\perp H_I(\mathbf{x}) \mid^2 \right] \right.$$

$$\left. +2\omega\mu(\mathbf{x}) H_R(\mathbf{x}) H_I(\mathbf{x}) \right\} d\mathbf{x} \qquad (91)$$

$$\mathrm{imag}(h, \Upsilon^\epsilon h)$$

$$= \min_{H_R|_{\partial\Omega}=h_R} \max_{H_I|_{\partial\Omega}=h_I} \int_\Omega \left\{ \omega\mu(\mathbf{x}) [H_R(\mathbf{x})]^2 - \omega\mu(\mathbf{x}) [H_I(\mathbf{x})]^2 \right.$$

$$\left. -2 \frac{1}{\sigma_0} e^{S(\mathbf{x})/\epsilon} \nabla^\perp H_R(\mathbf{x}) \cdot \nabla^\perp H_I(\mathbf{x}) \right\} d\mathbf{x} \quad (92)$$

In the next section we present an asymptotic analysis of $(h, \Upsilon^\epsilon h)$. We show that the extremal $\nabla^\perp H_R + i \, \nabla^\perp H_I$ in Ω can be approximated by current flow through a network. We begin with a local asymptotic analysis of (90) after which we solve the global problem by making use of variational principles (91) and (92).

4.1 Local Asymptotic Analysis

We rewrite equation (85) as

$$\nabla H(\mathbf{x}) + \frac{1}{\epsilon} \nabla^\perp S(\mathbf{x}) \cdot \nabla^\perp H(\mathbf{x}) = -i\omega\mu(\mathbf{x})\sigma(\mathbf{x})H(\mathbf{x}) \qquad (93)$$

and observe that the behavior of H in Ω is dictated by the magnitude of $\mu(\mathbf{x})\sigma(\mathbf{x})$ compared to $\frac{1}{\epsilon}$. We assume that equation (93) is in dimensionless form and $\omega = O(1)$. We divide the domain of the solution into "resistive" subdomains $\Omega_{r_j}, j = 1, \ldots N_r$, where $\mu(\mathbf{x})\sigma(\mathbf{x}) \ll \frac{1}{\epsilon}$ and "conductive" subdomains $\Omega_{c_j}, j = 1, \ldots N_c$, where $\mu(\mathbf{x})\sigma(\mathbf{x}) \gg \frac{1}{\epsilon}$. They are separated by interfaces $\mathcal{L}_j, j = 1, \ldots N_{\mathcal{L}}$, where $\mu(\mathbf{x})\sigma(\mathbf{x}) = \frac{1}{\epsilon}$. We study equation (93) in each of these subdomains separately.

1. In *resistive regions* with $\mathbf{x} \in \Omega_{r_j}$ for some $j = 1, \ldots N_r$, such that $\mu(\mathbf{x})\sigma(\mathbf{x}) \ll \frac{1}{\epsilon}$, we show that the current $\nabla^\perp H$ behaves essentially like static current in a high contrast medium with conductivity (5). We construct the static network in Ω_{r_j}, as shown in section 2.2, where the nodes are the maxima of σ and the branches go through the saddle points of the conductivity. Take a ridge of maximal σ in Ω_{r_j} as shown in figure 6. We introduce the curvilinear coordinates (ξ, η), where ξ is arclength along the ridge and η is orthogonal to it. We define a vicinity of the ridge by

$$|\eta| \le \delta \ll 1 , \qquad (94)$$

where $\frac{\delta}{\sqrt{\epsilon}} \to \infty$ as $\epsilon \to 0$. In this region the scaled logarithm of σ is given by

$$S(\xi, \eta) = S(\xi, 0) + \frac{k(\xi)}{2}\eta^2 + \frac{1}{6}\frac{\partial^3 S}{\partial \eta^3}(\xi, 0)\eta^3 + \cdots , \qquad (95)$$

where $k(\xi) > 0$. Suppose that at the point ξ_M of maximal conductivity along the ridge, $\mu(\xi_M, 0)\sigma(\xi_M, 0) = 1/\epsilon^{1-\alpha} \ll 1/\epsilon$, for some $\alpha > 0$. Then,

$$\mu(\xi, 0)\sigma(\xi, 0) = \frac{1}{\epsilon^{1-\alpha}} \frac{\mu(\xi, 0)}{\mu(\xi_M, 0)} \exp\left[\frac{S(\xi, 0) - S(\xi_M, 0)}{\epsilon}\right] \le \frac{1}{\epsilon^{1-\alpha}} . \qquad (96)$$

Now we write (93) in terms of η and ξ, for points in the vicinity (94) of a ridge of maximal conductivity. Take a point $\mathbf{x}(\xi)$ along the ridge, as shown in figure 6. Let $\hat{\xi}$ and $\hat{\eta}$ denote the unit tangent and normal to the ridge, respectively. In the vicinity (94), we have

$$\mathbf{x} = \mathbf{x}(\xi) + \eta\hat{\eta} \qquad (97)$$

and so,

$$d\mathbf{x} = \left[\hat{\xi} + \eta\hat{\eta}'\right]d\xi + \hat{\eta}d\eta, \quad \text{where } \hat{\eta}' = \gamma(\xi)\hat{\xi} \text{ and } \gamma(\xi) = O(1) . \qquad (98)$$

Then,

$$\left(\frac{\partial}{\partial x}, \frac{\partial}{\partial y}\right)H \approx \frac{\partial H}{\partial \xi} + \frac{\partial H}{\partial \eta},$$

$$\Delta H \approx \frac{\partial^2 H}{\partial \eta^2} + \gamma(\xi)\frac{\partial H}{\partial \eta} + \frac{\partial^2 H}{\partial \xi^2} \qquad (99)$$

and, with scaling $\eta = \sqrt{\epsilon}z$, equation (93) becomes

$$\frac{\partial^2 H}{\partial z^2} + \sqrt{\epsilon}\gamma(\xi)\frac{\partial H}{\partial z} + \epsilon\frac{\partial^2 H}{\partial \xi^2} + k(\xi)z\frac{\partial H}{\partial z} + \left[S'(\xi,0) + \epsilon z^2 k'(\xi)\right]\frac{\partial H}{\partial \xi}$$

$$\approx \omega\epsilon^{\alpha}\frac{\mu(\xi,0)}{\mu(\xi_M,0)}e^{\frac{S(\xi,0)-S(\xi_M,0)}{\epsilon} - \frac{k(\xi)z^2}{2}}H \ . \tag{100}$$

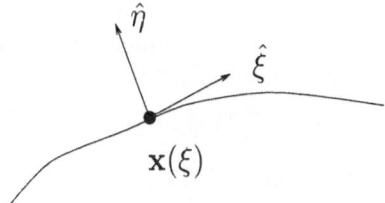

Fig. 6. Ridge of maximal $\sigma(\mathbf{x})$ and curvilinear coordinates (ξ, η).

Thus, $H(\xi, \eta) \approx H_0(\xi, \eta)$, where H_0 solves

$$\frac{\partial^2 H_0}{\partial z^2} + k(\xi)z\frac{\partial H_0}{\partial z} + S'(\xi,0)\frac{\partial H_0}{\partial \xi} = 0. \tag{101}$$

This is, to leading order, the static elliptic equation

$$\nabla^{\perp}\left\{\frac{1}{\sigma_0}e^{S(\xi,\eta)/\epsilon}\nabla^{\perp}\left[H_0(\xi, \frac{\eta}{\sqrt{\epsilon}}) + O(\sqrt{\epsilon})\right]\right\} = 0 \ , \tag{102}$$

studied in section 2.3. As we move away from the ridge of maximal conductivity, the coefficient of H on the right hand side of equation (93) becomes

$$\mu(\mathbf{x})\sigma(\mathbf{x}) = \frac{1}{\epsilon^{1-\alpha}}\frac{\mu(\mathbf{x})}{\mu(\xi_M,0)}\exp\left[\frac{S(\mathbf{x}) - S(\xi_M,0)}{\epsilon}\right]$$

$$= O\left(\frac{1}{\epsilon^{1-\alpha}}\exp\left[\frac{S(\mathbf{x}) - S(\xi_M,0)}{\epsilon}\right]\right) \ . \tag{103}$$

Here, $S(\mathbf{x}) < S(\xi_M,0)$ and $\sigma(\mathbf{x})$ is much smaller than $\sigma(\xi_M,0)$, the maximal conductivity along the ridge of maximal σ. Indeed, by assumption $S(\mathbf{x})$ is a smooth function with derivatives of order one, so outside the vicinity (94) of the ridge of maximal σ we have $|S(\mathbf{x}) - S(\xi_M,0)| \geq O(\sqrt{\epsilon})$ and $\mu(\mathbf{x})\sigma(\mathbf{x}) \ll 1$. Thus, $H(\mathbf{x}) \approx \tilde{H}(\mathbf{x})$, which is the solution of static equation

$$\nabla^{\perp}\left[\frac{1}{\sigma_0}e^{S(\mathbf{x})/\epsilon}\nabla^{\perp}\tilde{H}(\mathbf{x})\right] = 0 \ .$$

As shown in section 2.3, $\nabla^\perp H(\mathbf{x})$ for $\mathbf{x} \in \Omega_{r_j}, j = 1, \ldots N_r$ can be approximated by current flowing through a resistor network with its topology defined uniquely by the ridges of maximal σ contained in Ω_{r_j}.

2. In *conductive regions* $\mathbf{x} \in \Omega_{c_j}$ for some $j = 1, \ldots N_c$, so that $\mu(\mathbf{x})\sigma(\mathbf{x}) \gg \frac{1}{\epsilon}$, we see from equation (93), that $H \approx 0$ or, equivalently, the highly conductive regions Ω_{c_j} expell the magnetic and electric fields. This is a well known phenomenon, found in the literature under such names as skin depth penetration and the Meissner effect [31, 49]. Of course, the magnetic field H in Ω_{c_j} has to match the field in the nearby resistive regions Ω_{r_j}. The matching is done in the vicinity of interfaces $\mathcal{L}_j, j = 1, \ldots N_{\mathcal{L}}$, where $\mu(\mathbf{x})\sigma(\mathbf{x}) = \frac{1}{\epsilon}$.

3. At *interfaces* $\mathcal{L}_j$ we introduce curvilinear coordinates (ξ, η), as shown in figure 7. We define a neighborhood of $\mathcal{L}_j$ by

$$| \eta | \le \delta \ll 1 , \tag{104}$$

where $\frac{\delta}{\epsilon} \to \infty$ as $\epsilon \to 0$. Here, the scaled logarithm of σ is given by

$$S(\xi, \eta) = S(\xi, 0) - \lambda(\xi)\eta + \frac{1}{2}\frac{\partial^2 S}{\partial \eta^2}(\xi, 0)\eta^2 + \ldots , \tag{105}$$

where $\lambda(\xi) > 0$, so that σ increases with η. We also have that

$$\mu(\xi, 0)\frac{1}{\sigma_0}e^{S(\xi,0)/\epsilon} = \frac{1}{\epsilon} . \tag{106}$$

After a calculation similar to that in part 1 above, leading to equations (99), we have

$$\frac{\partial^2 H}{\partial \eta^2} - \gamma(\xi)\frac{\partial H}{\partial \eta} + \frac{\partial^2 H}{\partial \xi^2} - \lambda(\xi)\epsilon\frac{\partial H}{\partial \eta} + \frac{1}{\epsilon}\left[S'(\xi, 0) - \lambda'(\xi)\eta\right]\frac{\partial H}{\partial \xi} + i\frac{\omega}{\epsilon}e^{\lambda(\xi)\eta/\epsilon}H = 0 . \tag{107}$$

We introduce the stretching $\eta = \sqrt{\epsilon}z$ and look for a solution H of the form

$$H(\xi, z) \approx \exp\left[\frac{P(\xi, z)}{\sqrt{\epsilon}}\right] . \tag{108}$$

To leading order $\frac{1}{\epsilon^2}$ we have

$$\left(\frac{\partial P}{\partial z}\right)^2 - \lambda(\xi)\frac{\partial P}{\partial z} + i\,\omega\epsilon e^{\lambda(\xi)z/\sqrt{\epsilon}} = 0 \tag{109}$$

and

$$P(\xi, z) = P(\xi, 0) + \frac{1}{2}\lambda(\xi)z \pm \frac{1}{2}\lambda(\xi)\int_0^z \sqrt{1 - 4i\frac{\omega\epsilon}{\lambda^2(\xi)}e^{\lambda(\xi)t/\sqrt{\epsilon}}}dt . \tag{110}$$

From (110) we note that $H(\xi, z)$ to be bounded we must discard the solution with the plus sign in front of the integral. Thus, from (108) and (110) we have

$$H(\xi, \eta) \approx H(\xi, 0)\exp\left[\frac{\lambda(\xi)\eta}{2\epsilon} - \frac{\lambda(\xi)}{2}\int_0^{\eta/\epsilon}\sqrt{1 - 4i\frac{\omega\epsilon}{\lambda^2(\xi)}e^{\lambda(\xi)t}}dt\right] . \tag{111}$$

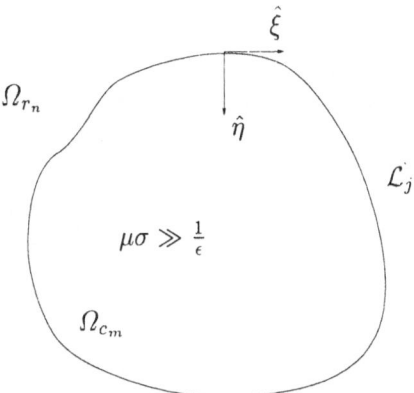

Fig. 7. Highly conductive region Ω_{c_m} separated from the resistive region Ω_{r_n} by interface $\mathcal{L}_j$.

From (111) we see that for $\eta < O(\epsilon)$, $H(\xi, \eta) \approx H(\xi, 0)$ and the magnetic field matches the field in the adjacent resistive region. However, for $\eta > O(\epsilon)$ (111) decays exponentially towards zero, the value of the magnetic field inside the conductive region. Because of the exponential decay of (111) there is a strong current $\nabla^\perp H(\xi, \eta) \approx -\frac{\partial H}{\partial \eta} \hat{\xi}$, along the interface $\mathcal{L}_j$.

4.2 Global Asymptotic Approximation

The local analysis of section 4.1 shows that $\nabla^\perp H$ is channeled in Ω either along ridges of maximal σ or along interfaces $\mathcal{L}_j$, $j = 1, \ldots N_{\mathcal{L}}$, that separate resistive subdomains Ω_{r_j}, $j = 1, \ldots N_r$ from highly conductive ones, Ω_{c_j}, $j = 1, \ldots N_c$. In fact, $\nabla^\perp H$ is like current through a network. First we establish the topology of the network, as follows.

We construct the static resistor network in the whole domain Ω as explained in section 2.2. Then we identify the highly conductive regions Ω_{c_j}, $j = 1, \ldots N_c$. We discard the pieces of the static network that fall inside Ω_{c_j} and we connect the loose ends through "wires" along the interfaces $\mathcal{L}_j$, $j = 1, \ldots N_{\mathcal{L}}$. The topology of the network is now uniquely defined for any σ of the form (5).

In figures 8 and 9 we illustrate the construction of the asymptotic network for an example of conductivity $\sigma_0 e^{-S(\mathbf{x})/\epsilon}$ with five maxima denoted by $\circ$, four minima denoted by $\bullet$ and saddle points counted from 1 to 4, respectively. Inside the dotted closed curve in figure 8 we assume that $\mu\sigma \gg \frac{1}{\epsilon}$. In the remainder of the domain $\mu\sigma \ll \frac{1}{\epsilon}$. The global network is shown in figure 9, where the assignment of impedances $Z_1, \ldots Z_8$ is discussed in the next section. The branches (e, f), (f, g), (g, h) and (h, e) are pieces of the dotted interface in figure 8 where $\mu\sigma = \frac{1}{\epsilon}$.

We will now analyze the quadratic forms $(h, \Upsilon^\epsilon h)$ up to $o(1)$ by making use of the variational principles (91) and (92). For simplicity we assume that h_R and h_I in (86) are constant along pieces of $\partial\Omega$ that lie between adjacent peripheral

224

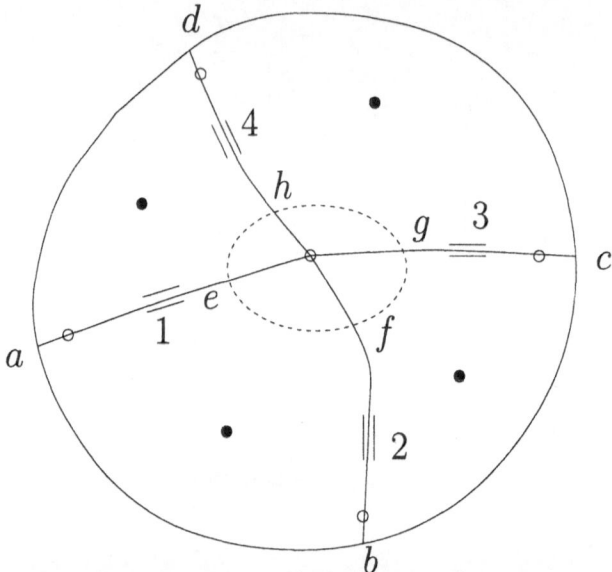

Fig. 8. Example of $\sigma(\mathbf{x}) = \sigma_0 e^{-S(\mathbf{x})/\epsilon}$. The topology of the static resistor network is shown with full line. Inside the dotted curve, $\mu\sigma \gg \frac{1}{\epsilon}$. In the remainder of the domain, $\mu\sigma \ll \frac{1}{\epsilon}$.

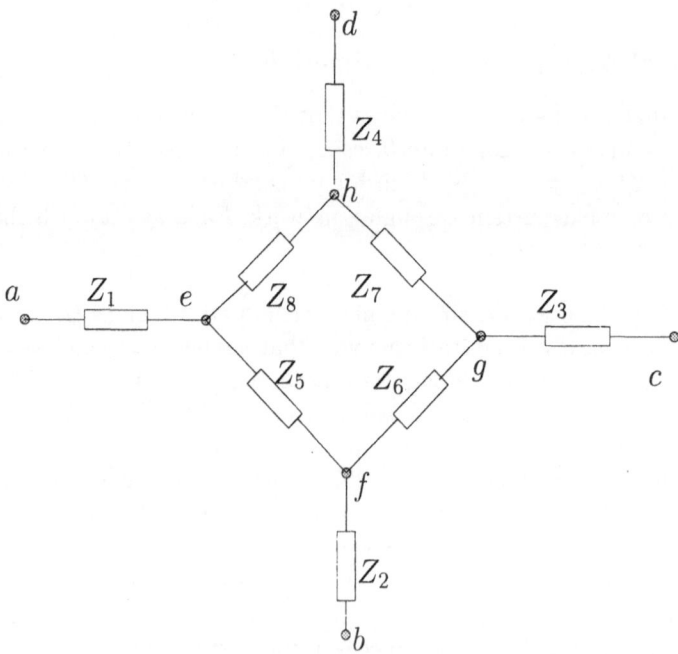

Fig. 9. A possible global network

nodes of the network in Ω. Equivalently, in the direction normal to the boundary, currents $\nabla^\perp H_R$ and $\nabla^\perp H_I$ are concentrated around the peripheral nodes of the network. In a general situation we have h_R and h_I arbitrary functions in $H^{\frac{1}{2}}(\partial\Omega)$ and, in thin boundary layers, $\nabla^\perp H_R$ and $\nabla^\perp H_I$ adjust to currents concentrated through the asymptotic network. The analysis is identical to that of section 2.3 and [10] and shall not be repeated here.

We begin with the calculation of an upper bound on $-\text{real}(h, \Upsilon^\epsilon h)$, obtained with a trial field H_R chosen as follows. For $\mathbf{x} \in \Omega_{r_j}$ for some $j = 1, \dots N_r$, such that $\mu(\mathbf{x})\sigma(\mathbf{x}) \ll \frac{1}{\epsilon}$ we take H_R to be a constant away from ridges of maximal σ. In the vicinity of a ridge of maximal σ consider coordinates (ξ, η) where ξ is arc length along the ridge. Here, the scaled logarithm of σ is given by

$$S(\xi, \eta) = S(\xi, 0) + \frac{k(\xi)}{2}\eta^2 + \frac{1}{6}\frac{\partial^3 S}{\partial \eta^3}(\xi, 0)\eta^3 + \dots , \tag{112}$$

where $k(\xi) > 0$ and H_R is chosen as

$$H_R(\xi, \eta) = -\frac{f_R}{2}\,\text{erf}\left(\frac{\eta}{\sqrt{\frac{2\epsilon}{k(\xi)}}}\right) + \text{constant} , \tag{113}$$

where f_R is the real flux along the ridge. By conservation of current, f_R is a constant along the ridge or changes only near maxima of σ. For $\mathbf{x} \in \Omega_{c_j}$ for some $j = 1, \dots N_c$, such that $\mu(\mathbf{x})\sigma(\mathbf{x}) \gg \frac{1}{\epsilon}$, we take $H_R = 0$.

For $\mathbf{x} \in \mathcal{L}_j$, $j = 1, \dots N_\mathcal{L}$, so that $\mu(\mathbf{x})\sigma(\mathbf{x}) = \frac{1}{\epsilon}$ we introduce curvilinear coordinates (ξ, η) as in section 4.1, part 2 and a vicinity of $\mathcal{L}_j$ as in (104). In this vicinity the scaled logarithm of σ is given by

$$S(\xi, \eta) = S(\xi, 0) - \lambda(\xi)\eta + \frac{1}{2}\frac{\partial^2 S}{\partial \eta^2}(\xi, 0)\eta^2 + \dots \tag{114}$$

and we take $H_R(\xi, \eta)$ to be

$$H_R(\xi, \eta) = \text{real}\left\{ \left[H_R(\xi, 0) + i\tilde{H}_I(\xi, 0)\right] \right.$$
$$\left. \exp\left[\frac{\lambda(\xi)\eta}{2\epsilon} - \frac{\lambda(\xi)}{2}\int_0^{\eta/\epsilon}\sqrt{1 - 4i\frac{\omega\epsilon}{\lambda^2(\xi)}e^{\lambda(\xi)t}}\,dt\right]\right\} \tag{115}$$

Here, $\tilde{H}_I = O(1)$, is a function of arclength along the interface, to be specified later. Note that for any $\tilde{H}_I$, when $\eta \leq O(\epsilon)$ equation (115) gives $H_R(\xi, \eta) \approx H_R(\xi, 0)$.

For $\eta \gg \epsilon$, $H_R(\xi, \eta) \approx 0$. Thus, the trial field H_R is continuous, as required, throughout Ω for an arbitrary $\tilde{H}_I$. In figure 10 we illustrate our choice of trial field H_R, for the example of figure 8.

We now write the integral in (91) as a sum of integrals over the resistive regions Ω_{r_j}, $j = 1, \dots N_r$, the conductive regions Ω_{c_j}, $j = 1, \dots N_c$ and the vicinities of interfaces $\mathcal{L}_j$, $j = 1, \dots N_\mathcal{L}$. We obtain an upper bound on $-\text{real}(h, \Upsilon^\epsilon h)$

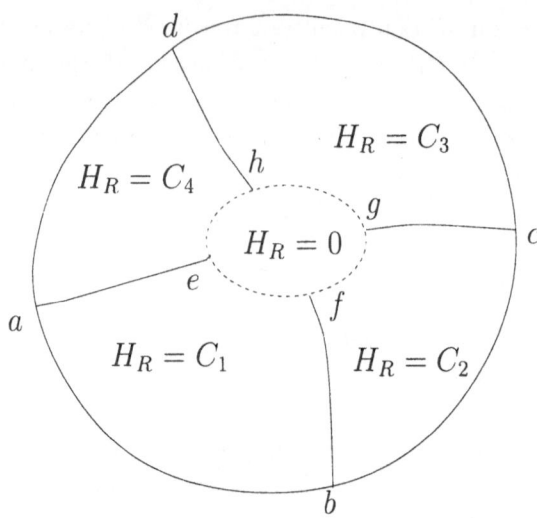

Fig. 10. Trial field H_R for the upper bound on $-\text{real}(h, \Upsilon^\epsilon h)$. In the resistive region, where $\mu\sigma \ll \frac{1}{\epsilon}$ and away from ridges of maximal σ we take H_R to be a constant. Across ridges of maximal σ, H_R is chosen as in (113). In the conductive region, where $\mu\sigma \gg \frac{1}{\epsilon}$, $H_R = 0$.

by using the trial field H_R defined above and maximizing the integral over each subdomain in Ω.

1. In resistive regions $\mathbf{x} \in \Omega_{r_j}$ for some $j = 1, \ldots N_r$, so that $\mu(\mathbf{x})\sigma(\mathbf{x}) \ll \frac{1}{\epsilon}$, we wish to calculate

$$\max_{H_I} \int_{\Omega_{r_j}} \left\{ \frac{1}{\sigma_0} e^{S(\mathbf{x})/\epsilon} \left[|\nabla^\perp H_R(\mathbf{x})|^2 - |\nabla^\perp H_I(\mathbf{x})|^2 \right] + 2\omega\mu(\mathbf{x})H_R(\mathbf{x})H_I(\mathbf{x}) \right\} .$$

$$(116)$$

The maximum is achieved by fields H_I that satisfy

$$\Delta H_I(\mathbf{x}) + \frac{1}{\epsilon}\nabla^\perp S(\mathbf{x}) \cdot \nabla^\perp H_I(\mathbf{x}) = -\omega\mu(\mathbf{x})\sigma(\mathbf{x})H_R(\mathbf{x}) . \qquad (117)$$

Clearly, this is very similar to equation (93), considered in section 4.1, part 1. We can state the result: $H_I(\mathbf{x}) \approx \tilde{H}_I(\mathbf{x})$, where

$$\nabla^\perp \left[\frac{1}{\sigma_0} e^{S(\mathbf{x})/\epsilon} \nabla^\perp \tilde{H}_I(\mathbf{x}) \right] = 0 . \qquad (118)$$

Suppose that the ridges of maximal σ divide Ω_{r_j} into regions $\mathcal{D}_l, l = 1, \ldots n_{r_j}$. In such a region, say $\mathcal{D}_l$, our trial field is a constant, say $H_R = C_l$. From (118) and the results of sections 2.2 and 2.3, $H_I \approx \tilde{H}_I = D_l$, where D_l is a constant. Near a ridge of maximal σ, both H_R and $\tilde{H}_I$ change very fast and we have strong currents $\nabla^\perp H_R$ and $\nabla^\perp \tilde{H}_I$ of order $1/\sqrt{\epsilon}$ along the ridge. From section 2.2 we also have

$$\int_{\Omega_{r_j}} \frac{1}{\sigma_0} e^{S(\mathbf{x})/\epsilon} \mid \nabla^\perp H_R(\mathbf{x}) \mid^2 dx = \sum_{\mathbf{x}_S \in \Omega_{r_j}} R(\mathbf{x}_S) \left[f_R(\mathbf{x}_S)\right]^2 [1 + o(1)] \qquad (119)$$

and

$$\int_{\Omega_{r_j}} \frac{1}{\sigma_0} e^{S(\mathbf{x})/\epsilon} \mid \nabla^\perp \tilde{H}_I(\mathbf{x}) \mid^2 dx = \sum_{\mathbf{x}_S \in \Omega_{r_j}} R(\mathbf{x}_S) \left[f_I(\mathbf{x}_S)\right]^2 [1 + o(1)] \quad , \qquad (120)$$

where $R(\mathbf{x}_S)$ is the effective resistance of the saddle point $\mathbf{x}_S$ and $f_R(\mathbf{x}_S)$ and $f_I(\mathbf{x}_S)$ are the real and imaginary fluxes through it, respectively. Note that f_R and f_I are completely determined by C_l and D_l, the values of H_R and $\tilde{H}_I$ in the regions $\mathcal{D}_l$, $l = 1, \ldots n_{r_j}$, of diffuse flow.

Collecting the contributions from the region Ω_{r_j} to the upper bound for $-\text{real}(h, \Upsilon^\epsilon h)$ it becomes

$$\max_{D_l} \left\{ \sum_{\mathbf{x}_S \in \Omega_{r_j}} R(\mathbf{x}_S) \left[(f_R(\mathbf{x}_S))^2 - (f_I(\mathbf{x}_S))^2\right] \right. \qquad (121)$$

$$\left. +2\omega \sum_{l=1}^{n_{r_j}} C_l D_l \int_{\mathcal{D}_l} \mu(\mathbf{x}) dx \right\} [1 + o(1)] \quad .$$

2. In transition regions $\mathbf{x} \in \mathcal{L}_j$ for some $j = 1, \ldots N_r$, such that $\mu(\mathbf{x})\sigma(\mathbf{x}) = \frac{1}{\epsilon}$ we calculate

$$\max_{H_I} \int d\xi \int_{-\delta}^{\delta} d\eta \left\{ \frac{1}{\sigma_0} e^{S(\xi,\eta)/\epsilon} \left[\mid \nabla^\perp H_R(\xi,\eta) \mid^2 - \mid \nabla^\perp H_I(\xi,\eta) \mid^2\right] \right.$$

$$\left. + 2\omega\mu(\xi,\eta) H_R(\xi,\eta) H_I(\xi,\eta) \right\} \quad , \qquad (122)$$

where we integrate in the vicinity (104) defined in section 4.1, part 2, such that $\omega\mu(\xi,\eta) \approx \mu(\xi,0)$. The maximum is achieved by fields H_I that satisfy

$$\Delta H_I(\xi,\eta) + \frac{1}{\epsilon} \nabla^\perp S(\xi,\eta) \cdot \nabla^\perp H_I(\xi,\eta)$$

$$\approx -\omega\mu(\xi,0)\sigma(\xi,\eta)\text{real}\left\{ \left[H_R(\xi,0) + i\tilde{H}_I(\xi,0)\right] \times \right.$$

$$\left. \exp\left[\frac{\lambda(\xi)\eta}{2\epsilon} - \frac{\lambda(\xi)}{2} \int_0^{\frac{\eta}{\epsilon}} \sqrt{1 - 4i\frac{\omega\epsilon}{\lambda^2(\xi)} e^{\lambda(\xi)t}} dt\right] \right\} \qquad (123)$$

Recall that any $\tilde{H}_I(\xi,0)$ in (115) gives an acceptable trial field H_R. We are therefore free to choose $\tilde{H}_I$ as the solution of equation (118) and since (123) has been solved in section 4.1, part 2, we have

$$H_I(\xi,\eta) \approx \text{imag}\left\{ \left[H_R(\xi,0) + i\tilde{H}_I(\xi,0)\right] \right.$$

$$\left. \exp\left[\frac{\lambda(\xi)\eta}{2\epsilon} - \frac{\lambda(\xi)}{2} \int_0^{\frac{\eta}{\epsilon}} \sqrt{1 - 4i\frac{\omega\epsilon}{\lambda^2(\xi)} e^{\lambda(\xi)t}} dt\right] \right\} \qquad (124)$$

Note that, for $\eta \lesssim O(\epsilon)$, we have the correct matching of the imaginary field, since $H_I(\xi, \eta) \approx \tilde{H}_I(\xi, 0)$. For $\eta \gg O(\epsilon)$, $H_I \approx 0$.

Because of the rapid variation with η of both H_R and H_I we have a strong current, of order $1/\sqrt{\epsilon}$, concentrated near the interface $\mathcal{L}_j$ and given by

$$\nabla^\perp \left[H_R(\xi, \eta) + i\, H_I(\xi, \eta) \right]$$

$$\approx -\frac{\partial}{\partial \eta} \left[H_R(\xi, \eta) + i\, H_I(\xi, \eta) \right] \hat{\xi}$$

$$= \left[H_R(\xi, 0) + i\tilde{H}_I(\xi, 0) \right] \frac{\lambda(\xi)}{\epsilon} \left[\sqrt{1 - 4i\, \frac{\omega\epsilon}{\lambda^2(\xi)} e^{\lambda(\xi)\eta/\epsilon}} - 1 \right]$$

$$\times \exp \left[\frac{\lambda(\xi)\eta}{2\epsilon} - \frac{\lambda(\xi)}{2} \int_0^{\eta/\epsilon} \sqrt{1 - 4i\, \frac{\omega\epsilon}{\lambda^2(\xi)} e^{\lambda(\xi)t}}\, dt \right] \hat{\xi} \ . \tag{125}$$

Here $\hat{\xi}$ is the unit vector tangential to $\mathcal{L}_j$. Thus, (122) becomes

$$\text{real} \int d\xi \frac{\lambda^2(\xi) \left[H_R(\xi, 0) + i\, \tilde{H}_I(\xi, 0) \right]^2}{4\epsilon^2 \sigma(\xi, 0)} \int_{-\delta}^{\delta} d\eta \left[\sqrt{1 - 4i\, \frac{\omega\epsilon}{\lambda^2(\xi)} e^{\frac{\lambda(\xi)\eta}{\epsilon}}} - 1 \right]^2$$

$$\cdot e^{-\lambda(\xi) \int_0^{\frac{\eta}{\epsilon}} \sqrt{1 - 4i\frac{\omega\epsilon}{\lambda^2(\xi)} e^{\lambda(\xi)t}}\, dt}$$

$$+ \text{imag} \int d\xi \omega \mu(\xi, 0) \left[H_R(\xi, 0) + i\, \tilde{H}_I(\xi, 0) \right]^2 \int_{-\delta}^{\delta} d\eta$$

$$\cdot e^{\frac{\lambda(\xi)\eta}{\epsilon} - \lambda(\xi) \int_0^{\frac{\eta}{\epsilon}} \sqrt{1 - 4i\frac{\omega\epsilon}{\lambda^2(\xi)} e^{\lambda(\xi)t}}\, dt} \ . \tag{126}$$

Note that the last term in (126) is of order $\delta\omega \int \mu(\xi, 0)d\xi$ and is much smaller than (122). We are left with

$$\text{real} \int d\xi \frac{\lambda^2(\xi) \left[H_R(\xi, 0) + i\, \tilde{H}_I(\xi, 0) \right]^2}{4\epsilon\sigma(\xi, 0)} \mathcal{F}(\xi) \ , \tag{127}$$

where with the change of variable $s = \frac{\eta}{\epsilon}$, $|\eta| < \delta$ and $\frac{\delta}{\epsilon} \to \infty$, we have

$$\mathcal{F}(\xi) = \int_{-\infty}^{\infty} ds \left[\sqrt{1 - 4i\, \frac{\omega\epsilon}{\lambda^2(\xi)} e^{\lambda(\xi)s}} - 1 \right]^2 e^{-\lambda(\xi) \int_0^s \sqrt{1 - 4i\frac{\omega\epsilon}{\lambda^2(\xi)} e^{\lambda(\xi)t}}\, dt} \ . \tag{128}$$

The integrand in (128) is a smooth complex function with compact support and $\mathcal{F}(\xi)$ is an absolutely convergent integral. We cannot evaluate explicitly $\mathcal{F}(\xi)$ but we can easily do so numerically. It has the form

$$\mathcal{F}(\xi) = -\frac{\epsilon^2\omega^2}{\lambda^5(\xi)} \mathcal{G}_1(\xi) - i\frac{\epsilon\omega}{\lambda^3(\xi)} \mathcal{G}_2(\xi) \ , \tag{129}$$

where $\mathcal{G}_1(\xi)$ and $\mathcal{G}_2(\xi)$ are two positive functions of order one that are independent of ϵ, ω and $\lambda(\xi)$. Thus, the contribution of the interface $\mathcal{L}_j$ to the upper bound on $-\text{real}(h, \Upsilon^\epsilon h)$ is

$$2\omega \int H_R(\xi, 0) \tilde{H}_I(\xi, 0) \frac{\mathcal{G}_2(\xi)}{\lambda(\xi)\sigma(\xi, 0)} d\xi . \tag{130}$$

Note also that along $\mathcal{L}_j$, $H_R(\xi, 0)$ and $\tilde{H}_I(\xi, 0)$ are approximately constant between two adjacent points of intersection of the interface and the ridges of maximal σ (see for example figure 10).

The upper bound for $-\text{real}(h, \Upsilon^\epsilon h)$, follows from (122) and (130) and is

$$-\text{real}(h, \Upsilon^\epsilon h) \le \min_{C_l} \max_{D_l} \sum_{j=1}^{N_r} \left\{ \sum_{x_S \in \Omega_{r_j}} R(x_S) \left[(f_R(x_S))^2 - (f_I(x_S))^2 \right] \right.$$

$$\left. + 2\omega \sum_{l=1}^{n_{r_j}} C_l D_l L_l \right\} [1 + o(1)] , \tag{131}$$

where L_l is the inductance associated with region $\mathcal{D}_l$, $l = 1, \dots n_{r_j}$, of diffuse flow. Suppose that the boundary of $\mathcal{D}_l$ consists only of ridges of maximal σ lying in the resistive region Ω_{r_j}. Then,

$$L_l = \int_{\mathcal{D}_l} \mu(x) dx . \tag{132}$$

If part of the boundary of $\mathcal{D}_l$ consists of a piece of interface $\mathcal{L}_k$ then the inductance L_l is given by

$$L_l = \int_{\mathcal{D}_l} \mu(x) dx + 2\omega \int_{\xi_1}^{\xi_2} \frac{\mathcal{G}_2(\xi)}{\lambda(\xi)\sigma(\xi, 0)} d\xi ,$$

where ξ_1 and ξ_2 are the endpoints of the piece of $\mathcal{L}_k$ that bounds $\mathcal{D}_l$. Note that here, $\frac{1}{\sigma(\xi, 0)} = \epsilon\mu(\xi, 0)$ and since μ has variations of order one equation (132) still holds, approximately.

We conclude our calculation of the upper bound (131) with the observation that we take the min-max over the real and imaginary constant values of the magnetic field, in regions $\mathcal{D}_l$ of diffuse flow. This min-max is constrained by the boundary conditions imposed on H_R and H_I, so that C_l and D_l are prescribed in regions $\mathcal{D}_l$ that touch $\partial\Omega$. Then, from the first order optimality condition in (131), we obtain a full-rank, linear system of equations which determines uniquely the upper bound for $-\text{real}(h, \Upsilon^\epsilon h)$.

The calculation of a lower bound for $-\text{real}(h, \Upsilon^\epsilon h)$ starts by choosing and imaginary trial field H_I. The technique is very similar to the one above and the resulting lower bound matches asymptotically the upper bound (131) so that

$$-\text{real}(h, \Upsilon^\epsilon h) = \min_{C_l} \max_{D_l} \sum_{j=1}^{N_r} \left\{ \sum_{x_S \in \Omega_{r_j}} R(x_S) \left[(f_R(x_S))^2 - (f_I(x_S))^2 \right] \right.$$

$$+2\omega \sum_{l=1}^{n_{r_j}} C_l D_l L_l \Bigg\} [1 + o(1)] \ . \tag{133}$$

A similar calculation gives

$$\mathrm{imag}(h, \Upsilon^\epsilon h) = \min_{C_l} \max_{D_l} \sum_{j=1}^{N_r} \Bigg\{ \omega \sum_{l=1}^{n_{r_j}} \left(C_l^2 - D_l^2 \right) L_l$$

$$- 2 \sum_{\mathbf{x}_S \in \Omega_{r_j}} R(\mathbf{x}_S) f_R(\mathbf{x}_S) f_I(\mathbf{x}_S) \Bigg\} [1 + o(1)] \ . \tag{134}$$

We have thus shown that the complex map Υ^ϵ has a discrete approximation in the asymptotic limit $\epsilon \to 0$, that comes from the strong channeling of currents in the domain of the solution. The topology of the flow is strongly dependent on frequency. For low ω, current is essentially like the static one and is concentrated along paths of maximal conductivity. However, as ω increases, the topology of the flow can change dramatically because of the development of regions such as Ω_{c_j}, $j = 1, \ldots N_C$, that expel the electric and magnetic fields.

4.3 Discussion

The results of section 4.2 lead us to expect a network approximation of quasistatic transport in high contrast media. The topology of the network is clear at first, as explained above. However, the problem of associating to each branch in the network an impedance is more complicated. To understand the difficulty, consider the example in figure 10. The asymptotic approximation (133) is

$$-\mathrm{real}(h, \Upsilon^\epsilon h) \approx R_1(C_1 - C_4)^2 - R_1(D_1 - D_4)^2 + R_2(C_1 - C_2)^2$$

$$- R_2(D_1 - D_2)^2 + R_3(C_2 - C_3)^2 - R_3(D_2 - D_3)^2$$

$$+ R_4(C_3 - C_4)^2 - R_4(D_3 - D_4)^2 + 2\omega \sum_{l=1}^{4} C_l D_l L_l \ , \tag{135}$$

and there is no min-max to take since all C_j and D_j, $j = 1, \ldots 4$, are fixed by the boundary conditions. In Ω we have strong currents concentrated along ridges of maximal σ that pass through the four saddle points and along the interface that surrounds the conducting region in the middle of the domain. Through the saddle points $1, 2, 3$ and 4 we have currents $(C_1 - C_4) + i(D_1 - D_4)$, $(C_2 - C_1) + i(D_2 - D_1)$, $(C_3 - C_2) + i(D_3 - D_2)$ and $(C_4 - C_3) + i(D_4 - D_3)$, respectively. Through the branches (e, h), (h, g), (g, f) and (f, e) we have counterclockwise currents $C_4 + iD_4$, $C_3 + iD_3$, $C_2 + iD_2$ and $C_1 + iD_1$, respectively. Thus, the conservation of current law (Kirchhoff's nodal law) is satisfied for arbitrary C_j and D_j, $j = 1, \ldots 4$. Nevertheless, in order to give to the approximation (135) a network interpretation, we must also have Kirchhoff's loop laws satisfied and they are not. Therefore we do not have a planar *network* approximation but we do have a finite dimensional approximation for the boundary impedance operator Υ^ϵ. In general, the discrete approximation may or may not correspond to a unique planar network. The fact that we do not have a network is not, however, essential in inversion. What really counts is that the map Υ^ϵ has a unique, discrete asymptotic approximation, as given by (133) and (134). The asymptotic approximation of the magnetic field H and current $\nabla^\perp H$ in Ω are also uniquely determined, as explained in section 4.2. Finally, we have strong channeling of currents which means that boundary data available in inversion contain information about a few important features of the conductivity, such as saddle points and distribution of ridges of maximal σ in Ω. The issue of possible network forms of the discrete approximation of Υ^ϵ needs further study.

Acknowledgement

Work supported by AFOSR grant F49620-98-1-0211 and by NSF grant DMS-9971972

References

1. Allers, A., Santosa, F.: Stability and resolution analysis of a linearized problem in electrical impedance tomography, Inverse Problems **7** (1990) 515–533.
2. Alessandrini, G.: Stable determination of conductivity by boundary measurements, App. Anal. **27** (1988) 153–172.
3. Alumbaugh, D.L., Morrison, H. F.: Electromagnetic conductivity imaging with an iterative Born inversion, IEEE Transactions on Geosciences and Remote Sensing, **31** (1993) 758–763.
4. Alumbaugh, L., Morrison, H. F.: Monitoring subsurface changes over time with cross-well electromagnetic tomography, Geophysical Prospecting, **43** (1995) 873–902.
5. Bender, C. M., Orszag, S. A.: Advanced Mathematical Methods for Scientists and Engineers, McGraw-Hill, Inc., New York, 1978.
6. Berryman, J. G., Kohn, R. V.: Variational constraints for electrical impedance tomography, Phys. Rev. Lett. **65** (1990) 325–328.
7. Berryman, J. G.: Convexity properties of inverse problems with variational constraints, J. Franklin Inst. **328** (1991) 1–13.
8. Borcea, L.: Asymptotic Analysis of Quasistatic Transport in High Contrast Conductive Media, SIAM J. Appl. Math. **59** (1999) 597–639.
9. Borcea, L., Berryman, J. G., Papanicolaou, G. C.: High Contrast Impedance Tomography, Inverse Problems **12** (1996) 935–958.
10. Borcea, L., Berryman, J. G., Papanicolaou, G. C.: Matching pursuit for imaging high contrast conductive media, Inverse Problems **15** (1999) 811–849.
11. Borcea, L., Papanicolaou, G. C.: Network approximation for transport properties of high contrast materials, SIAM J. Appl. Math **58** (1998) 501–539.
12. Borcea L., Papanicolaou, G.C.: A Hybrid Numerical Method for High Contrast Conductivity Problems, Journal of Computational and Applied Mathematics **87** (1997) 61–78.

13. Calderón, A. P.: On an inverse boundary value problem, Seminar on Numerical Analysis and its Applications to Continuum Physics, Soc. Brasileira de Matématicá Rínftyo de Janeiro (1980) 65–73.

14. Cheney, M., Isaacson, D.: Distinguishability in impedance imaging, IEEE Trans. Biomed. Eng. **39** (1992) 852–860.

15. Cherkaeva, E., Tripp, A.: Inverse conductivity problem for noisy measurements, Inverse Problems **12** (1996) 869–883.

16. Cherkaev, A. V., Gibiansky, L. V.: Variational principles for complex conductivity, viscoelasticity, and similar problems in media with complex moduli, J. Math. Phys. **35** (1994) 127–145.

17. Courant, R., Hilbert, D.: Methods of Mathematical Physics, vol. I, Wiley, New York (1953) pp. 240–242 (Dirichlet's principle) and pp. 267–268 (Thomson's principle).

18. Curtis, E. B., Morrow, J. A.: Determining the resistors in a network, SIAM J. Appl. Math. **50** (1990) 918–930.

19. Curtis, E.B., Ingerman, D., Morrow, J. A.: Circular planar graphs and resistor networks, Linear Algebra Appl. 283 (1998) 115–150.

20. Dines, K. A., Lytle, R. J.: Analysis of electrical conductivity imaging, Geophysics **46** (1981) 1025–1036.

21. Dobson, D. C., Santosa, F.: Resolution and stability analysis of an inverse problem in electrical impedance tomography: dependence on the input current patterns, SIAM J. Appl. Math. **54** (1994) 1542–1560.

22. Fannjiang, A., Papanicolaou, G. C.: Convection enhanced diffusion for periodic flows, SIAM J. Appl. Math. 54 (1994) 333–408.

23. Fucks, L. F., Cheney, M., Isaacson, D., Gisser, D. G., Newell, J. C.: Detection and imaging of electric conductivity and permittivity at low frequency, IEEE Trans. Biomed. Eng. **38** (1991) 1106-1110.

24. Grünbaum, F. A., Zubelli, J. P.: Diffuse tomography: computational aspects of the isotropic case, Inverse Problems **8** (1992) 421–433.

25. Habashy, T. M., Groom, R. W., Spies, B. R.: Beyond the Born and Rytov approximations: nonlinear approach to electromagnetic scattering, Journal of Geophysical Research **98** (1993) 1759-1775.

26. Isaacson, D.: Distinguishability of conductivities by electric current computed tomography, IEEE Trans. Med. Imag. **MI-5** (1986) 91–95.

27. Isaacson, D., Cheney, D.: Current problems in impedance imaging, Inverse problems in partial differential equations, editors: Colton, D., Ewing, R., Rundell, W., SIAM (1990) 141–149.

28. Isakov, V.: Uniqueness and stability in multi-dimensional inverse problems, Inverse Problems **9** (1993) 579–621.

29. Isakov, V.: Inverse problems for partial differential equations, Springer-Verlag, New York (1998).

30. Ito, K., Kunisch, K.: The augmented Lagrangian method for parameter estimation in elliptic systems, SIAM J. Control and Optimization **28** (1990) 113–136.

31. Jackson, J. D.: Classical Electrodynamics, second ed., Wiley, New York (1974).

32. Kallman, J. S., Berryman, J. G.: Weighted least-squares methods for electrical impedance tomography, IEEE Trans. Med. Imaging **11** (1992) 284–292.

33. Keller, G. V.: Rock and Mineral Properties, Electromagnetic Methods in Applied Geophysics, Vol. 1, Theory, (ed. Nabighian, M. N.) (1988) 13–52.

34. Keller, J. B.: Conductivity of a Medium Containing a Dense Array of Perfectly Conducting Spheres or Cylinders or Nonconducting Cylinders, J. Appl. Phys. **34** (1963) 991-993.

35. Keller, J. B.: Effective conductivity of periodic composites composed of two very unequal conductors, J. Math. Phys. **28** (1987) 2516-2520.

36. Kevorkian, J., Cole, J. D.: Multiple scale and singular perturbation methods, Springer Verlag, New York (1996).

37. Kohn, R. V., McKenney, A.: Numerical implementation of a variational method for electrical impedance tomography, Inverse Problems **6** (1990) 389–414.

38. Kohn, R., Vogelius, M.: Determining conductivity by boundary measurements, Comm. Pure App. Math. **38** (1985) 643–667.

39. Kohn, R., Vogelius, M.: Relaxation of a variational method for impedance computed tomography, Comm. Pure Appl. Math. **40** (1987) 745–777.

40. Kozlov, S. M.: Geometric aspects of averaging, Russian Math. Surveys **44** (1989) 91–144.

41. Mallat, S., Zhang, Z.: Matching pursuit in a time-frequency dictionary, IEEE Trans. Signal Processing **41** (1993) 3397–3415.

42. Nachman, A. I.: Global uniqueness for a two-dimensional inverse boundary problem, Annals of Mathematics **142** (1995) 71–96.

43. Santosa, F., Vogelius, M.: A backprojection algorithm for electrical impedance imaging, SIAM J. Appl. Math. **50** (1990) 216–243.

44. Sylvester, J., Uhlmann, G.: A global uniqueness theorem for an inverse boundary value problem, Ann. Math. **125** (1987) 153–169.

45. Sylvester, J., Uhlmann, G.: The Dirichlet to Neumann map and applications, Inverse problems in partial differential equations, editors: Colton, D., Ewing, R., Rundell, W., SIAM (1990) 101–139.

46. Wexler, A., Fry, B., Neuman, M.: Impedance-computed tomography algorithm and system, Appl. Opt. **24** (1985) 3985–3982.

47. Yorkey, T. J., Webster, J. G., Tompkins, W. J.: Comparing reconstruction algorithms for electrical impedance tomography, IEEE Trans. Biomed. Eng. **BME-34** (1987) 843–852.

48. Zhou, Q., Becker, A., Morrison, H. F.: Audio-frequency electromagnetic tomography in 2-D, Geophysics (1993) 482–495.

49. Ziman, J. M.: Principles of the Theory of Solids, Cambridge University Press (1971).

Inverse Scattering in Anisotropic Media

Gunther Uhlmann

Department of Mathematics
University of Washington
Seattle, WA 98195, USA

Abstract. We consider the inverse problem of determining a Riemannian metric in $\mathbf{R}^n$ which is euclidean outside a ball from scattering information. This is a basic inverse scattering problem in anisotropic media. By looking at the wave front set of the scattering operator we are led to consider the "classical" problem of determining a Riemannian metric by measuring the travel times of geodesics passing through the domain. We survey some recent developments on this problem.

1   The inverse scattering problem

Anisotropic materials include most crystals. A common case of aniso-tropy relevant to some Earth structures is transverse isotropy. The inverse scattering problem for this type of media is not very well understood at present. There are many mathematical difficulties associated with the inverse scattering problem for anisotropic Maxwell's equations or the system of elasticity for anisotropic materials. A more basic example of anisotropic media, which involves the study of a scalar partial differential equation, is the case of anisotropic conductors. In this case the electrical conductivity of the medium is represented by a positive definite, symmetric matrix. It is more convenient to look at the conductivity in geometric terms, thus we are going to think of it as a *Riemannian metric*. Notice that this equivalence between Riemannian metrics and conductivities is valid only in dimension $n \geq 3$ [L-U].

Let $g(x) = (g_{ij}(x))$ be a positive definite, symmetric matrix on $\mathbf{R}^n, n \geq 2$. We assume that the Riemannian metric g is smooth (many of the results in this paper are valid assuming finite smoothness). We also assume that the metric is euclidean outside a ball B of radius R centered at the origin, that is $g_{ij} = \delta_{ij}$ for $|x| > R$ where δ_{ij} denotes the Krönecker delta. The euclidean metric is denoted by $e = (\delta_{ij})$. We assume throughout this paper that there are no trapped rays in B, that is any geodesic for the metric g starting at a point in B leaves B in finite time.

We denote by Δ_g the Laplace-Beltrami operator associated to the metric g, i.e. in local coordinates

$$\Delta_g = (\det g)^{-\frac{1}{2}} \sum_{i,j=1}^{n} \frac{\partial}{\partial x_i} (\det g)^{\frac{1}{2}} g^{ij} \frac{\partial}{\partial x_j} \tag{1.1}$$

where $(g^{ij}) = (g_{ij})^{-1}$, $\det g = \det(g_{ij})$. Given $\lambda \in \mathbf{R} - 0, \omega \in S^{n-1}$, the *outgoing eigenfunctions*, $\psi_g(\lambda, x, \omega)$ are solutions of

$$\Delta_g \psi_g + \lambda^2 \psi_g = 0 \tag{1.2}$$

which have the asymptotic behavior

$$\psi_g = e^{i\lambda x \cdot \omega} + \frac{a_g(\lambda, \theta, \omega) e^{i\lambda |x|}}{|x|^{\frac{n-1}{2}}} + O(|x|^{-\frac{n-1}{2}-1}) \tag{1.3}$$

where $\theta = \frac{x}{|x|}$. The function $a_g(\lambda, \theta, \omega)$ is called the *scattering amplitude*. It measures, roughly speaking, the amplitude of the radial scattered wave which resulted from the interaction of the incident plane waves $e^{i\lambda x \cdot \omega}$ with the perturbation of the euclidean metric given by g.

The inverse scattering problem is whether one can determine the metric g from a_g, i.e. to study the non-linear map sending g to a_g. It is easy to see that it is not possible to determine the metric uniquely from this information. Let ψ be a smooth diffeomorphism of $\mathbf{R}^n$ which is the identity outside B. We define $v_g = \psi_g \circ \psi^{-1}$. A straightforward calculation shows that v_g satisfies

$$\Delta_{\psi * g} v_g + \lambda^2 v_g = 0 \tag{1.4}$$

where $\psi^* g$ denotes the pull back of the metric g under the diffeomorphism ψ that is

$$\psi^* g = (D\psi \circ g \circ {}^t D\psi) \circ \psi^{-1}.$$

Since the asymptotic behavior of v_g and the ψ_g at infinity is the same we conclude that

$$a_{\psi * g} = a_g. \tag{1.5}$$

The natural conjecture is that (1.5) is the only obstruction to uniqueness. This conjecture was proven recently. It is a consequence of the paper [B-K] which uses the boundary control method (BC) pioneered by Belishev (see [B] for a survey). In turn this method depends on a Holmgren type uniqueness theorem for hyperbolic equations which was proven by Tataru [T]. See also [R-Z].

The BC method has been greatly extended to solve the inverse scattering problem for any first order and zeroth order selfadjoint perturbation of the Laplace-Beltrami operator [K]. There are also recent results for the case of non-selfadjoint perturbations [K-L].

We remark that if two metrics g_1, g_2 are conformal to each other (i. e. $g_1 = \alpha(x)g_2$ with α a non-zero function) and $\psi * g_1 = g_2$ with ψ a diffeomorphism of $\mathbf{R}^n$ which is the identity outside B then ψ must be the identity and therefore $g_1 = g_2$.

The above mentioned results assume that we know the scattering amplitude for all frequencies and directions. Of course, this is too much information and we would like to measure the scattering amplitude for a more restricted set of angles and frequencies. An interesting physical problem is the *inverse backscattering problem* i.e. we measure $a_g(\lambda, \theta, -\theta)$ for all $\lambda \in \mathbf{R} - 0$ and all $\theta \in S^{n-1}$. The information given depends on n variables. The only known result about this problem is the following: if two metrics are conformal to each other and they are a priori close to the euclidean metric, then the two metrics are the same if their backscattering amplitudes are the same. This is not exactly the result stated in [S-U2] but the methods used there give this result.

Another inverse scattering problem which involves less data is the *fix energy problem*. In this case we measure the scattering amplitude at a fixed frequency λ_0 for all angles $\theta, \omega \in S^{n-1}$. The scattering amplitude $a_g(\lambda_0, \theta, \omega)$ depends on $2n - 2$ variables. It is well known (see for instance [U]) that knowledge of $a_g(\lambda_0, \theta, \omega)$ determines the set of Cauchy data for the Laplace-Beltrami operator on B. Namely we can recover from $a_g(\lambda_0, \theta, \omega)$

$$C_{g,\lambda_0} = \{(u|_{\partial B}, \frac{\partial u}{\partial \nu}|_{\partial \Omega}), \text{ with } u \in H^2(B) \text{ solution of (1.2) on } B.\} \qquad (1.6)$$

Notice that if λ_0 is not a Dirichlet eigenvalue for the Laplace-Beltrami operator then the set of Cauchy data is the graph of the Dirichlet to Neumann map Λ_{g,λ_0}. In the class of metrics conformal to the euclidean metric, it was proven in [Sy-U1] in dimension $n \geq 3$ that C_{g,λ_0} uniquely determines the metric g. In [L-U] it is shown in dimension $n \geq 3$ that C_{g,λ_0} uniquely determines g for real-analytic metrics. The smooth case remains open. The linearization of this problem is studied in [Sy-U2].

In the two dimensional case the anisotropic problem is in some sense easier since we can reduce it to the isotropic case by using isothermal coordinates [A]. In fact in this case the Laplace-Beltrami operator can be transformed, after a change of coordinates, to a conformal multiple of the standard Laplacian. Thus we can transform (1.2) into

$$c^2(x)\Delta + \lambda_0^2$$

with c positive and equal to 1 outside B. In this case it is not known at present for general smooth c whether we can recover c from the scattering amplitude at a non-zero fixed energy. It is known under the a priori assumption that c is small enough [Sy-U3] or for a generic set of $c's$ [Su-U1,2]. The anisotropic conductivity equation, which is the analog of (1.2), is given by

$$\sum_{i,j=1}^{2} \frac{\partial}{\partial x_i} \gamma^{ij} \frac{\partial \psi_\gamma}{\partial x_j} + \lambda_0^2 \psi_\gamma = 0$$

with $\gamma = (\gamma^{ij})$ a positive definite, symmetric smooth matrix which is the identity outside B. As before the inverse scattering problem at a fixed energy can be reduced to the question of whether the set of Cauchy data C_{γ,λ_0} determines γ uniquely up to conjugation by a group of diffeomorphism which is the identity on the boundary of B. A modification of the method of [A] allows us to reduce the problem to the case of an isotropic conductivity [S]. If $\lambda_0 = 0$ the isotropic problem was solved in [N] (see also [B-U] for another approach that allows for

238

less regular conductivities). For the case $\lambda_0 \neq 0$ uniqueness it is not known at present. Uniqueness has been proven for small enough conductivities or for generic conductivities [Su-U1,2].

In this paper we consider other information obtained from the scattering amplitude which involves less variables than the full scattering amplitude. Namely we will consider the singularities of the scattering operator whose kernel is, essentially, the distribution obtained by taking the Fourier transform of the scattering amplitude in the frequency variable. This leads to the problem of determining a metric from the *scattering relation,* which as we explain in the next section, can be considered as the "classical" analog of the inverse scattering problem. Knowledge of the scattering relation means that if we know the point of entry of the geodesic into B and its direction, we can determine the point of exit of the geodesic from B and the direction of exit. As we also show in section 2 the scattering relation determines, under some additional assumptions, the geodesic distance $d_g(x, y), x, y \in \partial B$ between points in the boundary of the ball. This function measures, roughly speaking, the travel time of geodesics passing through B. The inverse kinematic problem arising in seismology is to determine the metric g from these travel times. We discuss in section 3 this problem in detail. We make emphasis on a new identity which was derived in [S-U2] and played a fundamental role in proving that we can uniquely determine a metric sufficiently close to the euclidean metric (up to isometries) from its travel times. This is formula (3.19). We list in section 4 some open problems.

2   The scattering relation

To define the scattering operator and study its singularities we use the Lax-Phillips of scattering which uses the wave equation to define the scattering operator. It is quite natural in this context to use the wave equation since it is well understood how singularities propagate for solutions of this equation. For more details see [G].

Let $(u_0, u_1) \in C_0^\infty(\mathbf{R}^n) \times C_0^\infty(\mathbf{R}^n)$. We define the group of operators

$$U_g(t)(u_0, u_1) = (u(t), \frac{\partial u(t)}{\partial t}(t)) \tag{2.1}$$

where u solves the wave equation

$$\frac{\partial^2 u}{\partial t^2} - \Delta_g u = 0 \tag{2.2}$$

We denote by $U_e(t)$ the operator corresponding to the euclidean metric. The $U_g's$ are unitary groups associated to the energy space $\mathcal{H}_g$ defined as the completion of $C_0^\infty(\mathbf{R}^n) \times C_0^\infty(\mathbf{R}^n)$ under the norm defined by

$$\|(u_0, u_1)\|_g^2 = \int (|u_0|^2 + \sum_{i,j=1}^n g^{ij} \frac{\partial u}{\partial x_i} \frac{\partial u}{\partial x_j} \sqrt{det g}) \, dx. \tag{2.3}$$

We denote by $\mathcal{H}_e$ the energy space corresponding to the euclidean metric.

The *Wave Operators* are unitary operators from $\mathcal{H}_g$ to $\mathcal{H}_e$ which are defined by

$$W_{\pm} = \lim_{t \to \pm\infty} U_e(-t)U_g(t) \tag{2.4}$$

The scattering operator, which is a unitary operator from $\mathcal{H}_e$ to itself, is defined by

$$S_g = W_+ W_-^{-1} \tag{2.5}$$

It follows from finite speed of propagation of solutions of the wave equation that to compute $W_{\pm}$ acting on compactly supported data we don't need to take the limit in (2.4). Namely for $k \in \mathcal{H}_g$ compactly supported, $W_{\pm}k = U_e(-t) \circ U_g(t)k$ for $\pm t$ sufficiently large.

We now explain the connection between the scattering amplitude as defined in (1.3) and the scattering operator. In the context of the Lax-Phillips theory of scattering this is seen using a modification of the Radon transform to reduce the problem to a one dimensional problem depending on some parameters.

Let

$$R : \mathcal{E}'(\mathbf{R}^n) \longrightarrow \mathcal{E}'(\mathbf{R} \times S^{n-1})$$

be the Radon transform

$$Rf(s,\theta) = \int_{x \cdot \theta = s} f(x)d\sigma(x) \tag{2.6}$$

where $d\sigma$ is normalized Lebesgue measure on the hyperplane $\{x \cdot \theta = s\}$. Acting in the x-variable, R is defined on those elements of $D'(\mathbf{R}^n \times \mathbf{R} \times S^{n-1})$ having compact support in x for each t, ω.

It is well-known that the Radon transform intertwines the n-dimensional Laplacian with the one-dimensional Laplacian, i.e.,

$$R\Delta u = \frac{\partial^2}{\partial s^2} Ru, u \in \mathcal{E}'(\mathbf{R}^n). \tag{2.7}$$

The modified Lax-Phillips Radon transform [L-P] which maps $\mathbf{C}^2$- to $\mathbf{C}$- valued distributions, is defined by

$$R_{LP}(u_0, u_1) = C_n D_s^{\frac{n-1}{2}} (D_s Ru_0 - Ru_1), \quad n \text{ odd.} \tag{2.8}$$

For n even, in (2.8) one replaces $D_s^{\frac{n-1}{2}}$ by $|D_s|^{\frac{n-1}{2}}$. R_{LP} is a unitary isomorphism from the free energy space $\mathcal{H}_e$ to $L^2(\mathbf{R} \times S^{n-1})$. Furthermore the modified Radon transform has the key property that it intertwines the free group associated to solutions of the wave equation with the translation group on $\mathbf{R} \times S^{n-1}$. Namely we have

$$R_{LP}U_0(u_0, u_1) = T_t R_{LP}(u_0, u_1) \tag{2.9}$$

where T_t denotes the translation group to the right:

$$T_t f(s) = f(s - t), f \in \mathcal{E}'(\mathbf{R} \times S^{n-1}).$$

The scattering operator in "Radon transform land" is defined by

$$S_g = R_{LP} S_g R_{LP}^{-1}. \tag{2.10}$$

S_g is a unitary operator from $L^2(\mathbf{R} \times S^{n-1})$ to itself. It is easy to see that it is invariant under translation since the coefficients of the wave equation are independent of t. Thus, S_g is a convolution operator in the s-variable, which can be written as

$$S_g f(s, \theta) = I f(s, \theta) + \int_{S^{n-1}} k_g(s - s', \theta, \omega) f(\omega) d\omega \tag{2.11}$$

where I denotes the identity operator. The distribution $k_g(s, \theta, \omega)$ is called the *scattering kernel*. We have that

$$a_g(\lambda, \theta, \omega) = c_n \lambda^{n-3} \int_{\mathbf{R}} e^{-is\lambda} k_g(s, \theta, \omega) ds \tag{2.12}$$

where c_n is a constant.

This is a rough outline of the "quantum" picture using the wave equation approach. We describe now the "classical" picture in phase space by computing the singularities of the operators defined above.

It is a well-known result of Hörmander that singularities of solutions of the wave equation propagate along null-bicharacteristics. We consider the principal symbol of the wave equation

$$p(t, x, \tau, \xi) = \tau^2 - h_g(x, \xi) \tag{2.13}$$

with

$$h_g(x, \xi) = \sum_{i,j=1i}^{n} g^{ij} \xi_i \xi_j \tag{2.14}$$

The Hamiltonian vector field associated to p (resp. h_g) is defined by

$$H_p = 2\tau \frac{\partial}{\partial t} + \sum_{j=1}^{n} \frac{\partial p}{\partial \xi_j} \frac{\partial}{\partial x_j} - \sum_{j=1}^{n} \frac{\partial p}{\partial x_j} \frac{\partial}{\partial \xi_j},$$

$$(\text{resp. } H_{h_g} = \sum_{j=1}^{n} \frac{\partial h_g}{\partial \xi_j} \frac{\partial}{\partial x_j} - \sum_{j=1}^{n} \frac{\partial h_g}{\partial x_j} \frac{\partial}{\partial \xi_j}) \tag{2.15}$$

The bicharacteristics are integral curves of the Hamiltonian vector field H_p. The integral curves of H_{h_g} are tangent to the energy surface $h_g = 1$. We denote the bicharacteristic flow of H_p (resp. H_{h_g}) at time t by $\Phi(t)$ (resp. $\Theta_g(t)$.) We remark that the geodesics of the metric g are the projections of the bicharacteristics over x-space. In fact this is another way to define geodesics. $U_g(t)$ is a Fourier integral operator whose canonical relation is given by

$$C_g^\pm(t) = \{((x,\xi),(y,\eta)) \in T^*(\mathbf{R}^n) - 0 \times T^*(\mathbf{R}^n) - 0; \qquad (2.16)$$

$$(t,\tau,x,\xi) = \Phi^t(0,\tilde{\tau},y,\eta), \text{ with } \tilde{\tau} = \pm\sqrt{h_g(y,\eta)}\}$$

The "classical" free space is $T^*(\mathbf{R}^n - B) - 0$ together with vector field H_{h_e}. The perturbed "classical" space is $T^*(\mathbf{R}^n) - 0$ together with the vector field H_{h_g}. The natural "classical" analog of the wave operators (2.4) is given by the diffeomorphisms

$$\Psi_\pm = \lim_{t\to\pm\infty} \Theta_e(-t)\Theta_g(t) : T^*(\mathbf{R}^n) - 0 \longrightarrow T^*(\mathbf{R}^n - B) - 0 \qquad (2.17)$$

and the "classical" scattering diffeomorphism is given by

$$\Phi_g = \Psi_+ \circ \Psi_-^{-1} : T^*(\mathbf{R}^n - B) - 0 \longrightarrow T^*(\mathbf{R}^n - B) - 0 \qquad (2.18)$$

The scattering relation is the graph of Φ_g, that is, for some t

$$\mathcal{R}_g = \{((x,\xi),(y,\eta)) \in (\mathbf{R}^n - B) \times S^{n-1} \times (\mathbf{R}^n - B) \times S^{n-1};$$

$$(x,\xi) = \Theta_g(t)(y,\eta)\} \qquad (2.19)$$

To know the scattering diffeomorphism Φ_g is equivalent to knowing the scattering relation $\mathcal{R}_g$. Let $\Sigma = \{x \cdot \omega_0 = x_0 \cdot \omega_0\}$ be an hyperplane supported in $\mathbf{R}^n - B$ with normal $\omega_0 \in S^{n-1}$ and the point $x_0 \in \mathbf{R}^n - B$ near B.

Under the assumption of no conjugate points on the metric g near B (no caustics) the solution of the Hamilton-Jacobi equation near B

$$H_g(x, d_x S) = 0, \quad S = 0 \text{ on } \Sigma \qquad (2.20)$$

is given by $S(x) = d_g(x, \Sigma)$ where $d_g(x, \Sigma)$ denotes the geodesic distance from x to the hyperplane Σ. The Lagrangian manifold Λ obtained by the flow-out from Σ by the integral curves of H_{h_g} tangent to $h_g = 1$ is given by $\Lambda = (x, dS(x))$. To know the scattering relation is equivalent to knowing Λ in $T^*(\mathbf{R}^n - B)$. We then conclude that to know the scattering relation is equivalent to knowing this geodesic distance for all hyperplanes supported outside the ball. Since the metric is euclidean outside B we conclude that to know the scattering relation is equivalent to knowing $d_g(x, y), \forall x, y \in \partial B$. Physically this corresponds to knowing the travel times of geodesics passing through B.

3   The boundary distance function

In the last section we motivated the problem of determining a Riemannian metric on a bounded domain Ω in $\mathbf{R}^n$ from the geodesic distance function $d_g(x, y), x, y \in$

$\partial\Omega$. This function is also called *the hodograph* or the *boundary distance function*. This problem arose in geophysics for the case in which the metric is conformal to the euclidean metric, i.e. $g_{ij} = \frac{1}{c^2}\delta_{ij}$. In the literature this problem is referred to as the *inverse kinematic* problem. The function c models the sound speed of the medium. This problem has also attracted the interest of geometers because of rigidity questions in Riemannian geometry (see for instance [C1,2], [Gr], [M1], [O]). The problem can be formulated for general Riemannian manifolds with boundary.

We now state the problem more precisely. Let Ω be a bounded open set in $\mathbf{R}^n$ with smooth boundary Γ. We assume that $\bar{\Omega}$ is strictly convex with respect to g, i.e., for any two distinct points $x \in \bar{\Omega}$, $y \in \bar{\Omega}$ there is a unique geodesic joining x and y lying entirely in Ω with the possible exception of the endpoints x and y. Let $d_g(x, y)$ denote the geodesic distance between x and y. The inverse problem we address in this section is whether we can determine the Riemannian metric g knowing $d_g(x, y)$ for any $x \in \Gamma$, $y \in \Gamma$. As in (1.5) it is easy to see that g cannot be determined from this information. We have $d_{\psi^* g} = d_g$ for any diffeomorphism $\psi : \bar{\Omega} \to \bar{\Omega}$ that leaves the boundary pointwise fixed, i.e., $\psi|_\Gamma = Id$, where Id denotes the identity map and $\psi^* g$ is the pull-back of the metric g. R. Michel conjectured in [M1] that this is the only obstruction to uniqueness, namely if we have two Riemannian metrics g_1, g_2 with $\bar{\Omega}$ strictly convex with respect to both, and if

$$d_{g_1}(x, y) = d_{g_2}(x, y) \quad \forall (x, y) \in \Gamma^2, \tag{3.1}$$

there exists a diffeomorphism $\psi : \bar{\Omega} \to \bar{\Omega}$, $\psi|_\Gamma = Id$, so that

$$g_2 = \psi^* g_1. \tag{3.2}$$

As noted earlier in the case that the metrics g_1 and g_2 have the same boundary distance function and are in the same conformal class then the diffeomorphism must be the identity and therefore the conjecture in this case is that the metrics are the same.

In [G-M1,2] (see also [C-S]) the case of radial metric conformal to the euclidean is considered (i.e. the sound speed is assumed to be radial.) The first general result was proven by Mukhometov [Mu], who showed in two dimensions that if two metrics are conformal to the euclidean metric and the domain is geodesically convex for the two metrics then the metrics are the same. Moreover, he proved a stability estimate. The proof is very original and uses a form of an energy inequality for this problem. Energy inequalities are of standard use for hyperbolic equations but Mukhometov's energy inequality was, at the time, completely new. We also note that in two dimensions the problem is formally determined since the hodograph depends on two variables. Mukhometov's result was generalized to higher dimensions in [Mu-R] using a similar method. In [B-G] and [Be] it proved in all dimensions $n \geq 2$, again under the geodesically convex assumption on Ω, that if two metrics are conformal to each other and they have the same hodograph then they must be the same. Also in [B-G], [Be] stability estimates are proved in this case. Croke [C1] gave a nice geometric proof of the uniqueness part of the main result in [B-G], [Be] on Riemannian manifolds with boundary satisfying an additional assumption which is weaker than geodesically convex. As far as we know the stability estimates of [B-G] and [Be] don't follow from Croke's argument.

The conjecture (3.2) has been considered in [Gr], [M1] for general Riemannian manifolds with boundary under some assumptions on the curvature. In particular they have shown that if M is a Riemannian manifold with boundary and Riemannian metric g, geodesically convex with respect to g, the conjecture is valid if M is any compact subdomain of $\mathbf{R}^n$, any compact subdomain of an open n-dimensional hemisphere or any compact subdomain of the hyperbolic plane. In two dimensions in [G-N] the conjecture is proved under some restrictions on the behavior of an extension of the metric to $\mathbf{R}^2$ which are essentially a condition of negative curvature on the extension of the metric. The latter result was generalized in [C2],[O] in two dimensions for negatively curved surfaces with boundary under less stringent condition that geodesically convex and to n-dimensional negatively curved Riemannian manifolds with boundary under additional restrictions [C1].

3.1   The linearized problem

We consider the linearization of the map

$$g \longrightarrow d_g \tag{3.3}$$

in the direction of a $C_0^\infty(\Omega)$-tensor field $f_{ij}, i,j = 1, ..., n$. We consider the Hamiltonian vector field H_{h_g} and we denote by $(x(t), \xi(t))$ the integral curves of H_{h_g} on the energy level $h_g = 1$. We are going to use the following parameterization of those integral curves. Let us denote

$$ST^*\partial\Omega_- := \{(z,\omega) \in ST^*\partial\Omega; \ z \in \partial\Omega, \ \omega \in S^{n-1}, \ g^{-1}\omega \cdot \nu(z) \leq 0\}.$$

where $\nu(z)$ is the outer unit normal to $\partial\Omega$. Let us introduce the measure $d\mu(z,\omega) = g^{-1}\omega \cdot \nu(z)dS_z d\omega$ on $ST^*\partial\Omega_-$, where dS_z and $d\omega$ are the surface measures on $\partial\Omega$ and S^{n-1}, respectively. Then $(x(t), \xi(t)) = (x(t; z,\omega), \xi(t; z,\omega))$ is defined as the integral curve of H_{h_g} issued from $(z,\omega) \in ST^*\partial\Omega_-$.

Now we define the geodesic X-ray transform by

$$I_g(f)(z,\theta) = \int_\gamma \sum_{i,j=1}^3 f_{ij}(\gamma(t))\dot\gamma^i(t)\dot\gamma^j(t)\, dt, \tag{3.4}$$

where $\gamma(t) = \gamma(t; z,\theta)$ is the geodesic issued from, $(z,\theta) \in ST\partial\Omega_-$ parameterized by its arc-length. Here $ST\partial\Omega_-$ consists of all unit (with respect to the metric) vectors on the boundary pointing inside Ω. Since the tangent vector to the geodesic is related to the covector or ξ by the formula $g\dot\gamma = \xi$, we get

$$I_g(f)(z,\theta) = \int_\gamma \sum_{i,j=1}^3 m^{ij}(\gamma(t))\xi_i(t)\xi_j(t)\, dt \tag{3.5}$$

where $m = g^{-1}fg^{-1}$ or, in coordinates $m^{ij} = \sum_{i',j'} g^{ii'} f_{i'j'} g^{j'j}$. It is easy to see that if

$$d_{g+tf} = 0, \forall t$$

then

$$I_g(f) = 0. \tag{3.6}$$

Of course the transform $f \to I_g(f)$ is not injective because the distance function is invariant under change of variables which are the identity at the boundary. For the linearized problem this corresponds to $I_g(dv) = 0$ for any tensor field v satisfying $v|_\Gamma = 0$. Here dv denotes the symmetric covariant derivative of v.

By Theorem 3.3.2 of [Sh] we can uniquely decompose the tensor f_{ij} into its solenoidal and potential parts, i.e.

$$f = f^s + dv, \quad v|_\Gamma = 0 \tag{3.7}$$

The natural conjecture is that

$$I_g(f^s) = 0 \Longrightarrow f^s = 0. \tag{3.8}$$

This conjecture has been proved in [Sh] for Riemannian manifolds with boundary satisfying a positive bound on the sectional curvatures of g. In other words the curvature cannot be too big. We remark that integral geometry of tensor fields has also been extensively studied in the very nice book [Sh]

3.2 The local problem

As mentioned earlier conjecture (3.8) has only been proved under the assumption of constant or negative curvature on the metric g. The case of positive curvature remains open.

The first local result, for metrics sufficiently close to the euclidean metric, was proven in [S-U2]. We now state the result. We denote by $C^k_{(0)}(\Omega)$ the set of all $f \in C^k(\bar{\Omega})$ such that $\partial^\alpha f = 0$ on $\partial\Omega$ for $|\alpha| \leq k$. Then we have.

Theorem 1. *Suppose that g_1 and g_2 are two metrics satisfying (3.1). Then there exists $\varepsilon > 0$, such that if*

$$g_m - e \in C^{12}_{(0)}(\Omega), \quad \|g_m - e\|_{C^{12}(\bar{\Omega})} < \varepsilon, \quad m = 1, 2, \tag{3.9}$$

then there exists a C^{11} diffeomorphism $\psi : \bar{\Omega} \to \bar{\Omega}$ such that $\psi|_\Gamma = Id$ and $\psi^* g_1 = g_2$.

The proof of Theorem 1 relies on deriving a new identity (see (3.19)) for the difference of the metrics and working in suitable chosen coordinates. The linearized version of the identity at the euclidean metric gives, roughly speaking, that the integrals along the geodesics (lines in the linear case) of the difference of the two metrics is zero. Then one concludes that the metrics are the same in those coordinates by inverting the X-ray transform. In section 3 of [S-U2] Theorem 3.1 was proven by using the identity and a perturbation argument that leads to the inversion of a Fourier integral operator. It is likely that this identity will also give stability estimates and it will have other applications.

Recently Croke, Daiberkov and Sharafutdinov [C-D-S] have proven a different type of local result. In [C-D-S] it is assumed that the metric g satisfies the same curvature condition under which (3.8) is valid. Now assume that g_1 is a metric sufficiently close to g in an appropriate topology. It is shown in [C-D-S] that the two metrics are isometric. This result doesn't imply Theorem 3.1 since the neighborhood of g depends on the metric g. Eskin [E] wrote an article a few months earlier than the paper [C-D-S] was submitted proving a similar result but assuming that the curvature of g is sufficiently small.

3.3 The main identity

In this section we give complete details of the identity proved in [S-U2].

Assume that we have two metrics g_1 and g_2 satisfying

$$g - e \in C^k_{(0)}(\Omega), \quad \|g - e\|_{C^k(\bar{\Omega})} < \varepsilon \tag{3.10}$$

with some $k \geq 2$ and $\varepsilon > 0$. Assume also that they satisfy (3.9). By (3.10), g_1 and g_2 can be extended outside Ω as e and the so extended metrics belong to $C^k(\mathbf{R}^3)$. From now on we will denote by g_1 and g_2 the extended metrics.

Let $x^{(0)} \in \Gamma, \xi^{(0)} \in S^2$ such that $\nu(x^{(0)}) \cdot g^{-1} \xi^{(0)} < 0$. The integral curves of $H_{h_{g_j}}, j = 1, 2$ tangent to the energy $h_{g_j} = 1$ are denoted by $(x_{g_j}, \xi_{g_j}), j = 1, 2$. They solve the Hamiltonian system

$$\begin{cases} \frac{d}{ds} x_m = \sum_{j=1}^3 g^{mj} \xi_j, & \frac{d}{ds} \xi_m = -\frac{1}{2} \sum_{i,j=1}^3 \frac{\partial g^{ij}}{\partial x_m} \xi_i \xi_j, & m = 1, 2, 3, \\ x|_{s=0} = x^{(0)}, & \xi|_{s=0} = \xi^{(0)}. \end{cases} \tag{3.11}$$

Here g is either g_1 or g_2, while the initial conditions are the same for both metrics. We remark that if $\xi^{(0)} \cdot g^{-1} \xi^{(0)} = 1$, then s is the arc-length in (3.11). The assumption (3.1) implies the following property.

Lemma 1. *Let* g_1, g_2 *be two Riemannian metrics in* $\bar{\Omega}$ *with* $\bar{\Omega}$ *strictly convex with respect to anyone of them and assume* $g_1|_\Gamma = g_2|_\Gamma$. *Assume also (3.1). Let* $x_{g_m}, \xi_{g_m}, m = 1, 2,$ *be the solution of (3.11) with the same initial conditions*

$$x_{g_1}(0) = x_{g_2}(0) = x^{(0)}, \quad \xi_{g_1}(0) = \xi_{g_2}(0) = \xi^{(0)}.$$

Then

$$x_{g_1}(t) = x_{g_2}(t) \in \Gamma, \quad \xi_{g_1}(t) = \xi_{g_2}(t), \tag{3.12}$$

where t *is the common length of the corresponding geodesics joining* $x^{(0)}$ *and* $x_{g_1}(t) = x_{g_2}(t)$ *provided that* $\xi^{(0)} \cdot g^{-1} \xi^{(0)} = 1$.

Proof. This is a direct consequence of the discussion in section 2. Namely if the distance function for two metrics is the same then their scattering relation is the same. We give another proof due to Michel [M1]. Let x_{g_1} be the geodesics related to g_1 defined above. Denote by $s \mapsto y_{g_2}(s)$ the geodesics associated to g_2 joining $x_{g_1}(0)$ and $x_{g_1}(t) \in \Gamma$, where t is the length of $x_{g_1}x_{g_2}$. In other words, $y_{g_2}(0) = x_{g_1}(0)$, $y_{g_2}(t) = x_{g_1}(t)$. Note, that t is also the length of that geodesic. By [M1, Corollary 2.3], the geodesics x_{g_1} and y_{g_2} are tangent at the common endpoints. Since y_{g_2} solves (3.11) with $g = g_2$ and initial data $y_{g_2} = x^{(0)}$, $\xi(0) = \eta^{(0)}$ with some $\eta^{(0)}$, we get that $\eta^{(0)} = \xi^{(0)}$, because the two metrics coincide on the boundary. Therefore, y_{g_2} solves (3.11) with $g = g_2$ and by the uniqueness of that solution we get that $y_{g_2} = x_{g_2}$. This proves the lemma. $\square$

Consider the Hamiltonian system (3.11) with the following initial conditions

$$\begin{cases} \frac{d}{ds}x_m = \sum_{j=1}^{3} g^{mj}\xi_j, & \frac{d}{ds}\xi_m = -\frac{1}{2}\sum_{i,j=1}^{3} \frac{\partial g^{ij}}{\partial x_m}\xi_i\xi_j, \quad m = 1,2,3, \\ x|_{s=-\rho} = (-\rho, z), & \xi|_{s=-\rho} = (1,0,0). \end{cases}$$

$$(3.13)$$

Here $z \in \mathbf{R}^2$, $\rho > 0$ is such that $g = e$ for $|x| > \rho$ and the solution $x = x(s, z)$, $\xi = \xi(s, z)$ depends on the parameter z. If $g = e$, then $x = (s, z) = (s, z_1, z_2)$.

We now introduce as new coordinates $y = (s, z)$. Since the metrics are close to the euclidean metric it is easy to see that the map $\Omega \ni x \mapsto y$ is close to Id in the C^{k-1} topology for small $\varepsilon > 0$ and therefore is a diffeomorphism. In the new coordinates $g^{-1} = (g^{ij})$ will have the form

$$(g^{ij}) = \begin{pmatrix} 1 & 0 & 0 \\ 0 & g^{22} & g^{23} \\ 0 & g^{23} & g^{33} \end{pmatrix}.$$

$$(3.14)$$

Notice that g would have a similar form, too.

Denote by ψ_1, ψ_2 the maps $x \mapsto y$ related to g_1, g_2, respectively. Instead of g_1, g_2, consider $\tilde{g}_1 = \psi_1^* g_1$ and $\tilde{g}_2 = \psi_2^* g_2$, respectively. It is easy to see that s is the length parameter in (3.13) and therefore (3.1) implies $\psi_1(\Gamma) = \psi_2(\Gamma)$. So, both ψ_1 and ψ_2 map Ω to a new domain $\tilde{\Omega}$. We also have that $\psi_1 = \psi_2$ outside Ω. Therefore, (3.1) remains true for $\tilde{g}_1$, $\tilde{g}_2$ in $\tilde{\Omega}$ and instead of (3.10) we have

$$\tilde{g}_1 - \tilde{g}_2 \in C_{(0)}^{k-2}(\tilde{\Omega}), \quad \|\tilde{g}_m - e\|_{C^{k-2}(\tilde{\Omega})} < C\varepsilon, \quad m = 1,2$$

$$(3.15)$$

with some $C > 0$. We aim to prove that $\tilde{g}_1 = \tilde{g}_2$. This would prove Theorem 3.1 , because it would imply $\psi^* g_1 = g_2$ where $\psi := \psi_2^{-1}\psi_1$ would be a diffeomorphism in Ω fixing the boundary. For the sake of simplicity of notation, let us denote the new metrics again by g_1, g_2 and $\tilde{\Omega}$ by Ω.

Denote the solution of (3.11) by $x = x(s, x^{(0)}, \xi^{(0)})$, $\xi = \xi(s, x^{(0)}, \xi^{(0)})$. Let us introduce new notation

$$X := (x, \xi).$$

The solution to (3.11) related to g_1 and g_2, respectively, can therefore be written down as $X_{g_j} = X_{g_j}(s, X^{(0)}) = X_{g_j}(s, x^{(0)}, \xi^{(0)})$.

Set $F(s) := X_{g_2}(t - s, X_{g_1}(s, X^{(0)}))$. Here $t = t(X^{(0)})$ is the length of the geodesics issued from $X^{(0)}$ with endpoint on Γ and t is independent of $g = g_1$ or $g = g_2$. Notice that the x-component of $F(s)$ may not be in Ω but belongs to a neighborhood of Γ small with ε. By (3.12), $F(0) = X_{g_2}(t, X^{(0)}) = X_{g_1}(t, X^{(0)}) = F(t)$. Thus

$$\int_0^t F'(s)\,ds = 0. \tag{3.16}$$

Denote $V_{g_j} := (\partial H_{g_j}/\partial \xi, -\partial H_{g_j}/\partial x)$, $j = 1, 2$. Then

$$F'(s) = -V_{g_2}(X_{g_2}(t - s, X_{g_1}(s, X^{(0)}))) + \frac{\partial X_{g_2}}{\partial X^{(0)}}(t - s, X_{g_1}(s, X^{(0)}))$$

$$V_{g_1}(X_{g_1}(s, X^{(0)})). \tag{3.17}$$

We claim that

$$V_{g_2}(X_{g_2}(t - s, X_{g_1}(s, X^{(0)}))) = \frac{\partial X_{g_2}}{\partial X^{(0)}}(t - s, X_{g_1}(s, X^{(0)}))$$

$$V_{g_2}(X_{g_1}(s, X^{(0)})). \tag{3.18}$$

Indeed, (3.18) follows from

$$0 = \left.\frac{d}{ds}\right|_{s=0} X(T - s, X(s, X^{(0)})) = -V(X(T, X^{(0)}))$$

$$+ \frac{\partial X}{\partial X^{(0)}}(T, X^{(0)})V(X^{(0)}), \quad \forall T \tag{3.19}$$

after setting $T = t - s$. Therefore, (3.16), (3.17) and (3.18) combined together imply

$$\int_0^t \frac{\partial X_{g_2}}{\partial X^{(0)}}(t - s, X_{g_1}(s, X^{(0)}))(V_{g_1} - V_{g_2})(X_{g_1}(s, X^{(0)}))\,ds = 0. \tag{3.20}$$

Formula (3.20) is the main result used in [S-U2] to prove that the metrics coincide. This identity is a non-linear integral equation on the difference of the metrics g_1 and g_2. We formally linearize this identity at the euclidean metric to explain how to prove that the metrics coincide. In other words, we will formally replace X_{g_1} and X_{g_2} by X_e, where e is the euclidean metric, but we will keep V_{g_1} and V_{g_2}.

Suppose $g = e$. Then $X_e = (x_e, \xi_e)$ solves $x'_e = \xi_e$, $\xi'_e = 0$, therefore $V_e = (\xi, 0)$. It is easy to see that in this case

$$X_e = \begin{pmatrix} 1 & s \\ 0 & 1 \end{pmatrix} X^{(0)}, \quad \frac{\partial X_e}{\partial X^{(0)}} = \begin{pmatrix} 1 & s \\ 0 & 1 \end{pmatrix}. \tag{3.21}$$

Since $V = (g^{-1}\xi, -\frac{1}{2}\nabla_x(g^{-1}\xi)\cdot\xi)$ (recall that $g^{-1} = \{g^{ij}\}$), we get the following formal linearization formula for (3.20)

$$\int_0^t \left(m\xi - \frac{1}{2}(t - s)\nabla_x(m\xi)\cdot\xi, -\frac{1}{2}\nabla_x(m\xi)\cdot\xi\right)(x^{(0)} + s\xi)\,ds = 0, \tag{3.22}$$

where $\{m_{ij}\} := \{g_1^{ij}\} - \{g_2^{ij}\}$, $x^{(0)} \in \Gamma$, $\xi = \xi^{(0)} \in S^2$ and $\xi^{(0)} \cdot \nu(x^{(0)}) < 0$. By (3.14), m has the form

$$m = \begin{pmatrix} 0 & 0 & 0 \\ 0 & m_{22} & m_{23} \\ 0 & m_{23} & m_{33} \end{pmatrix}. \tag{3.23}$$

Equating the second components of both sides in (3.22), we get

$$\int_0^t \sum_{i,j=2}^3 \nabla_x m_{ij}(x^{(0)} + s\xi)\xi_i\xi_j \, ds = 0 \tag{3.24}$$

for $x^{(0)}$ and ξ as above. This equation easily implies

$$\sum_{i,j=2}^3 \eta \hat{m}_{ij}(\eta)\xi_i\xi_j = 0 \quad \text{for } \xi \cdot \eta = 0, \tag{3.25}$$

where $\hat{m}(\eta)$ is the Fourier transform of $m(x)$ extended as 0 outside Ω. Let $p = (0, p_2, p_3) \in S^2$ be a parameter. Picking

$$\xi = \xi_p(\eta) = \frac{\eta \times p}{|\eta \times p|} = \frac{(p_3\eta_2 - p_2\eta_3, -p_3\eta_1, p_2\eta_1)}{\sqrt{\eta_1^2 + (p_3\eta_2 - p_2\eta_3)^2}}, \tag{3.26}$$

we get

$$\eta \frac{p_2^2\eta_1^2\hat{m}_{33}(\eta) + p_3^2\eta_1^2\hat{m}_{22}(\eta) - 2p_2p_3\eta_1^2\hat{m}_{23}(\eta)}{\eta_1^2 + (p_3\eta_2 - p_2\eta_3)^2} = 0. \tag{3.27}$$

Choosing $p = (0, 1, 0)$ yields

$$\eta \frac{\eta_1^2}{\eta_1^2 + \eta_3^2} \hat{m}_{33}(\eta) = 0, \tag{3.28}$$

therefore $m_{33} = 0$. Next, setting $p = (0, 0, 1)$ in (3.27) leads to

$$\eta \frac{\eta_1^2}{\eta_1^2 + \eta_2^2} \hat{m}_{22}(\eta) = 0, \tag{3.29}$$

so $m_{22} = 0$. And finally, choosing $p = (0, 1, 1)/\sqrt{2}$, we obtain

$$\eta \frac{\eta_1^2\hat{m}_{33}(\eta) + \eta_1^2\hat{m}_{22}(\eta) - 2\eta_1^2\hat{m}_{23}(\eta)}{\eta_1^2 + (\eta_3 - \eta_2)^2/2} = 0, \tag{3.30}$$

thus $m_{23} = 0$.

4 Open problems

In this section we mention some open problems directly related to the conjecture (3.2).

- *Boundary determination.* Suppose we know d_g. Can we recover, in appropriate coordinates, all the derivatives of g at the boundary? This result was proven in the two dimensional case in [M2]. If the answer is affirmative it is likely that one can prove conjecture (3.2) for real-analytic metrics. Also, we wouldn't need to assume that the metrics coincide at the boundary in the statement of Theorem 3.1.
- *Compactness* Moding-out by the group of diffeomorphisms which are the identity on the boundary, is the set of metrics having the same boundary distance function compact in some appropriate topology? A result of this kind combined with the local results [C-D-S], [S-U2] would probably lead to a proof that, under appropriate restrictions on the curvature, there is only a finite number of metrics (up to isometry) with the same boundary distance function.
- *The two dimensional case* In this case we can use isothermal coordinates [A] to reduce the problem to the isotropic case. The problem is that the change of variables produced in this fashion is not the identity at the boundary and we cannot use Mukhometov's result. It is easy to see that it is enough to prove that the change of variables resulting at the boundary is the boundary value of a conformal map.
- *Caustics* Most of the results mentioned in this paper on the conjecture assume that the domain (or manifold) is geodesically convex. It is very easy to find counterexamples if the function d_g is multivalued [G-M1]. However, the scattering relation is well defined by just assuming that there are no trapped geodesics. Is it possible to generalize the known results about recovering the metric from the boundary distance function to recover the metric (up to isometry) from the scattering relation?
- *The Dirichlet to Neumann Map* It was proven in [Sy-U2] that from the hyperbolic Dirichlet to Neumann map we can recover the boundary distance function, assuming again that Ω is geodesically convex. Is there any connection between the elliptic Dirichlet to Neumann and the boundary distance function d_g? As mentioned above to know the elliptic DN map is the same as knowing the set of Cauchy data (1.6). This set is vaguely resemblant of the scattering relation (2.16).

Acknowledgement

The research for this paper was partly supported by NSF grant DMS-9705792, ONR grant N00014-93-1-0295 and a grant from the Royalty Research Fund at the University of Washington.

References

[A] L. AHLFORS, Quasiconformal mappings, Van Nostrand, 1966.

[B] M. BELISHEV Boundary control in reconstruction of manifolds and metrics (the BC method), *Inverse Problems* **13** (1997), no. 5, R1–R45.

[B-K] M. BELISHEV AND Y. KURYLEV, To the reconstruction of a Riemannian manifold via its spectral data (BC-method), *Comm. PDE* **17** (1992), 767-804.

[B-G] I. N. BERNSHTEIN AND M. GERVER, A problem of integral geometry for a family of geodesics and an inverse kinematic seismics problem. (Russian) *Dokl, Akad. Nauk SSSR* **243** (1978), no. 2, 302–305.

[Be] G. BEYLKIN, Stability and uniqueness of the solution of the inverse kinematic problem in the multidimensional case, *J. Soviet Math.* **21**(1983), 251–254.

[B-U] R. BROWN AND G. UHLMANN Uniqueness in the inverse conductivity problem for nonsmooth conductivities in two dimensions, *Comm. Partial Differential Equations* **22** (1997), no. 5-6, 1009–1027.

[C-S] K. CHADAN AND P. SABATIER, Inverse problems in Quantum scattering theory, Springer-Verlag, 1989.

[C1] C. CROKE, Rigidity and the distance between boundary points, *J. Differential Geom.* **33**(1991), no. 2, 445–464.

[C2] C. CROKE, Rigidity for surfaces of nonpositive curvature, Comment. Math. Helv. **65** (1990), no. 1, 150–169.

[C-D-S] C. CROKE, N. S. DAIRBEKOV AND V. A. SHARAFUTDINOV, Local boundary rigidity of a compact Riemannian manifold with curvature bounded above, to appear, Transactions AMS.

[E] G. ESKIN Inverse scattering problem in anisotropic media, preprint 1998.

[G-M1] M. GERBER AND V. MARKUSHEVICH, Determination of a seismic wave velocity from the travel time curve, *J. R. Astron. Soc. Geophysics*, **11** (1966), 165–173.

[G-M2] M. GERBER AND V. MARKUSHEVICH, On the characteristic properties of travel-time curves, *J. R. Astron. Soc. Geophysics*, **13** (1967), 241–246.

[G-N] M. L. GERVER AND N. S. NADIRASHVILI, An isometricity conditions for Riemannian metrics in a disk, *Soviet Math. Dokl.* **29** (1984), 199–203.

[Gr] M. GROMOV, Filling Riemannian manifolds, *J. Differential Geometry* **33** (1991), 445–464.

[G] V. GUILLEMIN, Sojourn times and asymptotic properties of the scattering matrix. Proceedings of the Oji Seminar on Algebraic Analysis and the RIMS Symposium on Algebraic Analysis (Kyoto Univ., Kyoto, 1976). *Publ. Res. Inst. Math. Sci.* **12** (1976/77), supplement, 69–88.

[K] Y. V. KURYLEV, Inverse boundary problems on Riemannian manifolds, *Contemp. Math.* **173,** (1994) 181–192

[K-L] Y. V. KURYLEV AND M. LASSAS, The multidimensional Gelfand inverse problem for non-self-adjoint operators, *Inverse Problems* **13** (1997), no. 6, 1495–1501.

[L-P] P. LAX AND R. PHILLIPS, Scattering Theory, revised edition, Academic Press, 1989.

[L-U] J. LEE AND G. UHLMANN Determining anisotropic real-analytic conductivities by boundary, *Comm. Pure Appl. Math.* **42** (1989), no. 8, 1097–1112

[M1] R. MICHEL, Sur la rigidité imposée par la longueur des géodésiques, *Invent. Math.* **65** (1981), 71-83.

[M2] R. MICHEL, Restriction de la distance géodésique à un arc et rigidité, *Bull. Soc. Math. France* **122** (1994), no. 3, 435–442.

[Mu] R. G. MUKHOMETOV, The reconstruction problem of a two-dimensional Riemannian metric, and integral geometry (Russian), *Dokl. Akad. Nauk SSSR* **232** (1977), no. 1, 32–35.

[Mu-R] R. G. MUKHOMETOV AND V. G. ROMANOV, On the problem of finding an isotropic Riemannian metric in an n-dimensional space (Russian), *Dokl. Akad. Nauk SSSR* **243** (1978), no. 1, 41–44.

[N] A. NACHMAN, Global uniqueness for a two-dimensional inverse boundary value problem, *Ann. of Math. (2)* **143** (1996), no. 1, 71–96.

[O] J. P. OTAL, Sur les longuer des géodésiques d'une métrique a courbure négative dans le disque, *Comment. Math. Helv.* **65** (1990), 334–347.

[R-Z] L. ROBIANNO AND, ZUILLY, Uniqueness in the Cauchy problem for operators with partially holomorphic coefficients, *Invent. Math.* **131** (1998), 493–539.

[Sh] V. A. SHARAFUTDINOV, Integral geometry of tensor fields, VSP, Utrech, the Netherlands (1994).

[S-U1] P. STEFANOV AND G. UHLMANN, Inverse backscattering for the acoustic equation, *SIAM J. Math. Anal.* **28**(5) (1997), 1191–1204.

[S-U2] P. STEFANOV AND G. UHLMANN, Rigidity for metrics with the same lengths of geodesics, *Math. Res. Lett.* **5** (1998), no. 1-2, 83–96.

[Su-U1] Z. SUN AND G. UHLMANN, Generic uniqueness for an inverse boundary value problem, Duke Math. J. **62** (1991), no. 1, 131–155.

[Su-U2] Z. SUN AND G. UHLMANN, Generic uniqueness for determined inverse problems in 2 dimensions, in K. Hayakawa, Y. Iso, M. Mori, T. Nishida, K. Tomoeda, M. Yamamoto (eds.), ICM-90 Satellite Conference Proceedings, Inverse Problems in Engineering Sciences, Springer-Verlag, (1991), 145–152.

[S] J. SYLVESTER An anisotropic inverse boundary value problem, *Comm. Pure Appl. Math.* **43** (1989), 20–232.

[Sy-U2] J. SYLVESTER AND G. UHLMANN, A global uniqueness theorem for an inverse boundary value problem, *Ann. of Math. (2)* **125** (1987), no. 1, 153–169.

[Sy-U] J. SYLVESTER AND G. UHLMANN, Inverse problems in anisotropic media, *Contemp. Math.* **122**(1991), 105–197.

[Sy-U3] J. SYLVESTER AND G. UHLMANN, A uniqueness theorem for an inverse boundary value problem in electrical prospection *Comm. Pure Appl. Math.* **39** (1986), no. 1, 91–112

[T] D. TATARU, Unique continuation for solutions to PDE's; between Hörmander's theorem and Holmgren's theorem, *Comm. Partial Differential Equations* **20** (1995), no. 5-6, 855–884.

[U] G. UHLMANN, Inverse boundary value problems and applications, *Astérisque* **207** (1992), 153–211.

Inverse Problems as Statistics

P.B. Stark

Department of Statistics
University of California
Berkeley, CA 94720-3860 USA
stark@stat.berkeley.edu,
WWW home page: http://www.stat.berkeley.edu/~stark

Abstract. What mathematicians, scientists, engineers, and statisticians mean by "inverse problem" differs. For a statistician, an inverse problem is an inference or estimation problem. The data are finite in number and contain errors, as they do in classical estimation or inference problems, and the unknown typically is infinite-dimensional, as it is in nonparametric regression. The additional complication in an inverse problem is that the data are only indirectly related to the unknown. Standard statistical concepts, questions, and considerations such as bias, variance, mean-squared error, identifiability, consistency, efficiency, and various forms of optimality apply to inverse problems. This article discusses inverse problems as statistical estimation and inference problems, and points to the literature for a variety of techniques and results.

1 Introduction

This paper casts inverse problems as statistical estimation and inference problems. Along the way, it presents some statistical ideas that I find helpful in thinking about inverse problems.

Inverse problems are formulated differently and raise different questions for a statistician and an applied mathematician. For example, to a statistician, the number of data is finite, and the data contain errors that are modeled at least in part as stochastic. To a statistician, bias, variance, identifiability, consistency, and similar notions figure prominently; emphasis is on estimation and inference. Applied mathematicians often are more interested in existence, uniqueness, and construction, given an infinite number of noise-free data, and stability given data contaminated by a deterministic disturbance.

These two viewpoints are related; for example, identifiability and uniqueness are similar. Moreover, there is a deep connection between statistical measures of the difficulty some estimation problems (estimating a linear functional of an element of a Hilbert space from linear observations observed with Gaussian errors) and the theory of optimal recovery of a linear functional from linear data with deterministic errors [17]. Moreover, many of the mathematical tools employed are the same: functional analysis, convex analysis, optimization theory, nonsmooth analysis, approximation theory, harmonic analysis, and measure theory.

Section 2 presents some probabilistic and statistical terminology in a sufficiently general framework that most inverse problems become special cases. Section 3 presents a canonical inverse problem, and translates it into the language of
section 2. Section 4 introduces some ideas and notation from statistical decision
theory. Section 5 applies the ideas of section 4 to estimation, and presents some
common loss functionals used to define optimal estimators. Section 6 does the
same thing for confidence sets. Section 7 examines some common estimators used
in statistics and inverse problems, including the Backus-Gilbert method, Bayes
estimation, maximum likelihood, regularization, shrinkage, and strict bounds.

2   Preliminaries

This section introduces notation used throughout the rest of the paper. We will
specialize to the case that the unknown model (parameter) is an element of
a separable Banach space, n real data are observed, and the joint probability
distribution of the data errors is known. These conditions can be relaxed in
various ways; in particular, the assumption that the parameter is in a Banach
space is often unnecessary. In most of the development, it suffices for the data
to take values in a separable Banach space, not necessarily $\Re^n$. We sometimes
consider what happens when $n \to \infty$. The development here is based largely on
Bass [11], Dunford and Schwartz [23] and Freedman [26]; see also Kolmogorov
[34].

For a finite set A, $\#A$ is the number of elements it contains. The intersection
of two sets A and B is $AB = A \cap B$. For $x \in \Re^n$, $\|x\|$ denotes the ordinary
Euclidean norm. For a positive semidefinite n by n matrix Σ,

$$\|x\|_\Sigma^2 = x \cdot \Sigma \cdot x \tag{1}$$

defines the seminorm $\|x\|_\Sigma$. The symbol $\|x\|_p$ denotes the ordinary ℓ_p norm of
x:

$$\|x\|_p = \begin{cases} \left(\sum_{j=1}^n |x_j|^p \right)^{1/p}, & 1 \le p < \infty \\ \max_{1 \le j \le n} |x_j|, & p = \infty. \end{cases} \tag{2}$$

A probability distribution $\mathbb{P}$ on a σ-algebra $\mathcal{F}$ of subsets of a set Ω is a count-
ably additive positive finite measure with total mass one. We refer to $(\mathbb{P}, \Omega, \mathcal{F})$
as a *probability triple*. The elements of $\mathcal{F}$ are called *events*. If a statement is true
on Ω except on some set F for which $\mathbb{P}(F) = 0$, the statement is said to hold
almost surely (a.s.). For any subset F of Ω, the *indicator function of the set F*
is

$$1_F : \Omega \to \{0,1\}$$
$$\omega \mapsto \begin{cases} 1, \omega \in F \\ 0, \omega \notin F. \end{cases} \tag{3}$$

Note that $1_{FG} = 1_F 1_G$. The function 1_F is a special case of a random variable. A
random variable X is a mapping from the set Ω of a probability triple $(\mathbb{P}, \Omega, \mathcal{F})$

into a separable Banach space $\mathcal{X}$. A random variable X is *measurable* if the pre-image under X of Borel sets in $\mathcal{X}$ are in $\mathcal{F}$. A random variable X is *integrable* if

$$\int_\Omega \|X(\omega)\| d\mathbb{P}(\omega) < \infty. \tag{4}$$

If X is integrable, one can define the *expected value of X*

$$\mathbb{E}X = \int_\Omega X(\omega) d\mathbb{P}(\omega) \tag{5}$$

as a limit of integrals of suitably defined "simple functions" that converge to X, where convergence is in a norm defined on functions from a measure space to a Banach space; see Dunford and Schwartz [23] volume I, chapter III for details. The norm Dunford and Schwartz use is

$$\|X\| = \inf_{\alpha > 0} \arctan\left(\alpha + \mathbb{P}\{\omega : \|X(\omega)\| > \alpha\}\right). \tag{6}$$

If there is more than one probability measure under consideration, for example a family of measures $\mathcal{P} = \{\mathbb{P}_\theta : \theta \in \Theta\}$, we write

$$\mathbb{E}_{\mathbb{P}_\theta} X \equiv \mathbb{E}_\theta X = \int_\Omega X(\omega) d\mathbb{P}_\theta. \tag{7}$$

If X is real-valued and integrable, the variance of X, $\mathbf{Var}(X)$ is

$$\mathbf{Var}(X) \equiv \mathbb{E}(X - \mathbb{E}X)^2. \tag{8}$$

The *σ-algebra generated by the random variable X, $\sigma(X)$*, is

$$\{(X \in B) : B \text{ is a Borel subset of } \mathcal{X}\}. \tag{9}$$

For any event $F \in \mathcal{F}$,

$$\mathbb{P}(F) \equiv \mathbb{E}1_F \equiv \int_F d\mathbb{P} \equiv \int_\Omega 1_F d\mathbb{P}. \tag{10}$$

Events F, $G \subset \mathcal{F}$ are *independent* if $\mathbb{P}(FG) = \mathbb{P}(F)\mathbb{P}(G)$. Two σ-algebras $\mathcal{F}$ and $\mathcal{G}$ are *independent* if every $F \in \mathcal{F}$ is independent of every $G \in \mathcal{G}$. If $\sigma(X)$ is independent of $\mathcal{F}$, we say X is independent of $\mathcal{F}$. We say that the random variables X and Y are independent if $\sigma(X)$ is independent of $\sigma(Y)$.

The *conditional probability of the event F given the event G* is $\mathbb{P}(F|G) \equiv \mathbb{P}(FG)/\mathbb{P}(G)$ for $\mathbb{P}G \neq 0$. *Bayes' Rule* is

$$\mathbb{P}(F|G) = \frac{\mathbb{P}(G|F)\mathbb{P}(F)}{\mathbb{P}(G)} \tag{11}$$

for $\mathbb{P}(F)$, $\mathbb{P}(G) > 0$.

Let $(\mathbb{P}, \Omega, \mathcal{F})$ be a probability triple, let X be a measurable, integrable random variable, and let $\mathcal{G} \subset \mathcal{F}$ be a sub-σ-algebra of $\mathcal{F}$. The *conditional expectation of X given $\mathcal{G}$, $\mathbb{E}(X|\mathcal{G})$*, is any $\mathcal{G}$-measurable function Y such that for every $G \in \mathcal{G}$, $\mathbb{E}(1_G Y) = \mathbb{E}(1_G X)$. If Y is a $\mathbb{P}$-measurable random variable, $\mathbb{E}(X|Y) = \mathbb{E}(X|\sigma(Y))$. If X is $\mathcal{G}$-measurable, it follows that $\mathbb{E}(X|\mathcal{G}) = X$. If X is independent of $\mathcal{G}$, it follows that $\mathbb{E}(X|\mathcal{G}) = \mathbb{E}(X)$. If X is $\mathcal{G}$-measurable, $\mathbb{E}(XY|\mathcal{G}) = X\mathbb{E}(Y|\mathcal{G})$. The *conditional probability of $B \in \mathcal{F}$ given the sub-σ-algebra $\mathcal{G}$ of $\mathcal{F}$* is $\mathbb{P}(B|\mathcal{G}) = \mathbb{E}(1_B|\mathcal{G})$.

A measurable random variable X taking values in a separable Banach space $\mathcal{X}$ induces a probability distribution on the Borel σ-algebra of $\mathcal{X}$ through $\mathbb{P}(B) = \mathbb{P}(X \in B)$ for all Borel sets $B \subset \mathcal{X}$. If Φ is a Borel measure on $\mathcal{X}$ such that $\mathbb{P}(X \in B) = \Phi(B)$ for all Borel sets B, we write $X \sim \Phi$. If two random variables have the same probability distribution (if they are defined on the same σ-algebra $\mathcal{F}$ on the same set Ω, take values in the same space $\mathcal{X}$, and $\mathbb{P}(X \in B) = \mathbb{P}(Y \in B)$ for all Borel sets B), we write $X \sim Y$ and say X *and Y are identically distributed*. If in addition X and Y are independent, then they are *independent and identically distributed (i.i.d.)*.

Suppose that the family $\mathcal{P} = \{\mathbb{P}_\theta : \theta \in \Theta\}$ on a measurable space $\mathcal{G}$ is dominated by a common measure μ. Let $p_\theta(g)$, $g \in \mathcal{G}$ denote the density of $\mathbb{P}_\theta$ with respect to μ. For fixed $g \in \mathcal{G}$, the function

$$\mathcal{L} : \Theta \to \Re^+$$
$$\theta \mapsto p_\theta(g) \tag{12}$$

is called the *likelihood function*. If $X \sim \mathbb{P}_\gamma$ for some $\gamma \in \Theta$ and the value $X = x$ is observed, $\mathcal{L}(\theta)$ is *the likelihood of θ given $X = x$*.

The family of probability distributions $\mathcal{P} = \{\mathbb{P}_\theta : \theta \in \Theta\}$ on a measurable linear vector space $\mathcal{G}$ is a *location family* if there is a mapping $L : \Theta \to \mathcal{G}$ such that $\mathbb{P}_\theta\{B\} = \mathbb{P}\{B - L\theta\}$ for all $B \in \mathcal{B}$, all $\theta \in \Theta$, and some measure $\mathbb{P}$ defined on $\mathcal{G}$.

A *statistic* is a measurable mapping of the data into a measurable space. Let $X \sim \mathbb{P}_\theta$ for some $\theta \in \Theta$. A statistic T is *sufficient for $\mathcal{P}$* if the conditional distribution of X given T does not depend on which "theory" $\mathbb{P}_\theta \in \mathcal{P}$ is true. It is trivially true that X is sufficient for $\mathcal{P}$.

A sequence of random variables $\{G_j\}$ taking values in a topological space $\mathcal{G}$ *converges in probability* to $g \in \mathcal{G}$ if for every neighborhood τ of g,

$$\lim_{j \to \infty} \mathbb{P}\{X_j \in \tau\} = 1. \tag{13}$$

One then writes $G_j \overset{P}{\to} g$.

If S is a bounded Borel subset of $\Re^n$, $U(S)$ is the uniform probability distribution on S. In the special case $n = 1$, $S = [0, 1]$, we write $U[0, 1]$ for the distribution defined by

$$\mathbb{P}(B) = \int_B d\mu, \tag{14}$$

where μ is Lebesgue measure, for all Borel subsets B of $[0, 1]$.

The d-variate normal distribution with mean $\theta \in \Re^d$ and d by d symmetric covariance matrix Σ is denoted $N(\theta, \Sigma)$; if Σ is positive-definite, the density of

$N(\theta, \Sigma)$ with respect to Lebesgue measure is

$$\phi(x) = |\Sigma^{1/2}|^{-1}(2\pi)^{-d/2} \exp\left(-\frac{1}{2}(x-\theta) \cdot \Sigma \cdot (x-\theta)\right), \qquad (15)$$

where $|\Sigma|$ is the determinant of Σ.

If X is a real-valued random variable, the *cumulative distribution function of X* is

$$F_X : \Re \to [0,1]$$
$$x \to \mathbb{P}(X \le x). \qquad (16)$$

The α quantile of a real-valued random variable X is

$$X_\alpha = \inf\{x \in \Re : F_X(x) \ge \alpha\}. \qquad (17)$$

Let Θ be a non-empty subset of a separable Banach space $\mathcal{T}$. The set Θ will represent possible "theories" about the state of nature—the competing models θ that might spawn the observations $X \in \Re^n$. Let $\mathcal{P} = \{\mathbb{P}_\theta : \theta \in \Theta\}$ be a collection of probability measures defined on a common σ-algebra $\mathcal{B}$ on $\Re^n$. If theory $\theta \in \Theta$ be true, the probability distribution of the data X is $\mathbb{P}_\theta$; that is, $X \sim \mathbb{P}_\theta$. We wish to learn something about θ from X. The quadruple $(\Theta, \mathcal{P}, \mathcal{X}, \mathcal{B})$ is a *statistical experiment indexed by Θ*.

Parameters are properties of θ we might wish to learn about. For the purposes of this paper, a parameter is the value at θ of a continuous mapping $g : \Theta \to \mathcal{G}$, where $\mathcal{G}$ is a Banach space. (Note that restricting attention to continuous mappings is not always necessary or desirable; see, e.g., [14].) We insist further that $g(\Theta)$ contain at least two points—g is not constant on Θ. If we sought to estimate θ in its entirety, we might take $\mathcal{G} = \mathcal{T}$. We might instead be interested in a low-dimensional projection of θ, the norm of θ, or some other function or functional.

A parameter $g(\theta)$ is *identifiable* if

$$g(\theta_1) \ne g(\theta_2) \implies \mathbb{P}_{\theta_1} \ne \mathbb{P}_{\theta_2}. \qquad (18)$$

That is, a parameter is identifiable if a change in the parameter always is accompanied by a change in the probability distribution of the data.

The things we can do with X to learn about θ are *decision rules*, measurable mappings from $\Re^n$ to a Banach space $\mathcal{A}$ of *actions*. An *estimator* δ of a parameter $g(\theta)$ is a decision rule whose range is $\mathcal{G}$. An estimator is *consistent* if it converges in probability to the value of the parameter it estimates $(\delta(X) \xrightarrow{P} g(\theta))$; in the norm topology on $\mathcal{G}$, if for every $\varepsilon > 0$,

$$\lim_{n \to \infty} \mathbb{P}_\theta\{\|g(\theta) - \delta(X)\| < \varepsilon\} = 1. \qquad (19)$$

If a parameter is not identifiable, there is no estimator that is consistent whatever be $\theta \in \Theta$. (Obviously, the estimator $\delta(X) = c$ is consistent when $g(\theta) = c$, but because $\#\{g(\Theta)\} \ge 2$, there is some $\theta \in \Theta$ for which $g(\theta) \ne c$.)

The *bias at θ* of the estimator $\delta : \Re^n \to \mathcal{G}$ of the parameter $g(\theta)$ is

$$\text{bias}(\delta) = \text{bias}(\delta; \theta, g) \equiv \mathbb{E}_\theta(\delta(X) - g(\theta)). \tag{20}$$

The *mean norm error at θ* of the estimator $\delta : \Re^n \to \mathcal{G}$ of the parameter $g(\theta)$ is

$$\text{MNE}(\delta) = \mathbb{E}_\theta \|\delta(X) - g(\theta)\|. \tag{21}$$

When $\mathcal{G}$ is a Hilbert space, the square of the MNE usually is called the mean squared error (MSE). When $\mathcal{G} = \Re$, the mean squared error can be decomposed into a sum of two terms, the variance of δ and the square of the bias of δ:

$$
\begin{aligned}
\mathbb{E}_\theta(\delta(X) - g(\theta))^2 &= \mathbb{E}_\theta((\delta(X) - \mathbb{E}_\theta\delta(X) - g(\theta) + \mathbb{E}_\theta\delta(X))^2) \\
&= \mathbb{E}_\theta(\delta(X) - \mathbb{E}_\theta\delta(X))^2 + (\mathbb{E}_\theta\delta(X) - g(\theta))^2 \\
&= \mathbf{Var}(\delta(X)) + \text{bias}^2(\delta(X)).
\end{aligned}
\tag{22}
$$

In most inverse problems, $g(\theta) = \theta$ is not identifiable: this is essentially the problem of nonuniqueness. Moreover, most linear functionals of θ are usually not identifiable in inverse problems either. One of the messages of Backus-Gilbert theory (see section 7.1) [2–5] is that in unconstrained linear inverse problems in infinite-dimensional spaces, the only linear functionals of the model θ that are identifiable are linear combinations of the representers (the components of the forward mapping).

A $1 - \alpha$ *confidence level confidence set* δ for the parameter $g(\theta)$ is a random subset of $\mathcal{G}$ that has probability at least $1 - \alpha$ of containing $g(\theta)$, whatever be $\theta \in \Theta$:

$$\mathbb{P}_\theta\{\delta(X) \ni g(\theta)\} \geq 1 - \alpha, \; \forall \theta \in \Theta. \tag{23}$$

3 A Canonical Inverse Problem

Let $L_2[0, 1]$ denote the Hilbert space of Lebesgue square-integrable functions on the interval $[0, 1]$. Let $(f|g)$ denote the inner product of the functions f and g:

$$(f|g) = \int_0^1 f(t)g(t)dt. \tag{24}$$

Let $\{\Delta_j\}_{j=1}^n$ be a fixed collection of closed, disjoint sub-intervals of $[0, 1]$, each of strictly positive length, and such that there exists an open set $\Delta_0 \subset [0, 1]$ for which

$$\Delta_0 \cap \{\cup_{j=1}^n \Delta_j\} = \emptyset. \tag{25}$$

For $f \in L_2[0, 1]$, define the continuous linear functionals K_j by

$$K_j f = \int_{\Delta_j} f d\mu = (f|1_{\Delta_j}). \tag{26}$$

The functions $1_{\Delta_j} \in L_2[0,1]$ are called the "representers" or "data kernels" for our problem. Note that $\{K_j\}_{j=1}^n$ are linearly independent. We observe data $X = (X_j)_{j=1}^n$, with

$$X_j = K_j f + Z_j, \quad j = 1, \ldots, n, \tag{27}$$

where the noise terms $\{Z_j\}$ are independent and identically distributed normal random variables with zero mean and variance σ^2 (we write $\{Z_j\}$ i.i.d. $N(0, \sigma^2)$). We abbreviate equation (27) by

$$X = Kf + Z. \tag{28}$$

Consider estimating the pair of values $(L_1 f, L_a f)$ from these data, where L_1 and L_a are given by

$$L_1(f) = \int_0^1 f d\mu, \tag{29}$$

and

$$L_a(f) = \int_0^1 \sum_{j=1}^n a_j 1_{\Delta_j} f d\mu, \tag{30}$$

with $(a_j)_{j=1}^n \in \Re^n$. Both L_1 and L_a are bounded linear functionals on $L_2[0,1]$.

We now translate this problem into the notation of the previous section. We identify $\Theta = L_2[0,1]$, $\theta = f$, $\mathcal{X} = \Re^n$, and $\mathcal{P}$ is the location family of n-dimensional normal distributions on $\Re^n$ with independent coordinates and common variance σ^2:

$$\mathbb{P}_\theta(B) = \int_B \prod_{j=1}^n \left(\frac{1}{\sqrt{2\pi}\sigma} \exp[-(x_j - (\theta|\Delta_j))^2/2\sigma^2] \right) d^n x \tag{31}$$

for all Borel sets $B \subset \Re^n$. The parameter space $\mathcal{G} = \Re^2$, and the mapping g is given by

$$g : \Theta \to \Re^2$$

$$\theta \to ((\theta|1), (\theta|\sum_{j=1}^n a_j 1_{\Delta_j})). \tag{32}$$

In this model, θ is not identifiable. Neither is $g(\theta)$, because its first component $(\theta|1)$ can be perturbed arbitrarily without changing the distribution of X (just change θ on Δ_0). The second component of $g(\theta)$, $(\theta|L_a)$, is identifiable. The estimator $\delta(X) = \sum_{j=1}^n a_j X_j = a \cdot X$ is unbiased for $(\theta|L_a)$; there is no estimator of $g(\theta)$ that is unbiased for all $\theta \in \Theta$.

Let $1 = (1, 1, \ldots, 1) \in \Re^n$. Suppose we estimate $g(\theta)$ by the linear estimator

$$\delta(X) = (1 \cdot X, a \cdot X). \tag{33}$$

The bias of $\delta(X)$ is

$$\mathbb{E}_\theta(\delta(X) - g(\theta)) = \left(\int_0^1 (1 - 1_{\bigcup_{j=1}^n \Delta_j}) \theta d\mu, 0 \right). \tag{34}$$

Let $b_1(\theta) = \int_0^1 (1 - 1_{\bigcup_{j=1}^n \Delta_j})\theta d\mu$. The mean squared error of $\delta(X)$ is

$$\mathbb{E}_\theta\|\delta(X) - g(\theta)\|^2 = b_1^2(\theta) + \sigma^2(n + \|a\|^2). \tag{35}$$

Note that the bias and the MSE of $\delta(X)$ are unbounded over Θ.

Suppose we seek a $1 - \alpha$ confidence set $S(X)$ for $g(\theta)$, valid whatever be $\theta \in \Theta$. We restrict attention to rectangular sets (Cartesian products of intervals) in $\mathcal{G} = \Re^2$. Let Z_α denote the α quantile of $N(0,1)$. There is no nontrivial confidence interval for the first component of $g(\theta)$; the interval

$$[a \cdot X - \sigma\|a\|Z_{1-\alpha/2}, a \cdot X + \sigma\|a\|Z_{1-\alpha/2}] \tag{36}$$

has probability $1 - \alpha$ of containing the second component of $g(\theta)$, so the rectangular set

$$\Re \times [a \cdot X - \sigma\|a\|Z_{1-\alpha/2}, a \cdot X + \sigma\|a\|Z_{1-\alpha/2}] \tag{37}$$

is a $1 - \alpha$ confidence set for $g(\theta)$.

Now suppose that in addition to the stochastic errors $\{Z_j\}$, each datum contains also a systematic error τ_j:

$$X_j = K_j f + Z_j + \tau_j, j = 1, \ldots, n. \tag{38}$$

Assume we know that every τ_j satisfies $|\tau_j| \le 1$. We can embed this case in the framework we have developed by appending to θ the n-vector $\tau = (\tau_j)_{j=1}^n \in \Re^n$. These additional parameters, which affect the probability distribution but are not the subject of the estimation problem, are called *nuisance parameters*. Let $\mathbf{T}$ be the hyperrectangular set $\prod_{j=1}^n \{|x| \le 1\} \subset \Re^n$. The parameter space Θ is now $L_2[0,1] \times \mathbf{T}$, which we endow with the following norm. If $\theta = (f, \gamma)$ with $f \in L_2[0,1]$ and $\gamma \in \mathbf{T}$,

$$\|\theta\|^2 = \|f\|^2 + \|\gamma\|^2, \tag{39}$$

where $\|\gamma\|$ is the ordinary Euclidean norm of $\gamma \in \Re^n$. For $\theta = (f, \gamma)$ and $\rho = (g, \delta)$ this corresponds to the inner product

$$(f|g) = \int_0^1 f g d\mu + \gamma \cdot \delta. \tag{40}$$

Let 1_j be the n vector that is zero in every component but j, for which it is 1. The probability distribution $\mathbb{P}_\theta$ on $\Re^n$ is

$$\mathbb{P}_\theta(B) = \int_B \prod_{j=1}^n \left(\frac{1}{\sqrt{2\pi}\sigma} \exp[-(x_j - (\theta|(\Delta_j, 1_j)))^2/2\sigma^2]\right) d^n x \tag{41}$$

for all Borel sets $B \subset \Re^n$. Now

$$g(\theta) = \left((\theta|(1,0)), (\theta|(\sum_{j=1}^n a_j 1_{\Delta_j}, 0)\right). \tag{42}$$

Neither component of $g(\theta)$ is identifiable once we have systematic errors. However, if we use the same estimator $\delta(X)$ as before, its bias is

$$\text{bias}(\delta) = \mathbb{E}_\theta(\delta(X) - g(\theta)) = (b_1(\theta) + 1 \cdot \tau, a \cdot \tau). \tag{43}$$

The first component of the bias is still unbounded for $\theta \in \Theta$, but the second is bounded by $\|a\|_1$. The mean squared error of $\delta(X)$ is

$$\mathbb{E}_\theta \|\delta(X) - g(\theta)\|^2 = ((b_1(\theta) + 1 \cdot \tau)^2 + (a \cdot \tau)^2 + \sigma^2(n + \|a\|^2). \tag{44}$$

As before, we cannot find a nontrivial confidence interval for the first component of $g(\theta)$, but we can still for the second, by widening the interval by the bound on the bias: the rectangular set

$$\Re \times [a \cdot X - \sigma\|a\|Z_{1-\alpha/2} - \|a\|_1, a \cdot X + \sigma\|a\|Z_{1-\alpha/2} + \|a\|_1] \tag{45}$$

is a $1 - \alpha$ confidence set for $g(\theta)$.

4 Statistical Decision Theory

This section presents a framework for comparing estimators and confidence sets in inverse problems: statistical decision theory [46]. Statistical decision theory can be developed in quite abstract settings (Le Cam [35] frames it in the context of mappings from an arbitrary set to an L-space); here we insist that the model set Θ is a subset of a separable Banach space, and that the observation is an element of a separable Banach space.

Decision theory frames statistical estimation and inference as a two-player game, nature versus statistician. Nature picks $\theta \in \Theta$; the value of θ is hidden from the statistician. The statistician picks a *decision rule* or *estimator* δ. Data X are generated according to the probability distribution $\mathbb{P}_\theta$. The statistician pays $\ell(\theta, \delta(X))$. In repeated play, the statistician would pay an average of $r(\theta, \delta) = \mathbb{E}_\theta \ell(\theta, \delta(X))$ per game; the risk r is the expected loss. The statistician seeks to make $r(\theta, \delta)$ small.

Because θ is unknown, different senses of "small" compete, and lead to different strategies for selecting the decision rule δ. The two most common strategies for picking an "optimal" decision function are *minimax* and *Bayes*. *Minimax* decision rules minimize over the statician's choice of decision functions $\delta \in \mathcal{D}$ the maximum risk over nature's possible choices of the parameter θ. This hedges against the possibility that nature plays the game aggressively, picking the value of θ that maximizes the statistician's guaranteed loss. *Bayes* decision rules minimize over decision functions a weighted (by the prior π) average risk over nature's possible choices of the parameter θ. This treats nature as if it draws θ at random from Θ according to the prior distribution π.

The choice of a loss function $\ell(\theta, \gamma)$ is essentially arbitrary; the most common loss function for point estimates of parameters in Euclidean spaces is squared error: $\ell(\theta, \gamma) = ||\gamma - \theta||^2$. For estimating a real-valued functional $L\theta$, the loss might be the absolute error

$$\ell(\theta, \gamma) = |\gamma - \theta|, \tag{46}$$

or a zero-one loss depending on the distance between γ and θ:

$$\ell(\theta, \gamma) = \begin{cases} 0, & |\theta - \gamma| \le c \\ 1, & \text{otherwise.} \end{cases} \tag{47}$$

For set-valued estimates $S(X)$, loss functions typically combine coverage of the parameter, and a measure of the size of the set. For example, we might take $\ell(S, \theta) = 1_{S \not\ni \theta} + \lambda |S|$, with $\lambda > 0$, where $|S|$ is the diameter of S if θ is an element of a metric space, or the Lebesgue measure of S, if $\theta \in \Re^n$. Another possibility is to combine coverage of the parameter and distance from the parameter to the closest point in the set. In particular problems, context can (and perhaps should) dictate the choice of loss functions.

Here is the set-up in a bit more detail. Associated with each possible state of nature $\theta \in \Theta$ and each possible action $a \in \mathcal{A}$ is a nonnegative *loss* $\ell : \Theta \times \mathcal{A} \to \Re^+$. It is helpful to require the loss ℓ to be convex in its second argument: for every $\theta \in \Theta$, for all $\gamma \in [0, 1]$, and for all a_1, a_2 in $\mathcal{A}$,

$$\ell(\theta, \gamma a_1 + (1 - \gamma)a_2) \le \gamma \ell(\theta, a_1) + (1 - \gamma)\ell(\theta, a_2). \tag{48}$$

This holds, for example, if we seek to estimate a parameter $g(\theta)$ using an estimator δ that takes values in $\mathcal{G}$, and the loss is the norm of the error of δ: $\ell(\theta, \delta(X)) = ||g(\theta) - \delta(X)||$.

The *risk at $\theta \in \Theta$ of a decision rule $\delta \in \mathcal{D}$* is

$$r(\theta, \delta) \equiv \mathbb{E}_\theta \ell(\theta, \delta(X)). \tag{49}$$

The *maximum risk of $\delta \in \mathcal{D}$ over Θ* is

$$\rho(\delta) \equiv \sup_{\theta \in \Theta} r(\theta, \delta). \tag{50}$$

The *minimax risk* is

$$\rho^* = \inf_{\delta \in \mathcal{D}} R(\delta). \tag{51}$$

If π is a probability measure on Θ, the *posterior risk of δ for prior π* is

$$\rho_\pi(\delta) = \int_\Omega r(\theta, \delta) d\pi(\theta). \tag{52}$$

An estimator δ is *admissible* for loss ℓ if there is no other estimator δ' such that

$$r(\theta, \delta') \le r(\theta, \delta), \quad \forall \theta \in \Theta \tag{53}$$

and $r(\theta, \delta') < r(\theta, \delta)$ for at least one $\theta \in \theta$. If such a δ' exists, it is said to *dominate* δ. If δ is not admissible it is *inadmissible*.

Theorem 1. *The Rao-Blackwell Theorem (see [37] Th. 1.7.8) Let X have probability distribution $\mathbb{P}_\theta \in \mathcal{P} = \{\mathbb{P}_\gamma : \gamma \in \Theta\}$, and let T be sufficient for $\mathcal{P}$. Let δ be an estimator of the parameter $g(\theta)$, and let the loss $\ell(\theta, a)$ be strictly convex in a. If $\delta(X)$ is integrable and*

$$R(\theta, \delta) = \mathbb{E}_\theta \ell(\theta, \delta(X)) < \infty \tag{54}$$

and

$$\eta(T) = E(\delta(X)|T), \tag{55}$$

then

$$R(\theta, \eta) < R(\theta, \delta) \tag{56}$$

unless $\delta(X) = \eta(T)$ a.s. $(\mathbb{P}_\theta)$.

Because X is sufficient for $\mathcal{P}$, the Rao-Blackwell theorem gives conditions under which randomized estimators are dominated by their non-randomized versions.

5   Estimation as a Decision Problem

We now specialize to the case of estimating a parameter $g(\theta)$ where $g : \Theta \to \mathcal{G}$, with $\mathcal{G}$ a Banach space with norm $\|\cdot\|$. Take the action space $\mathcal{A}$ to be $\mathcal{G}$ as well, and consider the set $\mathcal{D}$ of decision rules δ that are $\mathcal{P}$-measurable mappings from $\mathcal{X}$ to $\mathcal{G}$. A standard choice of $\ell(\theta, a)$ is then $\|g(\theta) - a\|$, which is convex. Then $r(\theta, \delta)$ is the average error in the estimator, measured in the norm of $\mathcal{G}$— the MNE defined in (21). A less common choice is $\ell(\theta, a) = 1_{g(\theta \notin B_c(a)}$, where $B_c(a) = \{g \in \mathcal{G} : \|g - a\| \le c\}$. When $\mathcal{G}$ is a Euclidean space, the most common loss function is $\|a - g(\theta)\|^2$. When $\mathcal{G} = \Re$ (estimating a single real parameter), common loss functions are $\ell(\theta, a) = |g(\theta) - a|^p$ and $\ell(\theta, a) = 1_{|g(\theta)-a|>c}$.

6   Confidence Sets as Decisions

A $1 - \alpha$ confidence set for the parameter $g(\theta) \in \mathcal{G}$ is a random set $\delta(X) \subset \mathcal{G}$ with the property that

$$\mathbb{P}_\theta\{\delta(X) \ni g(\theta)\} \ge 1 - \alpha, \ \forall \theta \in \Theta. \tag{57}$$

Take the action space $\mathcal{A}$ to be a collection of subsets of $\mathcal{G}$, and consider the set $\mathcal{D}$ of decision rules δ that are measurable in the following sense: Recall that $\mathcal{G}$ and $\mathcal{X}$ are separable Banach spaces. Let $f : \mathcal{G} \times \mathcal{X} \to \{0, 1\}$ be measurable in the product σ-algebra of $\mathcal{G}$ and $\mathcal{X}$; this implies that for fixed g, $f(g, \cdot)$ is measurable, and that for fixed x, $f(\cdot, x)$ is measurable. Let $\delta(X) = \{g : f(g, X) = 1\}$. Then $\delta(X)$ is a measurable confidence set; the class of subsets of $\mathcal{G}$ it produces is $\mathcal{A} = \{\{g \in \mathcal{G} : f(g, x) = 1\} : x \in \mathcal{X}\}$, the members of which are measurable subsets of $\mathcal{G}$.

We can choose f so that the elements of $\mathcal{A}$ have a given maximum diameter $|A| = \sup_{g,h \in A} \|g - h\|$, for example. In that case, a reasonable choice of loss functions is $\ell(\theta, a) = 1_{g(\theta) \notin a}$. Then $r(\theta, \delta)$ is the non-coverage probability: the probability that the set $\delta(X)$ does not include $g(\theta)$. One can seek small confidence sets with a given non-coverage probability whatever be $\theta \in \Theta$ by taking $\mathcal{A}$ to be a collection of Borel subsets of $\mathcal{G}$ with a given maximum diameter, and varying that diameter until the supremum of $R(\delta)$ over Θ is the target non-coverage probability.

The natural action space for confidence sets is a collection of subsets of $\mathcal{G}$. That is not a linear vector space, so convexity of the loss with respect to the action does not apply naturally. However, many results in decision theory depend on the assumption that the loss is convex in the action. One can make sense of convex combinations of actions in the problem of constructing confidence sets by considering randomized confidence sets, defined as follows. The basic idea is to replace the indicator function of a set, which takes values in $\{0, 1\}$, by a membership function that takes values in $[0, 1]$, much like the theory of "fuzzy sets" [47]; indicator functions are special cases. Convex combinations of membership functions are membership functions.

Let $\delta : \mathcal{G} \times \mathcal{X} \to [0, 1]$ be Borel measurable in the product topology. We augment the data X by an additional random variable $U \sim U[0, 1]$, independent of X; the random confidence set is

$$C(X, U) = \{g \in \mathcal{G} : \delta(X, g) \geq U\}. \tag{58}$$

The value of $\delta(X, g)$ is the probability that the confidence set contains the point g, with X held fixed. The special case that $\delta(\cdot, \cdot)$ takes only the values 0 and 1 corresponds to picking the indicator function of a subset of $\mathcal{G}$, as described above. The coverage probability of $C(X, U)$ is

$$\begin{aligned} \gamma(\theta, \delta) &= \mathbb{P}\{C(X, U) \ni g(\theta)\} \\ &= \mathbb{P}\{\delta(X, g(\theta)) \geq U\} \\ &= \mathbb{E}_\theta \delta(X, g(\theta)). \end{aligned} \tag{59}$$

The expected volume of $C(X, U)$ with respect to the measure μ on $\mathcal{G}$ is

$$\begin{aligned} \nu(\theta, \delta) &= \mathbb{E} \int_\mathcal{G} 1_{C(X,U)}(g) d\mu(g) \\ &= \mathbb{E}_\theta \int_\mathcal{G} \delta(X, g) d\mu(g) \\ &= \int_\mathcal{G} (\mathbb{E}_\theta \delta(X, g)) d\mu(g) \end{aligned} \tag{60}$$

by Fubini's theorem. With decisions δ defined this way, a convex combination of decisions is another element of the class. Moreover, the coverage probability and the expected volume are linear functionals of the decision rule, so they are convex, and we could consider a composite convex loss function

$$\ell(\theta, \delta) = -\gamma(\theta, \delta) + \lambda \nu(\theta, \delta) \tag{61}$$

for some $\lambda \in \Re^+$, to optimize a tradeoff between coverage probability and expected volume. This loss is convex, but not strictly convex.

7   The Things We Do to Data

There are a variety of "recipes" for constructing estimators; perhaps the most common in statistics are maximum likelihood (MLE) and Bayes, both of which have asymptotically optimal properties under certain assumptions. However, both can be inconsistent, even in finite-dimensional problems; see below. In the inverse problems literature, regularization (especially regularized least-squares, related to maximum penalized likelihood), truncation (related to the method of sieves) and the Backus-Gilbert method are among the most common procedures. In this section, $\theta \in \Theta \subset \mathcal{T}$, a separable Banach space, $K : \Theta \to \Re^n$ is linear, and $X = K\theta + Z$, where Z is usually a vector of i.i.d. Gaussian errors.

7.1   Backus-Gilbert Estimates

In a seminal series of papers [5, 2–4] George Backus and Freeman Gilbert introduced a rigorous basis to linear inverse theory in geophysics. Here, "a linear inverse problem" means that the data are linearly related to the unknown but for additive noise, and the estimators are linear in the data. In statistical terms, Backus and Gilbert showed that the only identifiable linear functionals in an unconstrained linear inverse problem ($\Theta = \mathcal{T}$) (meaning $\Theta = \mathcal{T}$) are linear combinations of the measurement functionals. That is, if the functional $L : \Theta \to \Re$, then $L\theta$ is identifiable iff $L = \gamma \cdot K$ for some $\gamma \in \Re^n$. In that case, $\gamma \cdot X$ is an unbiased estimate of $L\theta$, and if Σ is the covariance matrix of Z, the variance of the estimate is

$$E((\gamma \cdot X - \gamma \cdot K\theta)^2) = E((\gamma \cdot Z)^2) = \gamma \cdot \Sigma \cdot \gamma = \|\gamma\|_\Sigma^2. \qquad (62)$$

Backus and Gilbert focus on the case Θ is a Hilbert space of functions of position $r \in \mathcal{D} \subset \Re^n$, and develop a measure of "nearness" of L to the point-evaluator $\theta \mapsto \theta(r_0)$, which typically is not a member of Θ (Θ typically does not possess enough smoothness to be a reproducing kernel Hilbert space [1]). Backus and Gilbert develop a framework for trading off between the variance of the estimate and the nearness of the estimate to an estimate of $\theta(r_0)$.

When there are additional constraints on the unknown θ, for example, if Θ is a norm-bounded ellipsoid in a Hilbert space $\mathcal{T}$, much more is possible than Backus-Gilbert theory would suggest; see, e.g., [8–10, 40] and the sections below. Moreover, Backus-Gilbert estimates are not generally optimal for two-norm loss when three or more linear functionals are estimated; see section 7.5 below.

7.2   Maximum Likelihood Estimation and Its Variants

Recall the definition of the likelihood function $\mathcal{L}(\theta|X = x)$ from section 2. The basic idea behind maximum likelihood is to estimate θ by the value $\gamma \in \Theta$

for which the likelihood function $\mathcal{L}(\theta|X = x)$ for the observation $X = x$. The spirit of the approach is that the maximizing value of θ is "most likely" to be the correct one. In smooth finite-dimensional problems, this approach has some nice asymptotic properties; see Lehmann and Casella [37]. In many problems, however, it runs into trouble.

First of all, recall that in order to define the likelihood function, we need the set of probability distributions $\mathcal{P} = \{\mathcal{P}_\theta : \theta \in \times\}$ to be dominated by a common measure μ. Second, we need the likelihood to attain its maximum (this can be overcome by maximizing the likelihood approximately; see inequalities (63) and (64)). Third, we need the maximizer to be unique, or else we need a rule for choosing among maximizers. Fourth, we need $\arg\max_{\theta \in \Theta} \mathcal{L}(\theta|X)$ to be a measurable function of X; this requires additional assumptions. Even when these assumptions are met, maximum likelihood can have pathological properties, including being inconsistent even when a consistent estimator exists. See Le Cam [36] and examples in Lehmann and Casella [37].

One problem maximum likelihood faces even in quite regular inverse problems is the existence of infinitely many maximizers. A partial solution is to add a penalty term; this is quite analogous to regularization. Indeed, many regularization schemes can be viewed as maximum penalized likelihood estimators. An alternative is to limit the dimension of the parameter to obtain an overdetermined problem; the method of sieves implements this approach. See Shen [39] for a recent study of MLE, penalized MLE, and the method of sieves. Note that the likelihood function can be replaced by other functions of the parameter and the data, leading to more general M-estimates. For example, applying least-squares to data with non-Gaussian errors is a form of M-estimation.

Penalized maximum likelihood subtracts a nonnegative penalty term $J(\theta)$ from the the likelihood function before maximizing it. The penalty functional (or *regularization functional*) J is typically a quadratic norm or quadratic seminorm. For Gaussian errors, this leads to linear estimates of θ. The approximate penalized maximum likelihood estimator $\hat{\theta}_n$ is one that approximately maximizes the penalized likelihood: $\hat{\theta}_n$ such that

$$L(\hat{\theta}_n|X = x) - \lambda_n J(\hat{\theta}_n) \geq \sup_{\theta \in \Theta} L(\theta|X = x) - \lambda_n J(\theta) - O(\epsilon_n^2), \qquad (63)$$

where $\lambda_n \to 0$ as $n \to \infty$ and $\epsilon_n \to 0$ as $n \to \infty$. The constant λ_n is the *regularization parameter*; the constant ϵ_n allows us to find an approximate maximizer rather than the exact maximizer. If Θ and $\mathbb{P}_\theta$ satisfy various conditions and if λ_n vanishes at the right rate, this kind of estimator can be consistent and efficient in various senses.

Let $(\Theta_k)_{k=1}^\infty$ be a sequence of subsets of the Banach space $\mathcal{T}$ of which Θ is a subset. Let $d_k(\theta) = \inf_{\gamma \in \Theta_k} \|\gamma - \theta\|$. If for all $\theta \in \Theta$, $\lim_{k \to \infty} d_k(\theta) = 0$, (Θ_k) is a *sieve*. The collection of subspaces spanned by the first j elements of an ordered basis, $1 \leq j \leq k$, is a typical sieve. The idea is to maximize the likelihood approximately within the approximating set Θ_n: find $\hat{\theta}_n \in \Theta_n$ such that

$$L(\hat{\theta}_n|X = x) \geq \sup_{\theta \in \Theta_n} L(\theta|X = x) - O(\epsilon_n^2), \qquad (64)$$

where n is the number of data, and $\epsilon_k \to 0$ as $n \to \infty$.

Both penalization and sieves work essentially by forcing the estimator to lie in a compact subset of Θ, but allowing that subset to grow in a controlled way as $n \to \infty$. Consistency and optimality results for penalized maximum likelihood and the method of sieves depend on assumptions about θ to control the bias of the estimator; for example, if θ is a function, some smoothness is required. The choices of $\epsilon = \epsilon(n)$ and (Θ_n) or $J(\theta)$ and λ_n matter, and optimal rules tend to depend on assumptions about θ that cannot be verified empirically. Moreover, to my knowledge, the finite-sample behavior of approximate penalized likelihood and the method of sieves are not guaranteed to be good, even when all the technical conditions are met—the theorems are asymptotic as $n \to \infty$.

Choosing λ. From the point of view of stability, any strictly positive value of λ suffices; the error magnification decreases as λ increases. A common way to choose λ in geophysical inverse problems leads to "Occam's inversion," named for William of Occam by Constable *et al.* [12]. In paraphrase, Occam's razor is that when one choosing among competing hypotheses that explain the data adequately, one should pick the simplest. The quantitative prescription in [12] is to pick the model that attains the smallest value of the regularization functional among models that predict the data within a normalized sum of squared errors equal to one:

$$\frac{1}{n} \sum_{j=1}^{n} (X_j - K_j\theta)^2/\sigma_j^2 = 1. \tag{65}$$

For independent Gaussian errors, the expectation of the left hand side for the true value of θ is unity; the value on the right can be tuned more finely to be a quantile of the distribution of the sum of squares of the errors.

Genovese and Stark [28] give necessary conditions for such estimators to be consistent, and sufficient conditions, but not necessary and sufficient conditions. O'Sullivan [38] presents regularization of inverse problems from a statistical point of view, and gives various senses in which regularized estimates are optimal among linear estimates, along with a discussion of methods for selecting λ. Cross-validation and generalized cross-validation are popular in the statistical literature; see Wahba [45] for the nonparametric regression case and O'Sullivan [38] for applications to inverse problems and further references.

The method of sieves is quite close to a common numerical approach to inverse problems: approximating Θ by a finite-dimensional subspace and performing least-squares collocation within the subspace. Unfortunately, as mentioned above, the choice of the approximating space matters—it controls the bias/variance tradeoff—and good results depend critically on assumptions about θ that tend to be unverifiable.

7.3 Bayes Estimation

The description here is based in part on Lehmann and Casella [37], chapter 4; see Hartigan [30] for a more complete and technical exposition of the Bayesian approach, Gelman *et al.* [27] for Bayesian data analysis; and Le Cam [35] for more abstract and rigorous development. We shall need the assumption that Θ is a measurable space to use the Bayesian infrastructure; our assumption that $\mathcal{T}$ is a separable Banach space suffices. To be able to use likelihoods, we also assume that the probabilities $\mathcal{P} = \{\mathbb{P}_\theta : \theta \in \Theta\}$ are dominated by a common measure μ; the density of $\mathbb{P}_\theta$ with respect to μ is denoted $p_\theta(x)$. We shall assume that $p_\theta(x)$ is jointly measurable with respect to θ and x.

According to I.J. Good [29]

> ... the essential defining property of a Bayesian is that *he regards it as meaningful to talk about the probability $P(H|E)$ of a hypothesis H, given evidence E.*

One of the fundamental tenets of Bayesian inference is that uncertainty always can be represented as a probability distribution. Before any data are collected, a Bayesian has a *prior probability distribution* $\pi(\theta)$ for the unknown model θ—indeed, a Bayesian can estimate θ without data. (In addition to the probability distributions $\mathbb{P}_\theta$ on $\mathcal{X}$, we now have a probability distribution on another space, Θ; let $\mathbb{E}_\pi$ denote expectation with respect to π.) From the prior distribution and the law $\mathbb{P}_\theta$ of the data X given θ, we can find the joint distribution of θ and X. The *marginal distribution of X* or the *predictive distribution of X* is a mixture of the distributions $\mathbb{P}_\theta$ according to $\pi(\theta)$. When $\{\mathbb{P}_\theta\}$ is dominated by μ, as we have assumed, the density with respect to μ of the predictive distribution is

$$m(x) = \int p_\theta(x) d\pi(\theta). \tag{66}$$

Observing data allows one to update the prior probability distribution using the density of the observation given θ, $p_\theta(x)$, and Bayes' rule to find the *posterior distribution of θ given $X = x$*:

$$\pi(\theta|X = x) = \frac{p_\theta(x)\pi(\theta)}{m(x)}. \tag{67}$$

(Note that $m(x)$ can vanish; this happens with probability 0. It is not necessary that $\{\mathbb{P}_\theta\}$ be dominated for the posterior distribution to exist, but it makes the formula simple.) At this point, some Bayesians are done: the fundamental objects are probability distributions, and (given the prior), the posterior distribution $\pi(\theta|X = x)$ is all there is to know about θ. Computing posterior distributions can be quite difficult, and a considerable amount of computational machinery has been developed for that purpose; see Gelman *et al.* [27] for examples and references, and [44] for an example in an inverse problem in microwave cosmology.

Tarantola [43] gives a Bayesian treatment of inverse problems more-or-less along these lines, but truncates the problems to finite dimensions at the go; moreover, the distributions he treats computationally are limited primarily to Gaussians. Many of the difficulties of the Bayesian approach vanish if all the distributions are Gaussian and the dimension is finite. See below for a few of those and references. Backus [7] gives a Bayesian treatment of an infinite-dimensional inverse problem in geomagnetism; Backus [10] develops a framework for inference that defers the frequentist/Bayesian decision.

Some Bayesians and most frequentists are interested in estimating a parameter $g(\theta)$; the posterior distribution of θ induces a probability distribution on $g(\theta)$. We can also use the posterior to find point estimates that minimize the risk with respect to some loss function. Given a loss function $\ell(\theta, a) : \Theta \times \mathcal{A} \to \Re^+$, recall that the risk at θ of the estimator $\delta : \mathcal{X} \to \mathcal{A}$ is

$$r(\theta, \delta) = \mathbb{E}_\theta \ell(\theta, \delta(X)). \tag{68}$$

The *average risk of δ for prior π on Θ* is

$$\rho_\pi(\delta) = \int_\Theta r(\theta, \delta) d\pi(\theta). \tag{69}$$

An estimator δ_B that minimizes the average risk for prior π is called a *Bayes estimator*; its risk is the *Bayes risk for prior π*. Under various conditions, δ_B exists; for example, if there exists an estimator with finite risk, and if a minimizer $\delta_\pi(x)$ of

$$\mathbb{E}_\pi(\ell(\theta, \delta(x))|X = x) \tag{70}$$

exists for almost all x and depends measurably on x, then δ_π is a Bayes estimator for prior π and loss ℓ.

Under various conditions, every admissible estimator is either a Bayes estimator for some prior, or a limit of such estimators. See Le Cam [35] for details. The maximum risk over $\theta \in \Theta$ of a Bayes estimator can be quite large. Whatever be the probability measure π on Θ

$$\sup_{\theta \in \Theta} r(\theta, \delta) \geq \int_\Theta r(\theta, \delta) d\pi(\theta) : \tag{71}$$

the average risk lower-bounds the maximum risk. Because the Bayes estimator minimizes the right hand side, the maximum risk of any estimator δ is bounded below by the Bayes risk. In particular, the infimum over estimators δ of the maximum risk over θ, called the *minimax risk* (section 7.4) is bounded below by the Bayes risk for any prior π. In many circumstances, the minimax risk is equal to the maximum Bayes risk over all priors π on Θ; see Le Cam [35] for precise theorems. Both the Bayes risk and the Bayes estimator depend on the prior π; comparing the Bayes risk to the minimax risk is a way to quantify the sensitivity of the risk to the particular prior chosen.

Difficulties and Pathologies in Bayes Estimation If it is known that $\theta \in \Theta$, then the support of π should be a subset of θ: $\pi(\Theta) = 1$. Backus [6, 8, 7] points out the difficulty of capturing "hard" constraints, such as $\|\theta\| \leq 1$, using prior probability distributions in high- or infinite-dimensional spaces. Consider, for example, the case Θ is the subset $\{\theta : \|\theta\| \leq 1\}$ of a separable Hilbert space $\mathcal{T}$. Then Θ is rotationally invariant, so absent other information, it is reasonable to insist that π be rotationally invariant in $\mathcal{T}$ as well. Backus shows that any rotationally invariant prior on $\mathcal{T}$ assigns probability one either to the event $\{\|\theta\| = 0\}$ or to the event $\{\|\theta\| = \infty\}$: it is not possible to capture the constraint $\|\theta\| \leq 1$ as a prior probability distribution without injecting additional information (imposing a preference for directions in $\mathcal{T}$). Freedman [25] gives a complete characterization of infinite-dimensional probability distributions that are rotationally invariant in all finite-dimensional subspaces: they are mixtures of independent zero-mean Gaussian random variables.

A Bayesian estimate is *frequentist consistent* for $g(\theta)$ if for each $\theta \in \Theta$ and each neighborhood τ of $g(\theta)$, the posterior probability that $g(\theta) \in \tau$ given X converges to one almost surely ($\mathbb{P}_\theta$) as $n \to \infty$. Diaconis and Freedman [13] show that in nonparametric regression, whether or not a Bayes estimator is frequentist consistent depends on the prior. In particular, a hierarchical mixture of flat priors on nested finite-dimensional spaces leads to inconsistency in the problem they study.

7.4   Minimax Estimation

The minimax risk ρ^* is the smallest worst-case risk over $\theta \in \Theta$, over a class $\mathcal{D}$ of decisions:

$$\rho^* = \rho^*(\Theta, \mathcal{P}, \ell, \mathcal{D}) \equiv \inf_{\delta \in \mathcal{D}} \sup_{\theta \in \Theta} \mathbb{E}_\theta \ell(\theta, \delta(X)). \qquad (72)$$

If there exists an estimator $\delta^* \in \mathcal{D}$ that has maximum risk ρ^*, it is a *minimax estimator*.

The class of estimators $\mathcal{D}$ might be all measurable functions of X, or we might want to limit the complexity of the estimator, for example, by considering only linear, affine (inhomogeneous linear), or quadratic estimators.

Donoho [17] studies minimax estimation of a linear functional L of an element of a convex subset Θ of $\mathcal{T} = \ell_2$ from linear data $K\theta$ contaminated by additive i.i.d. zero-mean normally distributed errors. Donoho considers three loss functions: squared error, absolute error, and the length of a fixed length confidence interval with a specified minimum coverage probability. He shows for all these measures that the risk of the minimax affine estimator is within a fraction of the risk of the minimax measurable estimator, and shows how to construct the minimax affine estimator. He shows that the risk of the hardest one-dimensional subproblem is equal to the risk in the original infinite-dimensional problem, which allows him to use results about estimating the mean γ of a normal distribution subject to the constraint that $\gamma \in [-\tau, \tau]$ to find the risk in the original problem. The fundamental entity in the development is the *modulus of continuity of L*:

$$\omega(\epsilon; L, K, \Theta) \equiv \sup_{\theta_1, \theta_2 \in \Theta} \{|L\theta_1 - L\theta_2| : \|K\theta_1 - K\theta_2\| \leq \epsilon\}. \qquad (73)$$

This quantity also arises in the theory of optimal recovery, and Donoho establishes an equivalence between statistical estimation in Gaussian noise and optimal recovery in deterministic noise, through a re-calibration of the noise level. The results do not seem to translate to non-Gaussian noise, nor to estimating the whole object θ; however, Donoho [16] addresses minimax estimation of a function $\theta = \theta(t)$ in nonparametric regression and inverse problems, with L_∞ loss

$$\ell(\theta, g) = \sup_t |\theta(t) - g(t)|, \tag{74}$$

and Donoho and Nussbaum [22] study minimax estimation of a quadratic functional in a similar setting.

Donoho's [17] approach has been applied to inverse problems in geomagnetism [41] and microwave cosmology [44].

7.5   Shrinkage Estimators

In contrast with the problem of estimating a single linear functional, where, as noted above, affine estimators are nearly optimal, when one seeks to estimate three or more linear functionals, nonlinear estimators can reap large benefits.

Suppose that X has a d-variate normal distribution with independent coordinates: $X \sim N(\theta, I)$, with $d \geq 3$. Charles Stein proved in 1956 the surprising result that for squared-error loss, the maximum likelihood estimator of θ, namely X, is not admissible for θ [42]. He showed that estimators that "shrink" the observations nonlinearly towards the origin dominate the sample mean; in particular,

$$\delta_S(x) = \left(1 - \frac{\alpha}{\beta + \|x\|^2}\right) x \tag{75}$$

has uniformly smaller mean squared error than $\delta(x) = x$ when α is sufficiently small and β is sufficiently large. Stein's original result has been refined and extended in a variety of ways. James and Stein [33] showed that

$$\delta_{JS}(x) = \left(1 - \frac{\alpha}{\|x\|^2}\right) x \tag{76}$$

with $0 < \alpha \leq 2(d-2)$ suffices; $\alpha = d - 2$ is optimal in this family. For further generalizations (e.g., to distributions other than the normal) and references, see [24]. The variable $(1 - (d-2)/\|x\|^2)$ is called the *shrinkage factor*. The James-Stein estimator has the slightly unsavory feature that the shrinkage factor can be negative, yielding an estimator that both shrinks and reflects the data. Indeed, the *James-Stein positive part estimator*

$$\delta_{JS}^+(x) = \left(1 - \frac{d-2}{\|x\|^2}\right)^+ x \tag{77}$$

dominates the James-Stein estimator. The positive-part estimator is not minimax, but it is hard to improve upon [37].

Stein's result has implications for Backus-Gilbert theory: if there are three or more data, and one seeks to estimate three or more linear functionals of the model, a shrinkage estimator sometimes can do better in mean-squared error than the Backus-Gilbert unbiased estimates. The following theorem is relevant:

Theorem 2. *Lehmann and Casella [37], Theorem 5.7 Let $X \sim N(\theta, \Sigma)$ in dimension $d \geq 3$. Let $\mathrm{tr}(\Sigma)$ be the trace of Σ and let $\lambda_{\max}(\Sigma)$ be the largest eigenvalue of Σ. For squared error loss $\ell(\theta, a) = \|a - \theta\|^2$, the estimator*

$$\delta(x) = \left(1 - \frac{c(\|x\|^2)}{\|x\|^2}\right) x \tag{78}$$

is minimax provided

  1. $0 \leq c(\|x\|^2) \leq 2\mathrm{tr}(\Sigma)/\lambda_{\max}(\Sigma) - 4$, and
  2. $c(\cdot)$ is nondecreasing.

Because the risk of X is constant, it follows that this estimator dominates X in mean squared error; moreover, taking its positive part improves it further.

The intuition for the condition on the trace versus the maximum eigenvalue is as follows: suppose that there is one coordinate with very high variance compared with the rest. Then the risk is driven by that coordinate, and the problem is essentially one-dimensional—but shrinkage does not help in one dimension (in the absence of prior information).

In a series of interesting papers (e.g., [18, 15, 19, 21, 20]) Donoho, Johnstone, and co-authors have shown that in some problems with Gaussian errors, projecting the data onto an unconditional basis and shrinking the coefficients can attain essentially the minimax risk in some nonparametric regression and linear inverse problems with minimal constraints on the underlying model. This non-linear procedure out-performs any linear one in mean squared error. The key seems to be that the basis diagonalizes the prior information about the object to be recovered (in the case of inverse problems, there are two bases, a wavelet basis and a vaguelette basis, which nearly diagonalize the prior information and the forward mapping K), and that the basis is unconditional.

7.6 Confidence Set Inference and Strict Bounds

Backus [9] gives a method for constructing a conservative confidence set for a linear functional of a model in a separable Hilbert space $\mathcal{T}$ using a prior constraint on a quadratic functional of the model: $\Theta = \{\theta \in \mathcal{T} : Q(\theta, \theta) \leq 1\}$. The method decomposes the parameter θ into a part controlled by the prior constraint, and a part controlled by the data. The kinds of problems to which the method can be applied can also be treated using Donoho's minimax approach [17]; see [41] for a comparison of the methods on a problem in geomagnetism also treated by [9].

"Strict bounds" are a generic approach to constructing confidence sets in inverse problems. Let $g : \Theta \to \Re^k$, where k could be infinite. We seek a confidence interval for each component g_i of g. Using the knowledge of the family of distributions $\mathcal{P}$, we can find a $1 - \alpha$ confidence set $C(X)$ for $\theta \in \Theta$; i.e. a set such that whatever be $\theta \in \Theta$,

$$\mathbb{P}_\theta\{C(X) \ni \theta\} \geq 1 - \alpha. \tag{79}$$

(This is easiest when $\mathcal{P}$ is a translation family.) Report as the confidence interval for $g_i(\theta)$ the interval

$$[\inf_{\theta \in C(X)} g(\theta), \sup_{\theta \in C(X)} g(\theta)]. \tag{80}$$

The optimization problems to find the endpoints of the intervals sometimes can be solved exactly; when that is not possible, it is still sometimes possible to approximate them conservatively using conjugate duality. See [40] for examples and techniques, and an application to an inverse problem in seismology. See [31, 32] for an application to a problem in seismicity.

The strict bounds approach is a fairly general tool. For example, in many cases no metric or topology on Θ is needed, limiting the assumptions one needs to make [40]. It is straightforward to incorporate systematic errors into strict bounds, too; [40] gives an example in helioseismology that includes systematic uncertainty in the functionals measured, as well as the stochastic uncertainty in the measurements. However, strict bounds are typically far from optimal: the confidence intervals it generates often are longer than they need to be to attain the nominal confidence level. This seems to stem from choosing $C(X)$ generically, rather than tailoring $C(X)$ to the particular g of interest, and the geometry of the set Θ.

8 Conclusions

Couching inverse problems in statistical language permits a unified view and comparison of standard inversion techniques. It also makes it clearer how statistical tools can be brought to bear on inverse problems.

9 Acknowledgments.

This work was performed while the author was on appointment as Miller Research Professor in the Miller Institute for Basic Research in Science, and was supported by grants from the National Science Foundation and NASA. I thank D.A. Freedman for many helpful discussions.

References

1. N. Aronszajn. Theory of reproducing kernels. *Trans. Amer. Math. Soc.*, 68:337–404, 1950.

2. G. Backus. Inference from inadequate and inaccurate data, I. *Proc. Natl. Acad. Sci.*, 65:1–7, 1970.

3. G. Backus. Inference from inadequate and inaccurate data, II. *Proc. Natl. Acad. Sci.*, 65:281–287, 1970.

4. G. Backus. Inference from inadequate and inaccurate data, III. *Proc. Natl. Acad. Sci.*, 67:282–289, 1970.

5. G. Backus and F. Gilbert. The resolving power of gross Earth data. *Geophys. J. R. astron. Soc.*, 16:169–205, 1968.

6. G.E. Backus. Isotropic probability measures in infinite-dimensional spaces. *Proc. Natl. Acad. Sci.*, 84:8755–8757, 1987.

7. G.E. Backus. Bayesian inference in geomagnetism. *Geophys. J.*, 92:125–142, 1988.

8. G.E. Backus. Comparing hard and soft prior bounds in geophysical inverse problems. *Geophys. J.*, 94:249–261, 1988.

9. G.E. Backus. Confidence set inference with a prior quadratic bound. *Geophys. J.*, 97:119–150, 1989.

10. G.E. Backus. Trimming and procrastination as inversion techniques. *Phys. Earth Planet. Inter.*, 98:101–142, 1996.

11. R.F. Bass. *Probabilistic Techniques in Analysis.* Springer-Verlag, New York, 1995.

12. S.C. Constable, R.L. Parker, and C.G. Constable. Occam's inversion: A practical algorithm for generating smooth models from electromagnetic sounding data. *Geophysics*, 52:289–300, 1987.

13. P.W. Diaconis and D. Freedman. Consistency of Bayes estimates for nonparametric regression: normal theory. *Bernoulli*, 4:411–444, 1998.

14. D.L. Donoho. One-sided inference about functionals of a density. *Ann. Stat.*, 16:1390–1420, 1988.

15. D.L. Donoho. Nonlinear solution of linear inverse problems by wavelet-vaguelette decomposition. Technical report, Dept. of Statistics, Stanford Univ., 1991.

16. D.L. Donoho. Exact asymptotic minimax risk for sup norm loss via optimal recovery. *Probab. Theory and Rel. Fields*, 99:145–170, 1994.

17. D.L. Donoho. Statistical estimation and optimal recovery. *Ann. Stat.*, 22:238–270, 1994.

18. D.L. Donoho and I.M. Johnstone. Wavelets and optimal nonlinear function estimates. Technical Report 281, Dept. Statistics, Univ. of Calif., Berkeley, 1990.

19. D.L. Donoho and I.M. Johnstone. Ideal spatial adaptation by wavelet shrinkage. *Biometrika*, 81:425–455, 1994.

20. D.L. Donoho and I.M. Johnstone. Minimax estimation via wavelet shrinkage. *Ann. Stat.*, 26:879–921, 1998.

21. D.L. Donoho, I.M. Johnstone, G. Kerkyacharian, and D. Picard. Wavelet shrinkage: asymptopia? (with discussion). *J. Roy. Stat. Soc., Ser. B*, 57:301–369, 1995.

22. D.L. Donoho and M. Nussbaum. Minimax quadratic estimation of a quadratic functional. *J. Complexity*, 6:290–323, 1990.

23. N. Dunford and J.T. Schwartz. *Linear Operators.* John Wiley and Sons, New York, 1988.

24. S.N. Evans and P.B. Stark. Shrinkage estimators, Skorokhod's problem, and stochastic integration by parts. *Ann. Stat.*, 24:809–815, 1996.

25. D.A. Freedman. Invariants under mixing which generalize de Finetti's Theorem: continuous time parameter. *Ann. Math. Stat.*, 34:1194–1216, 1963.

26. D.A. Freedman. *Brownian Motion and Diffusion.* Springer-Verlag, New York, 1983.

27. A. Gelman, J. Carlin, H. Stern, and D.B. Rubin. *Bayesian Data Analysis.* Chapman & Hall, London, 1995.

28. C.R. Genovese and P.B. Stark. Data reduction and statistical consistency of ℓ_p misfit norms in linear inverse problems. *Phys. Earth Planet. Inter.*, 98:143–162, 1996.

29. I.J. Good. *The Estimation of Probabilities: An Essay on Modern Bayesian Methods.* MIT Press, Cambridge, MA, 1965.

30. J.A. Hartigan. *Bayes Theory.* Springer-Verlag, New York, 1983.

31. N.W. Hengartner and P.B. Stark. Confidence bounds on the probability density of aftershocks. Technical Report 352, Dept. Statistics, Univ. of Calif., Berkeley, 1992.

32. N.W. Hengartner and P.B. Stark. Finite-sample confidence envelopes for shape-restricted densities. *Ann. Stat.*, 23:525–550, 1995.

33. W. James and C. Stein. Estimation with quadratic loss. In *Proc. Fourth Berkeley Symposium on Mathematical Statistics and Probability*, volume 1, pages 361–380, Berkeley, 1961. Univ. California Press.

34. A.N. Kolmogorov. *Foundations of the Theory of Probability*. Chelsea Publishing Co., New York, 2nd edition, 1956.

35. L. Le Cam. *Asymptotic Methods in Statistical Decision Theory*. Springer-Verlag, New York, 1986.

36. L. Le Cam. Maximum likelihood: an introduction. *Intl. Stat. Rev.*, 58:153–171, 1990.

37. E.L. Lehmann and G. Casella. *Theory of Point Estimation*. Springer-Verlag, New York, 2nd edition, 1998.

38. F. O'Sullivan. A statistical perspective on ill-posed inverse problems. *Statistical Science*, 1:502–518, 1986.

39. X. Shen. On methods of sieves and penalization. *Ann. Stat.*, 25:2555–2591, 1997.

40. P.B. Stark. Inference in infinite-dimensional inverse problems: Discretization and duality. *J. Geophys. Res.*, 97:14 055–14 082, 1992.

41. P.B. Stark. Minimax confidence intervals in geomagnetism. *Geophys. J. Intl.*, 108:329–338, 1992.

42. C. Stein. Inadmissibility of the usual estimator of the mean of a multivariate normal distribution. In *Proc. Third Berkeley Symposium on Mathematical Statistics and Probability*, volume 1, pages 197–206, Berkeley, 1956. Univ. California Press.

43. A. Tarantola. *Inverse Problem Theory: methods for data fitting and model parameter estimation*. Elsevier Science Publishing Co., Amsterdam, 1987.

44. L. Tenorio, P.B. Stark, and C.H. Lineweaver. Bigger uncertainties and the Big Bang. *Inverse Problems*, 15:329–341, 1999.

45. G. Wahba. *Spline Models for Observational Data*. Soc. for Industrial and Appl. Math., Philadelphia, PA, 1990.

46. A. Wald. *Statistical Decision Functions*. John Wiley and Sons, New York, 1950.

47. L. Zadeh. Fuzzy sets. *Inf. Control*, 8:338–353, 1965.

SpringerMathematics

H. W. Engl, A. K. Louis,
W. Rundell (eds.)

Inverse Problems in Medical Imaging and Nondestructive Testing

Proceedings of the Conference in
Oberwolfach, Federal Republic of
Germany, February 4–10, 1996

1997. VII, 211 pages. 54 fig.
Softcover DM 98,–, öS 686,–
(recommended retail price)
ISBN 3-211-83015-4

14 contributions present mathematical
models for different imaging techniques
in medicine and nondestructive testing.
The underlying mathematical models are
presented in a way that also newcomers
in the field have a chance to understand
the relation between the special applications and the mathematics needed for
successfully treating these problems. The
reader gets an insight into a modern field
of scientific computing with applications
formerly not presented in such form, leading from the basics to actual research
activities.

H. W. Engl

Integralgleichungen

1997. V, 265 Seiten. 11 Abb.
Broschiert DM 53,–, öS 370,–
ISBN 3-211-83071-5

Dieses Lehrbuch behandelt zunächst den
klassischen Stoff wie Riesz- und Fred-
holmtheorie in funktionalanalytischer
Darstellung. Ein Schwerpunkt ist die
Anwendung von Methoden und Er-
gebnissen aus der Theorie der Integralgleichungen auf gewöhnliche und
partielle Differentialgleichungen. Neben
der Behandlung der analytischen
Aspekte wird auch auf die numerische
Lösung von Integralgleichungen einge-
gangen. Spezifisch für das Buch sind eine
ausführliche Behandlung von Integralgleichungen 1. Art, wie sie bei der
Modellierung inverser Probleme auftre-
ten, und ein Kapitel über nichtlineare
Integralgleichungen, das bis zu den
Grundlagen der Verzweigungstheorie
vordringt. Auch stark singuläre Glei-
chungen werden behandelt. Das Lehrbuch wendet sich an Studenten, die die
grundlegende Analysis-Ausbildung, in-
klusive der Grundlagen der linearen
Funktionalanalysis, absolviert haben.

 SpringerWienNewYork

A-1201 Wien, Sachsenplatz 4–6, P.O.Box 89, Fax +43.1.330 24 26, e-mail: books@springer.at, Internet: www.springer.at
D-69126 Heidelberg, Haberstraße 7, Fax +49.6221.345-229, e-mail: orders@springer.de
USA, Secaucus, NJ 07096-2485, P.O. Box 2485, Fax +1.201.348-4505, e-mail: orders@springer-ny.com
Eastern Book Service, Japan, Tokyo 113, 3–13, Hongo 3-chome, Bunkyo-ku, Fax +81.3.38 18 08 64, e-mail: orders@svt-ebs.co.jp

SpringerMathematics

Surveys on Mathematics
for Industry

Managing Editor
H. Engl, Linz

and an
International Editorial
and **Industrial Advisory Board**

Surveys on Mathematics for Industry is aiming to bridge the gap between university and industry by presenting mathematical methods relevant for industry and exposing industrial problems which are of interest to mathematicians.

To achieve this goal, the journal publishes surveys on new mathematical techniques, on established mathematical techniques with a new range of applications, on industrial problems for which appropriate mathematical models or methods are not yet available and broad historical surveys.

Furthermore, coverage includes comparisons of mathematical models or methods for particular industrial problems and descriptions of mathematical modelling techniques. Articles of general interest about the use of mathematics in industry will also be considered.

Subscription information
2000. Vol. 10 (4 issues) ISSN 0938-1953. Title No. 724
DM 386.–, ATS 2715.– plus carriage charges
approx. US $ 265.00 including carriage charges

View table of contents and abstracts online at: **www.springer.at/smi**

SpringerWienNewYork

A-1201 Wien, Sachsenplatz 4–6, P.O.Box 89, Fax +43.1.330 24 26, e-mail: journals@springer.at, Internet: **www.springer.at**
D-69126 Heidelberg, Haberstraße 7, Fax +49.6221.345-229, e-mail: orders@springer.de
USA, Secaucus, NJ 07096-2485, P.O. Box 2485, Fax +1.201.348-4505, e-mail: orders@springer-ny.com
Eastern Book Service, Japan, Tokyo 113, 3–13, Hongo 3-chome, Bunkyo-ku, Fax +81.3.38 18 08 64, e-mail: orders@svt-ebs.co.jp

Springer-Verlag
and the Environment

WE AT SPRINGER-VERLAG FIRMLY BELIEVE THAT AN international science publisher has a special obligation to the environment, and our corporate policies consistently reflect this conviction.

WE ALSO EXPECT OUR BUSINESS PARTNERS – PRINTERS, paper mills, packaging manufacturers, etc. – to commit themselves to using environmentally friendly materials and production processes.

THE PAPER IN THIS BOOK IS MADE FROM NO-CHLORINE pulp and is acid free, in conformance with international standards for paper permanency.